Plant Growth Promoting Rhizobacteria for Horticultural Crop Protection

P. Parvatha Reddy

Plant Growth Promoting Rhizobacteria for Horticultural Crop Protection

Springer

P. Parvatha Reddy
Indian Institute of Horticultural Research
Bangalore, Karnataka, India

ISBN 978-81-322-1972-9 ISBN 978-81-322-1973-6 (eBook)
DOI 10.1007/978-81-322-1973-6
Springer New Delhi Heidelberg New York Dordrecht London

Library of Congress Control Number: 2014947243

Printed on acid-free paper

Springer is part of Springer Science+Business Media (www.springer.com)

Preface

Over the last decades, world agriculture experienced high increase in crop yield achieved through high input of inorganic fertilizers and pesticides, and mechanization driven by fossil fuel. Over the years, this led to serious environmental problems such as depletion of soil quality and health, ocean and ground water pollution, and emergence of resistant pathogens. It is a big challenge to feed the increasing world population on decreasing farmland areas without damaging environment. Agriculture is plagued by two main constraints. The first is the depletion of nutrient supply in the agricultural soils, and consequently the sizable gap between achievable and actual yields in various crops. The second main constraint is crop protection. Devastating pathogens lead to 15–30 % average crop losses annually. In recent times, there has been a considerable pressure on consumers and farmers to reduce or eliminate the use of synthetic pesticides in horticulture, since fruits and vegetables are consumed afresh. This concern has encouraged looking for better alternatives to synthetic pesticides which are cheaper and eco-friendly.

There is a great need for environment friendly microbial technologies [including plant growth promoting rhizobacteria (PGPR)] in agriculture. There is a need for restructuring the crop rhizospheres for improving and sustaining the nutrient supply in the soils and enhancing the health and yield of crops through sustainable practices based on microbial technologies. Microbial technologies include the principles of microbial ecology, which encompass inoculation of crops with beneficial microorganisms and the use of cultural practices that enrich indigenous beneficial microorganisms in individual agricultural fields. It is well known that rhizosphere and soil microorganisms (PGPR) play an important role in maintaining crop and soil health through versatile mechanisms: nutrient cycling and uptake, suppression of plant pathogens, induction of resistance in plant host, and direct stimulation of plant growth. There are two main outcomes or effects from beneficial microorganisms: enhanced plant growth and crop protection, both of which represent the two main constraints to agriculture. Sustainable approaches are those that are not aimed solely at maximizing short-term production but rather those that consider long-term production gains, the ecology of agricultural systems, and profitability of farmers. Microbial inoculants have great potential to provide holistic health and sustainable crop yields. Maintaining biodiversity of PGPR in soil could be an important component of environment friendly sustainable agriculture strategies.

The information on the biomanagement of pests (insect and nematode pests, fungal, bacterial and viral/phytoplasma diseases) of horticultural crops (fruits, vegetables, plantation, spice, ornamental, medicinal and aromatic crops) using plant growth promoting rhizobacteria is very much scattered. There is no book at present which comprehensively and exclusively deals with the above aspects on horticultural crops. The present book deals with biomanagement of pests in horticultural crops in detail using PGPR. The present book is divided into six parts. The first part deals with the importance of PGPR including introduction, potential role of PGPR in agriculture, pest management, mechanism of biocontrol, mass production, formulation, delivery and commercialization. Pest management in tropical, sub-tropical and temperate fruit crops is dealt in Part II. The third part deals with pest management in solanaceous, bulbous, malvaceous, cruciferous, leguminous, cucurbitaceous, leafy, root and tuber vegetable crops. Pest management in plantation and spice crops is dealt in Part IV. Part V deals with pest management in ornamental, medicinal and aromatic crops. The last part deals with a road map ahead including challenges, future prospective and conclusions. The book is extensively illustrated with excellent quality photographs enhancing the quality of publication. The book is written in lucid style, easy to understand language along with adoptable recommendations involving eco-friendly components of IPM.

This book can serve as a useful reference to policy makers, research and extension workers, practicing farmers and students. The material can also be used for teaching post-graduate courses in crop protection. Suggestions to improve the contents of the book are most welcome (E-mail: reddy_parvatha@yahoo.com). The publisher, Springer India, deserves commendation for their professional contribution.

Bangalore, Karnataka, India P. Parvatha Reddy
February 20, 2014

Contents

Part V Pest Management in Ornamental, Medicinal and Aromatic Crops

About the Author

Dr. P. Parvatha Reddy obtained his doctorate degree jointly from the University of Florida, USA, and the University of Agricultural Sciences, Bangalore. He served as the Director of the prestigious Indian Institute of Horticultural Research (IIHR), Bangalore, from 1999 to 2002 during which period the Institute was honored with "ICAR Best Institution Award". Dr. Reddy also served as the Head, Division of Entomology and Nematology at IIHR and gave tremendous impetus and direction to research, extension and education in developing bio-intensive integrated pest management strategies in horticultural crops. These technologies are being practiced widely by the farmers across the country since they are effective, economical, eco-friendly and residue-free. Dr. Reddy has about 34 years of experience working with horticultural crops and involved in developing an F1 tomato hybrid "Arka Varadan" resistant to root-knot nematodes. Dr. Reddy has over 250 scientific publications to his credit, which also include 30 books. He has also guided two Ph.D. students at the University of Agricultural Sciences, Bangalore.

At present, Dr. Reddy is serving as Chairman, Research Advisory Committee (RAC) of the Indian Institute of Vegetable Research, Varanasi. He is a Member of the RAC of the National Centre for Integrated Pest Management, New Delhi; Member of the Expert Panel for monitoring the research program of National Initiative on Climate Resilient Agriculture (NICRA) in the theme of Horticulture including Pest Dynamics and Pollinators; and Member, QRT to review the progress of the Central Tuber Crops Research Institute, Trivandrum and AICRP on Tuber Crops.

Dr. Reddy has been awarded with the prestigious "Association for Advancement Pest Management in Horticultural Ecosystems Award", "Dr. G.I. D'souza Memorial Lecture Award", "Prof. H.M. Shah Memorial Award" and "Hexamar Agricultural Research and Development Foundation Award" for his unstinted efforts in developing sustainable, bio-intensive and eco-friendly integrated pest management strategies in horticultural crops.

Part I

Importance of Plant Growth Promoting Rhizobacteria

1 Introduction

Agriculture contributes to a major share of national income and export earnings in many developing countries while ensuring food security and employment. Of the 11 % of the world's land surface that is suitable for agriculture, 38 % has been degraded due to poor natural resource management practices. With no significant room to expand areas of cultivation, a good management of the available land is necessary to maintain agricultural productivity, ensure economic growth, protect biodiversity and meet the increasing food demands of a growing global population. Sustainable agriculture is vitally important in today's world because it offers the potential to meet our future agricultural needs, something that conventional agriculture will not be able to do. It integrates environmental health, economic viability and social equity to ensure long-term productivity of natural resources and improved livelihoods. It helps reduce the risks in developing countries of complex problems like climate variability and climate change.

Conventional agriculture, which involves high-yielding plants, mechanized tillage, inorganic fertilizers and chemical pesticides, is detrimental to the environment. For instance, fertilizer run-off from conventional agriculture is the chief culprit in creating dead zones with low oxygen where marine life cannot survive. Once fertile soils have become acidic due to heavy chemical fertilizer application. In the same way, reliability on chemical pesticides to manage pest problems has aggravated environmental ruins. The challenge of enhancing productivity while maintaining environmental soundness calls for educating farmers, emphasizing the long-term consequences of their traditional methods of agriculture and helping them develop and implement innovative, appropriate farming practices.

Pathogenic microorganisms affecting plant health are a major and chronic threat to food production and ecosystem stability worldwide. At the present time, among the most important factors limiting production of different crops are insect pests and diseases caused by fungi, bacteria, viruses and nematodes. By this reason, different methods have been used to control these pests (Ristaino 1991). Cultural practices and chemical control using synthetic chemicals are the most used control methods (Parra and Ristaino 2001). However, use of some of these synthetic products has caused various problems due to environmental pollution, with consequences such as toxicity to humans, as well as resistance of certain pests to these pesticides (Hernández-Castillo et al. 2005). Furthermore, the growing cost of pesticides, particularly in less-affluent regions of the world, and consumer demand for pesticide-free food have led to a search for substitutes for these products. There are also a number of fastidious pests for which chemical solutions are few, ineffective or nonexistent (Hoffland et al. 1995). Despite the use of available means of plant protection, about one third of the crops produced are destroyed by pests and diseases.

P.P. Reddy, *Plant Growth Promoting Rhizobacteria for Horticultural Crop Protection*,
DOI 10.1007/978-81-322-1973-6_1, © Springer India 2014

Recently, there has been a great interest in eco-friendly and sustainable agriculture. There is a need for gradually shifting to biological control methods from total dependence on chemical methods. It is in this context that bio-alternatives or bioagents, such as biopesticides including plant growth-promoting rhizobacteria (PGPR), are considered as viable options for pest management. Scientists across the world have been researching on suitable and effective PGPR as among the strategies to raising crop productivity levels.

The application of microbial products for management of biotic and abiotic stresses is less common. An exception to this is the use of rhizobial inoculants for legumes to ensure efficient nitrogen fixation, a practice that has been occurring in North America for over 100 years (Smith 1997). The region around the root, the rhizosphere, is relatively rich in nutrients, due to the loss of as much as 40 % of plant photosynthates from the roots (Lynch and Whipps 1991). Consequently, the rhizosphere supports large and active microbial populations capable of exerting beneficial, neutral or detrimental effects on plant growth. The importance of rhizosphere microbial populations for maintenance of root health, nutrient uptake and tolerance of environmental stress is now recognized (Bowen and Rovira 1999; Cook 2002). These beneficial microorganisms can be a significant component of management practices to achieve the attainable yield, which has been defined as crop yield limited only by the natural physical environment of the crop and its innate genetic potential (Cook 2002).

The prospect of manipulating crop rhizosphere microbial populations by inoculation of beneficial bacteria to increase plant growth and management of pests has shown considerable promise in laboratory and greenhouse studies, but responses have been variable in the field (Bowen and Rovira 1999). The potential environmental benefits of this approach, leading to a reduction in the use of agricultural chemicals and the fit with sustainable management practices, are driving this technology. Recent progress in our understanding of the biological interactions that occur in the rhizosphere and of the practical requirements for inoculants formulation and delivery should increase the technology's reliability in the field and facilitate its commercial development.

1.1 Plant Growth-Promoting Rhizobacteria (PGPR)

The thin layer of soil surrounding crop roots and the volume of soil occupied by roots is known as the rhizosphere. The shear extent of crop roots in soil dictates that a significant portion of soil is actually within the rhizosphere (about 5–40 % of soil in the rooting zone depending upon crop root architecture). The rhizosphere is the region of soil that is immediately near to the root surface and that is affected by root exudates. This zone is where the majority of soil microorganisms (bacteria and fungi) reside. They reside in the rhizosphere to utilize compounds and materials released from crop roots providing microorganisms with energy. The majority of microorganisms present in the rhizosphere are thought to have no direct consequence on plant growth and vigour. However, there are many present that are deleterious or beneficial to plant growth. Farmers have exploited to great success the use of inoculants of nitrogen-fixing bacteria to limit the need for costly fertilizers in legume crops. As we strive to optimize the performance of all crops, increasingly the value of inoculating soil with other microorganisms or promoting the activity of residing beneficial microorganisms through management practices is being explored.

The characteristics of an ideal PGPR (Jeyarajan and Nakkeeran 2000) are as follows:

- High rhizosphere competence
- High competitive saprophytic ability
- Enhancement of plant growth
- Ease for mass multiplication
- Broad spectrum of action
- Excellent and reliable pest control
- Safe to environment
- Compatible with other rhizobacteria
- Tolerant to desiccation, heat, oxidizing agents and UV radiations

Based on their experiments on radishes, Kloepper and Schroth (1978) introduced the term 'rhizobacteria' to the soil bacterial community that competitively colonized plant roots and stimulated growth, thereby reducing the incidence of plant diseases. Kloepper and Schroth (1981) termed these beneficial rhizobacteria as plant growth-promoting rhizobacteria (PGPR). PGPR can be defined as the indispensable part of rhizosphere biota that when grown in association with the host plants can stimulate the growth of the host. PGPR seemed as successful rhizobacteria in getting established in soil ecosystem due to their high adaptability in a wide variety of environments, faster growth rate and biochemical versatility to metabolize a wide range of natural and xenobiotic compounds. Cook (2002) considered PGPR as the significant component in the management of agricultural practices with innate genetic potential. The concept of PGPR has now been confined to the bacterial strains that can fulfil at least two of the three criteria such as aggressive colonization, plant growth stimulation and biocontrol (Weller et al. 2002).

According to Whipps (2001) there are three basic categories of interactions (neutral, negative or positive) that generally exist between the rhizobacteria and growing plants. Most rhizobacteria associated with plants are commensals in which the bacteria establish an innocuous interaction with the host plants exhibiting no visible effect on the growth and overall physiology of the host (Beattie 2006). In negative interactions, the phytopathogenic rhizobacteria produces phytotoxic substances such as hydrogen cyanide or ethylene, thus negatively influencing the growth and physiology of the plants. Counter to these deleterious bacteria, there are some PGPR that can exert a positive plant growth by direct mechanisms such as solubilization of nutrients, nitrogen fixation, production of growth regulators, etc. or by indirect mechanisms such as stimulation of mycorrhizae development, competitive exclusion of pathogens or removal of phytotoxic substances. However, in accordance with their degree of association with the plant root cells, PGPR can be classified into extracellular plant growth-promoting rhizobacteria (ePGPR) and intracellular plant growth-promoting rhizobacteria (iPGPR). The ePGPR may exist in the rhizosphere, on the rhizoplane or in the spaces between the cells of the root cortex, while iPGPR are located generally inside the specialized nodular structures of root cells. The bacterial genera such as *Agrobacterium*, *Arthrobacter*, *Azotobacter*, *Azospirillum*, *Bacillus*, *Burkholderia*, *Caulobacter*, *Chromobacterium*, *Erwinia*, *Flavobacterium*, *Micrococcus*, *Pseudomonas* and *Serratia* belong to ePGPR (Gray and Smith 2005). The iPGPR includes the endophytes and *Frankia* species, both of which can symbiotically fix atmospheric N_2 with the higher plants. Endophytes include a wide range of soil bacterial genera such as *Allorhizobium*, *Azorhizobium*, *Bradyrhizobium*, *Mesorhizobium* and *Rhizobium* of the family *Rhizobiaceae* that generally invades the root systems in crop plants to form nodules and stimulates growth either directly or indirectly. This group of rhizobacteria is mostly Gram negative and rod shaped with a lower proportion being Gram-positive rods, cocci and pleomorphic. Some of the important plant species forming symbiotic association with these rhizobial species include *Cajanus cajan*, *Phaseolus vulgaris* and *Pisum sativum*.

Several actinomycetes, one of the major components of rhizosphere microbial populations, are also useful because of their significant ecological roles in soil nutrient cycling (Elliot and Lynch 1995) as well as in plant growth-promoting activities (Merzaeva and Shirokikh 2006). A number of reports are available on the potential of actinomycetes as plant growth-promoting agents. Actinomycetes strains like *Micromonospora* spp., *Streptomyces* spp., *Streptosporangium* spp. and *Thermobifida* spp. are recorded as best to colonize the plant rhizosphere, showing an immense potentiality as biocontrol agents against a range of root pathogenic fungi (Franco-Correa et al. 2010). Rhizosphere streptomycetes as potential biocontrol agent of *Fusarium* and *Armillaria* and as PGPR was reported. Evidences are now available on actinobacteria used in the control of *Rhizoctonia solani* and *Ralstonia solanacearum* in tomato (Sabaratnam and Traquair 2002) and

Colletotrichum musae in banana (Taechowisan et al. 2003). Soil actinomycetes are also an important source of diverse antimicrobial metabolites. The metabolites of antagonistic actinobacteria in *Araucaria angustifolia* rhizosphere[especially indole acetic acid (IAA) and chitinase] are responsible for the degradation of different complex and relatively recalcitrant organic compounds present in soil. Similar antagonistic activity of endophytic *Streptomyces griseorubiginosus* against *Fusarium oxysporum* f. sp. *cubense* has been recorded by Cao et al. (2004).

The potential role of PGPR in conferring resistance to water stress in tomatoes and peppers has been investigated (Mayak et al. 2004). Fluorescent pseudomonads and species of *Bacillus* were reported with very high efficiency in host root colonization and production of growth metabolites resulting in improved strategic crop yield. The various modes of action of a *Bacillus subtilis* strain, FZB24, against phytopathogens suggest the role of the bacterium in plant vitality. Soil rhizobacterial populations are capable of exerting beneficial effects on many plants like potato, pea and cucumber by colonizing rhizosphere. Thus, it has been established that the inoculation of PGPR can increase nodulation, nitrogen uptake, growth and yield response of crop plants. In addition to this, employing microorganisms as co-culture in biotization is also another important area of research in recent decade. Biotization is a metabolic response of in vitro grown plant material to microbial inoculants leading to developmental and physiological changes of the derived propagules, causing enhancement in the biotic and abiotic stress resistance in plants. Here, plantlets are usually co-cultured with PGPR to produce more biomass and secondary metabolites. The importance of PGPR in maintaining root health, nutrient uptake and tolerance to environmental stress is well recognized. The specific traits of promoting plant growth and development are limited at a given environment of plant–microbe interactions. Several PGPR formulations are currently available as commercial products for agricultural production of beneficial crops. Our understanding on PGPR is now advancing at the cellular, genomic and proteomic level. Large numbers of PGPR strains of different bacterial classes and genera with multifunctional traits have, therefore, been described for their potent application in boosting plant activities in modern agriculture. However, it is equally important to study in detail the potentiality of this group of rhizospheric microbiota along with their mechanism of action involved in sustainable crop production. There is also a need to improve our knowledge for the selection of potent microbial strains colonizing rhizosphere of growing plants for specific restoration programmes. PGPR can promote growth and yield of crop plants by direct and indirect mechanisms. In some PGPR species, plant growth promotion dominates with nitrogen fixation, phosphate solubilization and production of phytohormones like auxin and cytokinin and volatile growth stimulants such as ethylene and 2,3-butanediol.

The ways that PGPR promote plant growth are as follows:

- Increasing nitrogen fixation in legumes
- Promoting free-living nitrogen-fixing bacteria
- Increasing supply of other nutrients, such as phosphorus, sulphur, iron and copper
- Producing plant hormones
- Enhancing other beneficial bacteria or fungi
- Controlling diseases, nematodes and insect pests

Siderophore production for rhizosphere colonization has also been recorded as one of the important mechanisms by certain PGPR (*Bradyrhizobium japonicum, Rhizobium leguminosarum* and *Sinorhizobium meliloti*) with plant growth-promoting activity. Besides iron-chelating siderophores, antibiotics (Weller 1988) and hydrogen cyanides are also likely to be produced by PGPR strains, participating tremendously in the reduction of phytopathogens and deleterious rhizobacteria with a corresponding improvement in plant health. However, regardless of the mechanism of plant growth promotion, PGPR must colonize the rhizosphere or root itself (Glick 1995). There is a need to examine our current understanding on the known, putative and speculative mechanism of plant growth promotion by the rhizobacteria along with their potential emergence on overall plant growth and development in agriculture.

In the context of increasing international concern for food and environmental quality, the use of PGPR for reducing chemical inputs in agriculture is a potentially important issue. PGPR have gained worldwide importance and acceptance for sustainable agricultural benefits. PGPR are the potential tools for the future of sustainable agriculture. Currently, there is an active and growing group of researchers working on fundamental and applied aspects of PGPR. The application and commercialization of PGPR for sustainable agriculture is a growing and demanding market worldwide.

1.2 Historical Background

Inoculation of plants with beneficial bacteria can be traced back for centuries. From experience, farmers knew that when they mixed soil taken from a previous legume crop with soil in which non-legumes were to be grown, yields often improved. By the end of the nineteenth century, the practice of mixing 'naturally inoculated' soil with seeds became a recommended method of legume inoculation in the USA. A decade later, the first patent ('Nitragin') was registered for plant inoculation with *Rhizobium* sp. Eventually, the practice of legume inoculation with rhizobia became a common practice.

For almost 100 years, *Rhizobium* inoculants have been produced around the world, primarily by small companies. Significant contributions to the production of legumes by inoculation of *Rhizobium* were made in Australia, North America, Eastern Europe, Egypt, Israel, South Africa, New Zealand and, to a lesser extent, Southeast Asia. For the large majority of less developed countries in Asia, Africa, and Central and South America, inoculant technology has had no impact on productivity of the family farm, because inoculants are not used or are of poor quality.

Inoculation with non-symbiotic, associative rhizosphere bacteria, like *Azotobacter*, was used on a large scale in Russia in the 1930s and 1940s. The practice had inconclusive results and was later abandoned. Interest in *Azotobacter* as an inoculant for agriculture has only recently been revived. An attempt to use *Bacillus megaterium* for phosphate solubilization in the 1930s on large scale in Eastern Europe apparently failed. Two major breakthroughs in plant inoculation technology occurred in the late 1970s: (i) *Azospirillum* was found to enhance non-legume plant growth by directly affecting plant metabolism, and (ii) Biocontrol agents, mainly of the *Pseudomonas fluorescens* and *P. putida* groups, began to be intensively investigated.

In recent, years, various other bacterial genera, such as *Bacillus*, *Flavobacterium*, *Acetobacter* and several *Azospirillum*-related microorganisms have also been evaluated. The immediate response to soil inoculation with associative, non-symbiotic PGPR (but also for rhizobia) varies considerably depending on the bacteria, plant species, soil type, inoculant density and environmental conditions. A major role of inoculant formulation is to provide a more suitable microenvironment (even temporarily) to prevent the rapid decline of introduced bacteria in the soil. Although much is known about the survival of bacteria within the protective environment of an inoculant carrier, little is known about the stresses that bacteria must endure upon transfer to the competitive and often harsh soil environment. Inoculants have to be designed to provide a dependable source of beneficial bacteria that survive in the soil and become available to the plant.

1.3 Genera of PGPR

PGPR generally include the strains in the genera *Serratia*, *Pseudomonas*, *Burkholderia*, *Agrobacterium*, *Erwinia*, *Xanthomonas*, *Azospirillum*, *Bacillus*, *Enterobacter*, *Rhizobium*, *Alcaligenes*, *Arthrobacter*, *Acetobacter*, *Acinetobacter*, *Achromobacter*, *Aerobacter*, *Azotobacter*, *Clostridium*, *Klebsiella*, *Micrococcus*, *Rhodobacter*, *Rhodospirillum*, *Flavobacterium*, *Bradyrhizobium*, *Frankia*, *Pseudomonas*, *Thiobacillus*, *Paenibacillus*,

Table 1.1 Plant growth-promoting rhizobacteria (Glick et al. 1999)

Azorhizobium caulinodans	*B. polymyxa*	*Klebsiella planticola*
Azospirillum amazonense	*B. pumilus*	*Kluyvera ascorbata*
A. halopraeferens	*B. sphaericus*	*K. cryocrescens*
A. irakense	*B. subtilis*	*Phyllobacterium rubiacearum*
A. lipoferum	*Burkholderia cepacia*	*Pseudomonas aeruginosa*
A. brasilense	*B. gladioli*	*P. aureofaciens*
Azotobacter chroococcum	*B. graminis*	*P. corrugata*
Bacillus cereus	*B. vietnamensis*	*P. fluorescens*
B. coagulans	*Citrobacter freundii*	*P. marginalis*
B. lateroporus	*Curtobacterium flaccumfaciens*	*P. putida*
B. licheniformis	*Enterobacter agglomerans*	*P. rubrilineans*
B. macerans	*E. cloacae*	*Rathayibacter rathayi*
B. megaterium	*Erwinia herbicola*	*Serratia marcescens*
B. mycoides	*Flavimonas oryzihabitans*	*Stenotrophomonas* sp.
B. pasteurii	*Hydrogenophaga pseudoflava*	*Streptomyces griseoviridis*

Alicyclobacillus, *Aneurinibacillus*, *Virgibacillus*, *Solibacillus*, *Gracilibacillus* and others (Table 1.1).

PGPR for which evidence exists that their plant stimulation effect is related to their ability to fix N_2 include the endophytes *Azospirillum* sp., *Azoarcus* sp., *Burkholderia* sp., *Gluconacetobacter diazotrophicus* and *Herbaspirillum* sp. and the rhizospheric bacteria *Azotobacter* sp. and *Paenibacillus* (*Bacillus*) *polymyxa*. Notable nitrogen-fixing bacteria associated with legumes include *Allorhizobium*, *Azorhizobium*, *Bradyrhizobium* and *Rhizobium*.

Various species of bacteria like *Pseudomonas*, *Azospirillum*, *Azotobacter*, *Bacillus*, *Burkholderia*, *Enterobacter*, *Rhizobium*, *Erwinia*, *Mycobacterium*, *Mesorhizobium*, *Flavobacterium*, *Klebsiella*, *Alcaligenes*, *Acinetobacter*, *Erwinia*, *Arthrobacter* and *Serratia* are associated with the plant rhizosphere and are able to exert a beneficial effect on plant growth.

The biological control agents include the genera *Bacillus*, *Streptomyces*, *Pseudomonas*, *Burkholderia* and *Agrobacterium*. Actinomycetes strains like *Micromonospora* sp., *Streptomyces* spp., *Streptosporangium* sp., and *Thermobifida* sp. are recorded as best to colonize the plant rhizosphere, showing an immense potentiality as biocontrol agents against a range of root pathogenic fungi.

1.3.1 Pseudomonads

All species and strains of *Pseudomonas* are Gram-negative rods (Fig. 1.1). A significant number of cells can produce exopolysaccharides known as biofilms. The role of fluorescent pseudomonads in agriculture has been a matter of interest due to their potential for plant growth promotion and pathogen suppression. Specific strains of fluorescent pseudomonads have the potential to suppress plant pathogens, enhance plant growth and participate in carbon and nitrogen cycling in nature. A large number of fluorescent pseudomonad species such as *P. putida*, *P. fluorescens*, *P. aeruginosa*, *P. aureofaciens* (now considered as *P. chlororaphis*) and *P. pyrocinia* have been well documented for their antagonistic potential.

The pathogen-suppressing ability of these bacteria is mainly due to their potential to produce an array of antibiotics, viz., phenazines, pysrole-type compounds, polyketides and peptides. Sigma factor genes, rpoD and rpoS, have been reported to control these antibiotics' production and enhance the antagonistic activities of fluorescent pseudomonads. The plant growth-promoting ability of these bacteria is due to the production of siderophore and phosphatase, phytohormones and 1-aminocyclopropane-1-carboxylate (ACC) deaminase.

Pseudomonads are well known for their involvement in the biological control of several plant

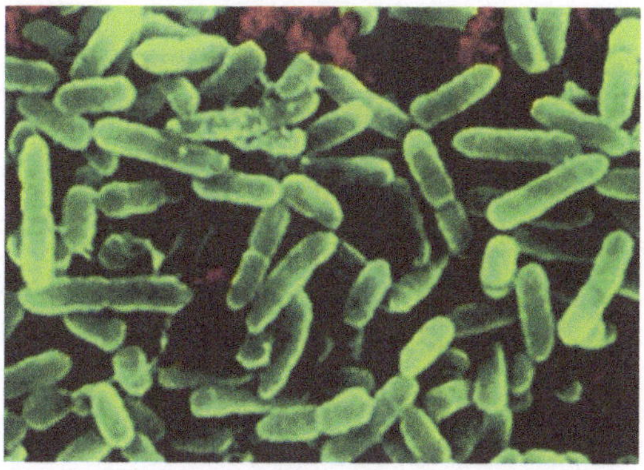

Fig. 1.1 *Pseudomonas* sp.

pathogens. Alabouvette et al. (1993) showed that in addition to nonpathogenic *Fusarium oxysporum*, *P. fluorescens* and *P. putida* are the main candidates for the biological control of *Fusarium* wilts. The fluorescent pseudomonads are involved in the natural suppressiveness of some soils to *Fusarium* wilts, and they have been applied successfully to suppress *Fusarium* wilts of various plant species (Lemanceau and Alabouvette 1993). For many pseudomonads, production of metabolites such as antibiotics, siderophores and hydrogen cyanide (HCN) is the primary mechanism of biocontrol (Weller and Thomashow 1993). Several evidences indicate that siderophore production when iron is limited is responsible for the antagonism of some strains of *P. aeruginosa* against *Pythium* spp., the causal agents of damping off and root rot of many crops (Charest et al. 2005). The antibiotics produced by bacterial biocontrol agents and their role in microbial interaction were reviewed by Raaijmakers et al. (2002). *P. fluorescens* CHAO isolated and intensively studied by the group of Défago in Switzerland produces several bioactive compounds (antibiotics, siderophores, HCN, indole acetic acid), giving it one of the broadest spectra of potential biocontrol and growth-promoting mechanisms of known PGPR (Weller and Tomashow 1993).

In Canada, *Pseudomonas* species were developed for the biological control of *Pythium* diseases in hydroponic systems for greenhouses (Paulitz and Bélanger 2001). In a spring cucumber crop, *P. corrugata* strain 13 and *P. fluorescens* strain 15 produced 88 % more marketable fruit, while in a fall crop with severe disease pressure due to higher slab temperatures, both strains significantly increased marketable fruit yield by 600 % of the marketable fruit. Strain 15 also increased fruit production in treatments not inoculated with pathogen (Paulitz and Bélanger 2001). Several reports show the critical role played by fluorescent *Pseudomonas* spp. in naturally occurring soils that are suppressive to *Fusarium* wilt (Mazzola 2002).

1.3.1.1 *Pseudomonas fluorescens*

It plays a major role in the plant growth promotion, induced systemic resistance, biological control of pathogens, etc. PGPR are known to enhance plant growth promotion and reduce severity of many nematode diseases.

P fluorescens produce the antibiotics 2, 4-diacetylphloroglucinol (DAPG), phenazine-1-carboxylic acid (PCA) and phenazine (Phe) and antimicrobial metabolites such as pyoluteorin (Plt), hydrogen cyanide (HCN), siderophores pyoverdine (Pvd), salicylic acid (Sal) and pyochelin (Pch) which inhibit a broad spectrum of plant pathogenic fungi, bacteria and nematodes and control a variety of root and seedling diseases. There are many evidences substantiating the importance of DAPG in biological control. The production of DAPG is governed by the *Phl* gene. The population size of DAPG producers in the rhizosphere is correlated with the disease suppressiveness of the soil and in situ antibiotic production. The diverse DAPG-producing *Pseudomonas* spp. have been isolated from the rhizosphere of various crop plants, and their role in promoting plant growth and inhibiting root diseases is the subject of ongoing investigations worldwide.

P. fluorescens strain WCS417r could elicit systemic disease resistance in plants through a variety of signal translocation pathways like SA-independent JA–ethylene-dependent signalling, ISR-related gene expression, NPR1-dependent signalling, etc.

1.3.1.2 *P. putida*

It has demonstrated potential biocontrol properties, as an effective antagonist of damping off (*Pythium*) and wilt (*Fusarium*) diseases. The diverse metabolism of *P. putida* may be exploited for bioremediation;

for example, it is used as a soil inoculant to remedy naphthalene contaminated soils.

1.3.1.3 *P. chitinolytica*

P. chitinolytica isolated from crustacean shell-amended soil was evaluated for the control of *Meloidogyne javanica* on tomatoes. Greenhouse, screen house and microplot tests all indicated that the bacterium improved the growth and yield and reduced nematode population in treated plants.

1.3.1.4 *P. chlororaphis*

They are rod-shaped, non-spore-forming, Gram-negative bacteria with one or more polar flagella. *P. chlororaphis* is a heterotrophic, soil bacterium that can be found in the rhizosphere, phyllosphere and surrounding sediments. Many applications of *P. chlororaphis* are in agriculture and horticulture by acting against various fungal plant pathogens by creating phenazine (antibiotic to root rot of cucumbers, peppers, tomatoes and many other agricultural crops). *P. chlororaphis* is an effective biocontrol agent against *Pythium aphanidermatum*, a causal agent of damping off of hot pepper seedlings in greenhouse vegetable production.

Several strains of *P. chlororaphis* have been shown to be efficient root colonizers (Fig. 1.2)

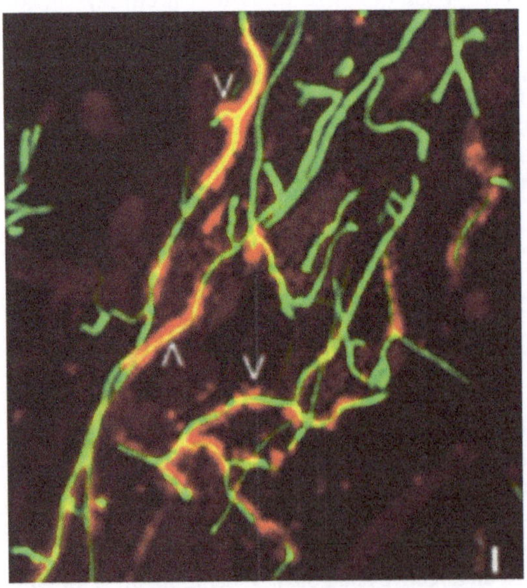

Fig. 1.2 Colonization of *Pseudomonas chlororaphis* on *Fusarium oxysporum* of tomato roots

capable of producing a variety of antifungal substances including phenazine-1-carboxamide (PCN), hydrogen cyanide, chitinases and proteases. The effectiveness of *P. chlororaphis* as a biocontrol agent has made it an organism of interest for agriculture.

1.3.2 Bacilli

Bacillales is the order to which Bacillaceae family belongs within the genus *Bacillus*. This genus is characterized by having a rod shape within the group of Gram positive and is therefore classified as strict aerobes or facultative anaerobes and integrated by 88 species. A feature associated with this genus is that it forms a type of cell called endospore as a response to adverse growth conditions which distorts the structure of the cell. This spore form is resistant to high temperatures and current chemical disinfectants. Generally, *Bacillus* species used in biocontrol are mobile, with peritrichous flagella.

Many species of *Bacillus*, including *B. subtilis*, *B. licheniformis*, *B. pumilus*, *B. amyloliquefaciens*, *B. cereus* and *B. mycoides*, are known to suppress growth of several fungal pathogens such as *Rhizoctonia*, *Fusarium*, *Sclerotinia*, *Sclerotium*, *Gaeumannomyces*, *Nectria*, *Pythium*, *Phytophthora* and *Verticillium*. The main property of antagonist bacterial strains is production of antifungal antibiotics, which seem to play a major role in biological control of plant pathogens and postharvest spoilage fungi. The main antibiotic producers of this genus are *B. brevis* (gramicidin, tyrothricin), *B. licheniformis* (bacitracin), *B. polymyxa* (polymyxin, colistin), *B. pumilus* (pumulin), *B. subtilis* (polymyxin, difficidin, subtilin, mycobacillin, bacteracin), *B. cereus* (cerexin, zwittermicin), *B. circulans* (circulin) and *B. laterosporus* (laterosporin). Many of these antifungal substances have been characterized and identified as peptide antibiotics. Antifungal peptides (iturins) produced by *Bacillus* species include mycosubtilins, bacillomycins, surfactins, fungistatins and subsporins. *Bacillus* spp. produce also a range of other metabolites including phytoalexins, chitinases and other cell wall-degrading enzymes and volatile compounds which elicit plant resistance mechanisms against attack by fungi, bacteria and pathogenic nematodes.

1.3.2.1 *Bacillus subtilis*

It is the most studied and has been reported as growth promoter and antagonistic to a variety of pathogens, such as *Phytophthora cactorum*, *Sclerotium cepivorum*, *Fusarium oxysporum*, *Rhizoctonia solani*, *Alternaria carthami*, *Phytophthora capsici* and *Fusarium solani*, among others, in different cultures and evaluated in vitro, greenhouse and field level, so that they can be used as an alternative in biological control for the management of plant diseases.

B. subtilis strain RB14 produces the cyclic lipopeptide antibiotics iturin A and surfactin active against several phytopathogens. The antifungal activity of the selected isolates was associated with their ability to produce inhibitory volatile substances and diverse and complex lytic chitinases.

This prevalent inhabitant of soil is widely recognized as a powerful biocontrol agent. In addition, it has a broad host range and is able to form endospores and produce different biologically active compounds with a broad spectrum of activity.

1.3.2.2 *B. pumilus*

It is a Gram-positive, aerobic, spore-forming bacillus commonly found in soil. *B. pumilus* strain GB34 is used as an active ingredient in agricultural fungicides. Growth of the bacterium on plant roots prevents *Rhizoctonia* and *Fusarium* spores from germinating. It induces resistance to a number of plant diseases.

1.3.2.3 *B. megaterium*

Bacillus megaterium from tea rhizosphere is able to solubilize phosphate and produce IAA, siderophore and antifungal metabolites, and thus it helps in the plant growth and reduces disease intensity (Chakraborty et al. 2006).

1.3.2.4 *B. licheniformis*

It is effective against a variety of pathogens, both as a preventive and curative, particularly on leaf spots and blights, and acts as growth-promoting bacteria with likely production of gibberellins.

1.3.2.5 *B. amyloliquefaciens*

It is effective against *Colletotrichum dematium*, *C. lagenarium*, *Rosellinia necatrix*, *Pyricularia oryzae*, *Agrobacterium tumefaciens*, *Xanthomonas campestris* pv. *campestris* and *X. campestris* pv. *vesicatoria* under in vitro and in vivo conditions; *Botrytis elliptica* under greenhouse conditions; *Botrytis cinerea* under postharvest conditions; and *Rhizoctonia solani*, *Fusarium* spp. and *Pythium* spp. under field conditions and also induces resistance mechanism in plants.

1.3.2.6 Diazotrophic PGPR

Presently, diazotrophic PGPR for which evidence exists that their plant stimulation effect is related to their ability to fix N_2 include the endophytes *Azospirillum* sp., *Azoarcus* sp., *Burkholderia* sp., *Gluconacetobacter diazotrophicus* and *Herbaspirillum* sp. and the rhizospheric bacteria *Azotobacter* sp. and *Paenibacillus* (*Bacillus*) *polymyxa* (Vessey 2003).

1.3.2.7 *Azospirillum* spp.

They (Fig. 1.3) are not considered classic biocontrol agents of soilborne plant pathogens. However, *A. brasilense* has moderate capabilities of biocontrolling crown gall-producing *Agrobacterium* (Bakanchikova et al. 1993), bacterial leaf blight of mulberry (Sudhakar et al. 2000) and bacterial leaf and/or vascular tomato diseases (Bashan and

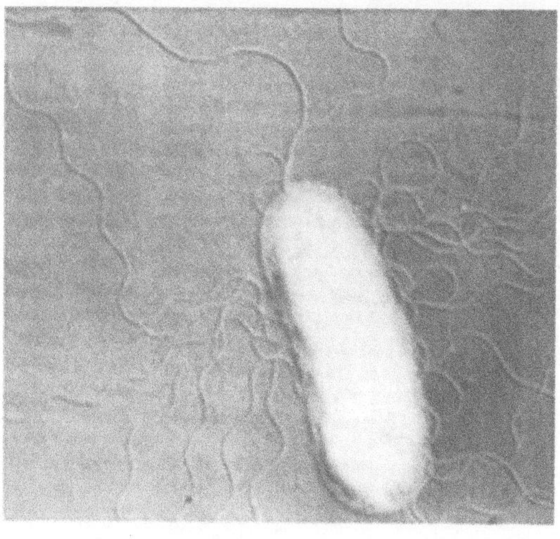

Fig. 1.3 *Azospirillum* sp.

de-Bashan 2002a, b). In addition, the proliferation of other nonpathogenic rhizosphere bacteria can be restricted by *A. brasilense* (Holguin and Bashan 1996). These *Azospirillum* antibacterial activities could be related to its already known ability to produce bacteriocins (Oliveira and Drozdowicz 1987) and siderophores (Shah et al. 1992). In addition, *A. brasilense* was recently reported to synthesize phenylacetic acid (PAA), an auxin-like molecule with antimicrobial activity (Somers et al. 2005).

1.3.2.8 *Azotobacter* spp.

They (Fig. 1.4) free-living nitrogen-fixing bacteria and normally fix molecular nitrogen from the atmosphere in plant roots. Nitrogenase is the most important enzyme involved in nitrogen fixation. *Azotobacter* also synthesizes some biologically active substances, including some phytohormones such as auxins, thereby stimulating plant growth. They also facilitate the mobility of heavy metals in the soil and thus enhance bioremediation of soil from heavy metals, such as cadmium, mercury and lead.

The heterotrophic bacterium *A. chroococcum*, antagonistic to *M. incognita*, inhibited the hatching of egg masses and prevented the larvae to penetrate into the roots of brinjal to form galls (Chahal and Chahal 1988).

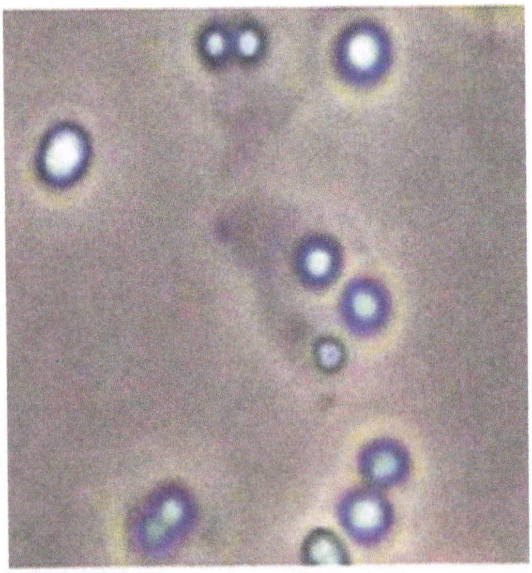

Fig. 1.4 *Azotobacter* sp.

Chahal and Chahal (2003) reported significant reduction in the gall formation and soil and root population of *M. incognita* both in tomato and brinjal by *A. chroococcum*.

1.3.3 Actinomycetes

A number of reports are available on the potential of actinomycetes as plant growth-promoting agents. Actinomycetes strains like *Micromonospora* spp., *Streptomyces* spp., *Streptosporangium* spp. and *Thermobifida* spp. are recorded as best to colonize the plant rhizosphere, showing an immense potentiality as biocontrol agents against a range of root pathogenic fungi (Franco-Correa et al. 2010).

1.3.3.1 *Streptomyces griseoviridis*

Streptomyces griseoviridis, the active ingredient of Mycostop, effectively inhibits the growth of a wide range of fungal pathogens. The antagonistic effect seems to be based on several modes of action (rhizosphere competence, hyperparasitism, antifungal metabolites, auxin excretion), and it is possible that the antagonist uses different mechanisms in diverse situations and against various pathogens.

As a seed dressing or soil treatment the organism secretes antibiotic substances which inhibit seed-borne and soilborne fungal pathogens. In greenhouse tests *S. griseoviridis* controlled *Alternaria brassicicola* on cauliflower and cabbage.

1.3.4 Enterobacteria

1.3.4.1 *Serratia marcescens*

They are Gram-negative, rod-shaped bacteria, which synthesize different types of enzymes, nuclease pigments, chitinase and extracellular polysaccharides. Various species of *Serratia* have been reported to show antimicrobial activity against different plant pathogenic bacteria, fungi and viruses. The extracellular products from *S. marcescens* are mainly responsible for antagonism.

S. marcescens has been reported to secrete extracellular chitinase, several proteases, a nuclease and a lipase. It is an important bacterium that is reported to induce systemic resistance to various pathogens. *S. marcescens* NBRI1213 protects betel vine against foot and root rot disease caused by *Phytophthora nicotianae. S. marcescens* strain 90-166 is known to exhibit antifungal activity against *Fusarium oxysorum* f. sp. *cucumerinum*, *Phytophthora parasitica* and *P. capsici*. In addition, *S. marcescens* strain Gsm01 protects cucumber from cucumber mosaic virus.

1.3.5 Rhizobia

Rhizobia and bradyrhizobia are well known as the microbial symbiotic partners of legumes, forming N_2-fixing nodules. However, these bacteria also share many characteristics with other PGPR. In fact rhizobia can produce phytohormones, siderophores and HCN; they can solubilize sparingly soluble organic and inorganic phosphates, and they can colonize the roots of many non-legume plants (Antoun et al. 1998). Rhizobia have a good potential to be used as biological control agents against some plant pathogens.

Strains of *Sinorhizobium meliloti* are antagonistic to *Fusarium oxysporum* (Antoun et al. 1998), and rhizobia antagonistic to *F. solani* f. sp. *phaseoli* isolated from commercial snap bean appeared to have a good potential for controlling *Fusarium* rot (Buonassisi et al. 1986). Ehteshamul-Haque and Ghaffar (1993) observed under field conditions that *S. meliloti*, *R. leguminosarum* bv. *viciae* and *B. japonicum* used either as seed dressing or as soil drench reduced infection of *Macrophomina phaseolina*, *Rhizoctonia solani* and *Fusarium* spp. in okra plants. In a field naturally infested with *Pythium* spp., inoculation of pea with strain R12 of *R. leguminosarum* bv. *viciae*, isolated from lentil in Alberta, Canada, significantly increased seedling emergence 4 weeks after planting (Bardin et al. 2004). This strain was as effective as *Pseudomonas fluorescens* 708, a biological control agent of *Pythium* sp. (Bardin et al. 2003). Two other strains R20 and R21 isolated from pea showed comparable results and are potentially good biocontrol agents against *Pythium* diseases in pea. Reitz et al. (2000) showed that the lipopolysaccharides of *R. etli* G12 induce the systemic resistance to infection by the cyst nematode *Globodera pallida* in potato roots.

1.4 PGPR as Biocontrol Agents

The introduction of PGPR for increasing plant growth promotion during the 1950s from research findings opened new vistas to use PGPR as an alternate to chemical pesticides for the management of soilborne pathogens (Kloepper 1993). PGPR are indigenous to soil and the plant rhizosphere and play a major role in the biocontrol of plant pathogens. They can suppress a broad spectrum of bacterial, fungal and nematode diseases. PGPR can also provide protection against viral diseases. The use of PGPR has become a common practice in many regions of the world. Although significant control of plant pathogens has been demonstrated by PGPR in laboratory and greenhouse studies, results in the field have been inconsistent. Recent progress in our understanding of their diversity, colonizing ability and mechanism of action, formulation and application should facilitate their development as reliable biocontrol agents against plant pathogens. Some of these rhizobacteria may also be used in integrated pest management programmes. The major groups of rhizobacteria with potential for biological control include *Pseudomonas* spp. and *Bacillus* spp. which are ubiquitous bacteria in agricultural soils.

References

Alabouvette C, Lemanceau P, Steinberg C (1993) Recent advances in the biological control of fusarium wilts. Pestic Sci 37:365–373

Antoun H, Beauchamp CJ, Goussard N, Chabot R, Lalande R (1998) Potential of *Rhizobium* and *Bradyrhizobium* species as plant growth promoting rhizobacteria on nonlegumes: effect on radishes (*Raphanus sativus* L.). Plant Soil 204:57–67

Bakanchikova TI, Lobanok EV, Pavlova-Ivanova LK, Redkina TV, Nagapetyan ZA, Majsuryan AN (1993) Inhibition of tumor formation process in dicotyledonous plants by *Azospirillum brasilense* strains. Mikrobiologiya 62(Suppl 3):515–523

Bardin SD, Huang HC, Liu L, Yanke LJ (2003) Control of *Pythium* damping off of canola, safflower, dry pea and sugar beet by microbial seed treatment. Can J Plant Pathol 25:268–275

Bardin SD, Huang HC, Pinto J, Amundsen EJ, Erickson RS (2004) Biological control of *Pythium* damping-off of pea and sugar beet by *Rhizobium leguminosarum* bv. *viceae*. Can J Bot 82:291–296

Bashan Y, de-Bashan LE (2002a) Reduction of bacterial speck (*Pseudomonas syringae* pv. *tomato*) of tomato by combined treatments of plant growth-promoting bacterium, *Azospirillum brasilense*, streptomycin sulfate, and chemothermal seed treatment. Eur J Plant Pathol 108:821–829

Bashan Y, de-Bashan LE (2002b) Protection of tomato seedlings against infection by *Pseudomonas syringae* pv. *tomato* by using the plant growth-promoting bacterium *Azospirillum brasilense. Appl. Environ.* Microbiology 68(Suppl 6):2637–2643

Beattie GA (2006) Plant-associated bacteria: survey, molecular phylogeny, genomics and recent advances. In: Gnanamanickam SS (ed) Plant-associated bacteria. Springer, Dordrecht, pp 1–56

Bowen GD, Rovira AD (1999) The rhizosphere and its management to improve plant growth. Adv Agron 66:1–102

Buonassisi AJ, Copeman RJ, Pepin HS, Eaton GW (1986) Effect of *Rhizobium* spp. on *Fusarium solani* f. sp. *phaseoli*. Can J Phytopathol 8:140–146

Cao L, Qiu Z, Dai X, Tan H, Lin Y, Zhou S (2004) Isolation of endophytic actinomycetes from roots and leaves of banana (*Musa acuminata*) plants and their activities against *Fusarium oxysporum* f. sp. *cubense*. World J Microbiol Biotechnol 20:501–504

Chahal PPK, Chahal VPS (1988) Biological control of root-knot nematode on brinjal (*Solanum melongena* L.) with *Azotobacter chroococcum*. In: Maqbool MA, Golden AM, Gaffar A, Krusberg LQ (eds) Advances in plant nematology. National Nematological Research Centre, Univ. of Karachi, Karachi, pp 257–263

Chahal VPS, Chahal PPK (2003) *Bacillus thuringiensis* for the control of *Meloidogyne incognita*. In: Trivedi PC (ed) Nematode management in plants. Scientific Publishers (India), Jodhpur, pp 251–257

Chakraborty U, Chakraborty B, Basnet M (2006) Plant growth promotion and induction of resistance in *Camellia sinensis* by *Bacillus megaterium*. J Basic Microbiol 46(Suppl 3):186–195

Charest MH, Beauchamp CJ, Antoun H (2005) Effects of the humic substances of de-inking paper sludge on the antagonism between two compost bacteria and *Pythium ultimum*. FEMS Microbiol Ecol 52:219–227

Cook RJ (2002) Advances in plant health management in the twentieth century. Annu Rev Phytopathol 38:95–116

Ehteshamul-Haque S, Ghaffar A (1993) Use of rhizobia in the control of root rot diseases of sunflower, okra, soybean and mungbean. J Phytopathol 138:157–163

Elliot LF, Lynch JM (1995) The international workshop on establishment of microbial inocula in soils: cooperative research project on biological resource management of the Organization for Economic Cooperation and Development (OECD). Am J Altern Agric 10:50–73

Franco-Correa M, Quintana A, Duque C et al (2010) Evaluation of actinomycete strains for key traits related with plant growth promotion and mycorrhiza helping activities. Appl Soil Ecol 45:209–217

Glick BR (1995) The enhancement of plant growth by free-living bacteria. Can J Microbiol 41:109–117

Glick BR, Patten CL, Holguin G, Penrose DM (1999) Biochemical and genetic mechanisms used by plant growth-promoting bacteria. Imperial College Press, London

Gray EJ, Smith DL (2005) Intracellular and extracellular PGPR: commonalities and distinctions in the plant-bacterium signaling processes. Soil Biol Biochem 37:395–412

Hernández-Castillo FD, Carvajal CR, Guerrero E, Sánchez A, Gallegos G, Lira-Saldivar RH (2005) Susceptibilidad a fungicidas de grupos de anastomosis del hongo Rhizoctonia solani Khün colectados en zonas paperas de Chihuahua. Méx Int J Expt Bot 74:259–269

Hoffland E, Pieterse C, Bik L, van Pelt JA (1995) Induced systemic resistance in radish is not associated with accumulation of pathogenesis-related proteins. Physiol Mol Plant Pathol 46:309–320

Holguin G, Bashan Y (1996) Nitrogen-fixation by Azospirillum brasilense Cd is promoted when co-cultured with a mangrove rhizosphere bacterium (Staphylococcus sp). Soil Biol Biochem 28(Suppl 12):1651-1660

Jeyarajan R, Nakkeeran S (2000) Exploitation of microorganisms and viruses as biocontrol agents for crop disease management. In: Upadhyay RK et al (eds) Biocontrol potential and their exploitation in sustainable agriculture. Kluwer Academic/Plenum Publishers, New York, pp 95–116

Kloepper JW (1993) Plant growth promoting rhizobacteria as biological control agents. In: Metting FB Jr (ed) Soil microbial ecology- applications in agricultural and environmental management. Marcel Dekker, New York, pp 255–274

Kloepper JW, Schroth MN (1978) Plant growth-promoting rhizobacteria on radishes. In: Proc. of the 4th Int. Conf. on Plant Pathogenic Bacteria, vol 2, Station de Pathologie Vegetale et Phytobacteriologie, INRA, Angers, pp 879–882

Kloepper JW, Schroth MN (1981) Plant growth promoting rhizobacteria and plant growth under gnotobiotic conditions. Phytopathology 71:642–644

Lemanceau P, Alabouvette C (1993) Suppression of Fusarium wilts by fluorescent *Pseudomonas*: mechanisms and applications. Biocontrol Sci Tech 3:219–234

Lynch JM, Whipps JM (1991) Substrate flow in the rhizosphere. In: Keister DL, Cregan B (eds) The rhizosphere and plant growth, vol 14, Beltsville symposium. in agricultural research. Kluwer, Dordrecht, pp 15–24

Mayak S, Tirosh T, Glick BR (2004) Plant growth-promoting bacteria confer resistance in tomato plants to salt stress. Plant Physiol Biochem 42:565–572

Mazzola M (2002) Mechanisms of natural soil suppressiveness to soil-borne diseases. Antonie Van Leeuwenhoek 81:557–564

Merzaeva OV, Shirokikh IG (2006) Colonization of plant rhizosphere by actinomycetes of different genera. Microbiology 75:226–230

Oliveira RGB, Drozdowicz A (1987) Inhibition of bacteriocin producing strains of *Azospirillum lipoferum* by their own bacteriocin. Zentralbl Mikrobiol 142:387–391

Parra G, Ristaino J (2001) Resistance to Mefenoxam and Metalaxyl among field isolates of Phytophthora capsici causing Phytophthora blight of bell pepper. Plant Dis 85:1069–1075

Paulitz TC, Belanger RB (2001) Biological control in greenhouse systems. Annu Rev Phytopathol 39:103–133

Raaijmakers JM, Vlami M, de Souza JT (2002) Antibiotic production by bacterial biocontrol agents. Antonie Van Leeuwenhoek 81:537–547

Reitz M, Rudolph K, Schröder I, Hoffmann-Hergarten S, Hallmann J, Sikora RA (2000) Lipopolysaccharides of Rhizobium etli strain G12 act in potato roots as an inducing agent of systemic resistance to infection by the cyst nematode Globodera pallida. Appl Environ Microbiol 66:3515–3518

Ristaino JB (1991) Influence of rainfall, drip irrigation, and inoculum density on the development of Phytophthora root and crown rot epidemics and yield in bell pepper. Phytopathology 81:922–929

Sabaratnam S, Traquair JA (2002) Formulation of a *Streptomyces* biocontrol agent for the suppression of *Rhizoctonia* damping-off in tomato transplants. Biol Control 23:245–253

Shah S, Karkhanis V, Desai A (1992) Isolation and characterization of siderophore, with antimicrobial activity, from *Azospirillum lipoferum* M. Curr Microbiol 25(Suppl 6):347–351

Smith RS (1997) New inoculant technology to meet changing legume management. In: Elmerich C, Kondorosi A, Newton WE (eds) Biological nitrogen fixation for the 21st century. Kluwer, Dordrecht, pp 621–622

Somers E, Ptacek D, Gysegom P, Srinivasan M, Vanderleyden J (2005) *Azospirillum brasilense* produces the auxin-like phenylacetic acid by using the key enzyme for indole-3-acetic acid biosynthesis. Appl Environ Microbiol 71(Suppl 4):1803–1810

Sudhakar P, Gangwar SK, Satpathy B, Sahu PK, Ghosh JK, Saratchandra B (2000) Evaluation of some nitrogen fixing bacteria for control of foliar diseases of mulberry (*Morus alba*). Indian J Seric 39(Suppl 1):9–11

Taechowisan T, Peberdy JF, Lumyong S (2003) Isolation of endophytic actinomycetes from selected plants and their antifungal activity. World J Microbiol Biotechnol 19:381–385

Vessey JK (2003) Plant growth promoting rhizobacteria as biofertilizers. Plant Soil 255:571–586

Weller DM (1988) Biological control of soil borne plant pathogens in the rhizosphere with bacteria. Annu Rev Phytopathol 26:379–407

Weller DM, Thomashow LS (1993) Use of rhizobacteria for biocontrol. Curr Opin Biotechnol 4:306–311

Weller DM, Raaijmakers JM, McSpadden Gardener BB, Thomashow LS (2002) Microbial populations responsible for specific soil suppressiveness to plant pathogens. Annu Rev Phytopathol 40:309–348

Whipps JM (2001) Microbial interactions and biocontrol in the rhizosphere. J Exp Bot 52:487–511

2 Potential Role of PGPR in Agriculture

PGPR strains have diverse applications in agriculture, horticulture and forestry. There are several reports in the literature indicating that PGPR could be proved as a boon in sustainable agriculture. PGPR are used as bioinoculants, biofertilizers and biocontrol agents, with practical potential in improved agriculture (Table 2.1). Their beneficial events could be biological control of diseases and pests, plant growth promotion, increasing crop yields and quality improvement that can take place simultaneously and sequentially (Fig. 2.1). Genera of PGPR include *Azotobacter*, *Azospirillum*, *Pseudomonas*, *Acetobacter*, *Burkholderia*, *Bacillus* and *Paenibacillus*, and some are members of the *Enterobacteriaceae*. Direct use of microorganisms to promote plant growth and to control plant pests continues to be an area of rapidly expanding research.

2.1 Plant Growth Promotion

PGPR strains can promote plant growth and development mainly by the following means:

- Production of phytohormones
- Biological nitrogen fixation
- Solubilization of mineral phosphates and other nutrients
- Production of siderophores that chelate iron and make it available to plant root
- PGPR as biofertilizer
- Promotion of mycorrhizal functioning

PGPR may use more than one of these mechanisms to enhance plant growth as experimental evidence suggests that the plant growth stimulation is the net result of multiple mechanisms that may be activated simultaneously.

2.1.1 Production of Phytohormones

PGPR produce phytohormones that are believed to be related to their ability to stimulate plant growth. Hormones are organic compounds that are effective at very low concentration; they are usually synthesized in one part of the plant and are transported to another location. They interact with specific target tissues to cause physiological responses, such as growth or fruit ripening. Botanists recognize five major groups of hormones: auxins, gibberellins, ethylene, cytokinins and abscisic acid.

IAA is a phytohormone which is known to be involved in root initiation, cell division and cell enlargement. This hormone is very commonly produced by PGPR and implicated in the growth promotion by PGPR. However, the effect of IAA on plants depends on the plant sensitivity to IAA and the amount of IAA produced from plant-associated bacteria and induction of other phytohormones. The bacterial IAA from *Pseudomonas putida* played a major role in the development of host plant root system. Most commonly, IAA-producing PGPR are believed to increase root growth and root length, resulting in greater root

P.P. Reddy, *Plant Growth Promoting Rhizobacteria for Horticultural Crop Protection*, DOI 10.1007/978-81-322-1973-6_2, © Springer India 2014

Table 2.1 Forms of PGPR and their mechanism of action stimulating plant growth

PGPR forms	Definition	Mechanism of action
Biofertilizer	A substance that contains live microorganisms which, when applied on the seed, plant surface or soil, colonizes the rhizosphere and promote plant growth through increased supply of primary nutrients for the host plant	Biological nitrogen fixation. Utilization of insoluble phosphorus
Phytostimulator	Microorganism, with the ability to produce phytohormones such as indole acetic acid, gibberellic acid, cytokinins and ethylene	Production of phytohormones
Biopesticide	Microorganisms that promote plant growth by controlling phytopathogenic agents	Production of antibiotics, siderophores, HCN
		Production of hydrolytic enzymes
		Acquired and induced systemic resistance

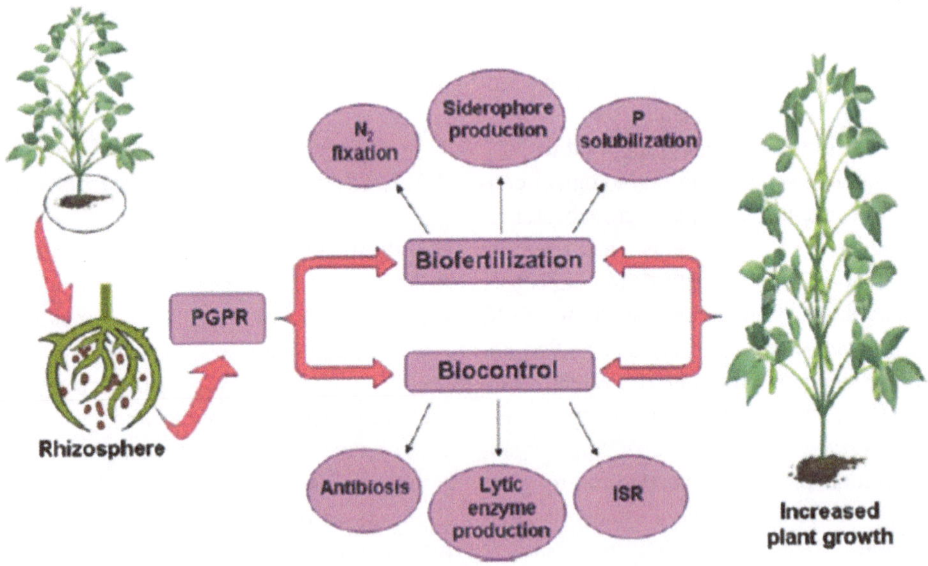

Fig. 2.1 Schematic illustration of important mechanisms (biofertilization and biocontrol of pathogens) known for plant growth promotion by PGPR

surface area which enables the plant to access more nutrients from soil.

Cytokinins are a class of phytohormones which are known to promote cell divisions, cell enlargement, and tissue expansion in certain plant parts (Salisbury 1994). Cytokinin is produced by *Pseudomonas fluorescens* isolated from the rhizosphere of the soybean (De Salamone et al. 2001).

Gibberellins are a class of phytohormones most commonly associated with modifying plant morphology by the extension of plant tissue, particularly stem tissue (Salisbury 1994). Evidence of GA production by PGPR is rare; however, Gutierrez-Manero et al. (2001) provide evidence that four different forms of GA are produced by *Bacillus pumilus* and *B. licheniformis*.

Ethylene is the only gaseous phytohormone. It is also known as the wounding hormone because its production in the plant can be induced by physical or chemical perturbation of plant tissues (Salisbury 1994). Glick and Pasternak (2003) put forward the theory that the mode of action of

Table 2.2 Examples of different phytohormone-producing PGPR

Phytohormones	PGPR
Indole-3-acetic acid (IAA)	*Acetobacter diazotrophicus*, *Herbaspirillum seropedicae*, *Aeromonas veronii*, *Enterobacter cloacae*, *Azospirillum brasilense*, *Enterobacter* spp., *Agrobacterium* spp., *Alcaligenes piechaudii*, *Bradyrhizobium* spp., *Comamonas acidovorans*, *Rhizobium leguminosarum*
Cytokinin	*Paenibacillus polymyxa*, *Pseudomonas fluorescens*, *Rhizobium leguminosarum*
Zeatin and ethylene	*Azospirillum* spp.
Gibberellic acid (GA_3)	*Azospirillum lipoferum*, *Bacillus* spp.
Abscisic acid (ABA)	*Azospirillum brasilense*
ACC deaminase	*Bacillus pumilus*, *Burkholderia cepacia*

Table 2.3 Efficacy of PGPR formulations on growth promotion

Formulation	Crop	Results	Reference
Talc-based *P. fluorescens*	Potato	Significant plant growth promotion	Kloepper and Schroth (1981)
	Pigeon pea	Significant increase in grain yield	Vidhyasekaran et al. (1997)
B. subtilis strain LS213 (commercial product)	Muskmelon, watermelon	Increased plant growth and improved yield	Vavrina (1999)
Vermiculite and kaolin based *B. subtilis*	Lettuce	Increased fresh weight	Amer and Utkhede (2000)
Chitin based formulation of *B. subtilis* strain GBO3 + *B. pumilus* strain INR7 (LS256) and *B. subtilis* strain GBO3 + *B. subtilis* strain IN937b	Tomato, pepper	Increased yield of pepper and tomato	Burelle et al. (2002)
Talc-based formulation of *B. subtilis* and *P. chlororaphis* (PA23)	Tomato	Increased growth promotion	Kavitha et al. (2003)
Talc-based *P. fluorescens* FP7 supplemented with chitin	Mango	Increased fruit yield and quality	Vivekananthan et al. (2004)
Talc-based *B. subtilis* (BSCBE4) and *P. chlororaphis* (PA23)	Turmeric	Significant yield increase of rhizomes	Nakkeeran et al. (2004)

some PGPR was the production of 1-carboxylate deaminase, an enzyme which could cleave ACC, the immediate precursor to ethylene in the biosynthetic pathway for ethylene in plant. The signalling pathway that is activated in this case depends on ethylene but is independent of salicylic acid (SA) and jasmonic acid (JA) signalling (Ryu et al. 2004). It would be interesting to investigate the capacity of plant growth-promoting *Pseudomonas* spp. to produce 2, 3-butanediol and its possible involvement in ISR.

Abscisic acid (ABA), also known as abscisin II and dormin, is a plant hormone. ABA is the major hormone that controls plants' ability to survive in a harsh, changing environment. It is a signalling pathway that is conserved in all types of plants and is considered a very early adaptation to the terrestrial environment. ABA participates in the control of root growth, seed desiccation and dormancy, guard cell responses and cellular osmoprotection. ABA functions in many plant developmental processes, including bud dormancy. It can stimulate root growth in plants that need to increase their ability to extract water from the soil.

Tables 2.2 and 2.3 represent some of the efficient PGPR strains as the producers of different plant growth regulators. The enhancement in various agronomic yields due to PGPR has been reported because of the production of growth stimulating phytohormones such as indole-3-acetic acid (IAA), gibberellic acid (GA_3), zeatin, ethylene and abscisic acid (ABA).

2.1.2 Biological Nitrogen Fixation

PGPR facilitate plant growth and development directly by supply of nitrogen to plants through nitrogen fixation. Nitrogen (N_2) is one of the principal plant nutrients, becoming a limiting factor in agricultural ecosystems due to heavy losses by rainfall or mineral leaching. Nitrogen is an essential element for all forms of life – a basic requisite for synthesizing nucleic acids, proteins and other organic nitrogenous compounds. Regrettably no plant species are capable of fixing atmospheric dinitrogen into ammonia and expend it directly for its growth. Thus the plants depend on biological nitrogen fixation (BNF). PGPR can fix atmospheric N_2 either symbiotically or non-symbiotically.

2.1.2.1 Symbiotic Nitrogen Fixation

Symbiotic N_2 fixation to legume crops with the inoculation of effective PGPR is well known (Esitken et al. 2006). These PGPR live inside the plant cells and produce nodules. Examples include *Rhizobium*, *Mesorhizobium*, *Bradyrhizobium*, *Azorhizobium*, *Allorhizobium* and *Sinorhizobium* species. The process of symbiotic N_2 fixation is limited only to legume crops and various trees and shrubs that form actinorhizal roots with *Frankia*.

i) *Rhizobium* Inoculants

These help in establishing efficient symbiotic association with leguminous crops (Fig. 2.2) and thus can fix 50–100 kg N/ha. A 10–70 % increase in yield of crops due to inoculation with *Rhizobium* inoculants over uninoculated has been reported. *Rhizobium* is specific to each legume crop and only the recommended inoculant should be used for each leguminous crop such as cowpea, pigeon pea, peas, etc.

Fig. 2.2 Root nodules colonized with rhizobacteria

2.1.2.2 Non-symbiotic Nitrogen Fixation

Non-symbiotic biological N_2 fixation is basically carried out by free-living diazotrophs that belong to the genera like *Azoarcus* (Reinhold-Hurek et al. 1993), *Azospirillum* (Bashan and de-Bashan 2010), *Burkholderia* (Estrada de los Santos et al. 2001), *Gluconacetobacter* (Fuentes-Ramirez et al. 2001), *Pseudomonas* (Mirza et al. 2006), *Azotobacter*, *Arthrobacter*, *Acinetobacter*, *Bacillus*, *Enterobacter*, *Erwinia*, *Flavobacterium*, *Klebsiella*, *Acetobacter*, etc. associated with the plant rhizosphere and fix atmospheric N_2 into form which is taken up by the plants. These are free-living rhizobacteria and live outside the plant cells and do not produce nodules. These rhizobacteria are reported to fix atmospheric N_2 in soil (Riggs et al. 2001) and make it available to plants.

i) *Azotobacter* Inoculants

Azotobacter is a free-living aerobic nitrogen fixer which is recommended for non-leguminous crop like potato, tomato, mustard etc. *Azotobacter* fixes atmospheric nitrogen in soil and helps in saving chemical fertilizers by 15–20 kg N/ha. The family Azotobacteriaceae includes species *A. agilis*, *A. insignis*, *A. macrocytogenes*, *A. chroococcum*, *A. vinelandii*, *A. beijerinckii*, and *A. paspali*.

ii) *Azospirillum* Inoculants

Azospirillum fix nitrogen under microaerophilic conditions and are frequently associated with root and rhizosphere of a large number of agriculturally important non-leguminous crops. *Azospirillum* fix atmospheric N_2 in soil and help to save chemical fertilizers by 15–20 kg N/ha and include species like *A. lipoferum*, *A. brasilense*, *A. amazonense*, *A. halopraeferens*, *A. irakense*, *A. largimobile*, *A. doebereinerae*, *A. oryzae*, *A. melini* and *A. canadensis*.

Recently, Minorsky (2008) reported a PGPR strain, *Pseudomonas fluorescens* B16, exhibiting vigorous colonization in the roots of tomatoes, causing enhancement in plant height, flower number and total fruit weight. A similar investigation on rhizobia to replace the use of nitrogen fertilizer was made by Vessey (2003) and thereby demonstrated a clear picture of improvement in crop yield after the inoculation of available nutrients by rhizobacteria in agricultural soil.

Biological nitrogen fixation contributes 180×10^6 metric tons/year globally, out of which symbiotic associations contribute 80 % and the rest comes from free-living or associative systems. The use of biofertilizer and bio-enhancer such as N_2-fixing bacteria and beneficial microorganisms can reduce chemical fertilizer applications and consequently lower production cost. PGPR retain more soil organic N and other nutrients in the plant soil system, thus reducing the need for fertilizer N and P and enhancing release of the nutrients. Besides, combined inoculations of rhizobacterial species to improve the quality of soil also seemed to be a potent area of research in present-day agriculture.

2.1.3 Solubilization of Mineral Phosphates and Other Nutrients

PGPR can change the plant physiology and certain nutritional and physical properties of rhizospheric soil and indirectly influence on the colonization patterns of soil microorganisms in that particular region. Inoculation of rhizobacteria increased uptake of nutrient elements like Ca, K, Fe, Cu, Mn and Zn by plants through stimulation of proton pump ATPase (Mantelin and Touraine 2004). Reports are available on the combinations of *Bacillus* and *Microbacterium* inoculants to improve the uptake of the mineral elements by crop plants (Karlidag et al. 2007). This increase in nutrient uptake by plants might be explained through organic acid production by the plants and PGPR, decreasing the soil pH in rhizosphere. There is an ample evidence (Glass et al. 2002) for the maintenance of soil fertility by the rhizobacterial isolates to increase the availability of nutrients to plants. Solubilization of unavailable forms of nutrients is one of the essential criteria in facilitating the transport of most of these nutrients (Glick 1995).

Plant growth-promoting rhizobacteria are actively involved in the solubilization of important minerals such as phosphorous and iron, thereby enhancing the availability of these essential nutrients to plants (Glick 1995). The positive role of PGPR in stimulating the plant growth by improving solubilization (releasing siderophores or organic acid) and nutrient uptake by the plants has been well documented in the literature (Glick 1995).

2.1.3.1 Phosphorous Solubilization

Phosphorus (P) is the second most important essential plant nutrient but most of P remains fixed in soil which is not available to plants. Phosphorus is necessary for plant growth and is taken by the plants from soil as phosphate anions. Phosphate anions are highly reactive and may be trapped via precipitation with cations such as Mg_2^+, Ca_2^+, Al_3^+ and Fe_3^+ depending on the quality of the soil. Phosphorus is extremely insoluble and unavailable to plants in these forms. As a result, the amount available to plants is usually a small proportion of this total. This low availability of phosphorus to plants is because the vast majority of soil P is found in insoluble forms. Many scientists have reported the ability of different bacterial species to solubilize insoluble inorganic phosphate compounds

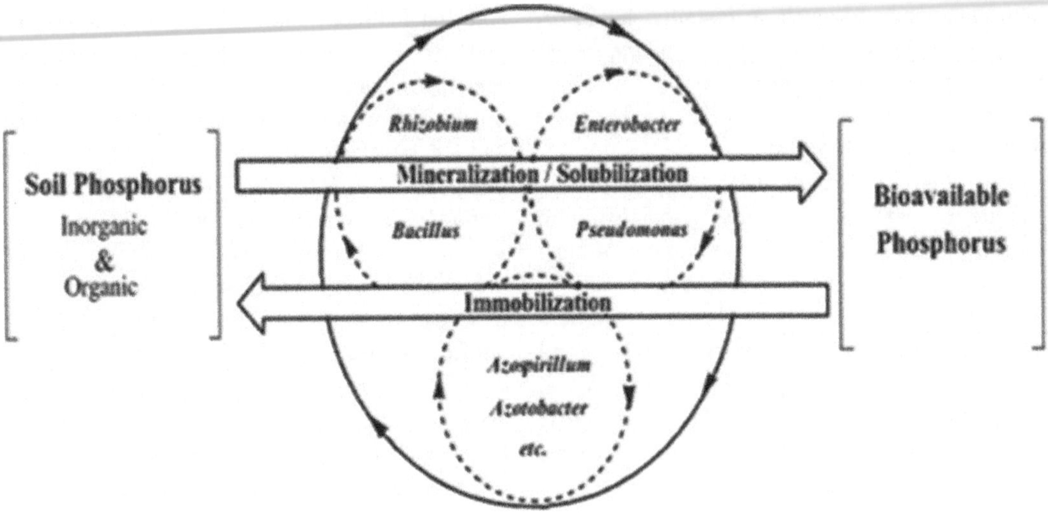

Fig. 2.3 Schematic representation of solubilization of soil phosphorus by rhizobacteria (Khan et al. 2009)

such as dicalcium phosphate, tricalcium phosphate, rock phosphate and hydroxyapatite. These bacteria solubilize phosphate through the production of acids and by some other mechanism and are termed as phosphate-solubilizing bacteria (PSB). Several researchers consequently have isolated PSB from various soils and proved that inoculations of these bacteria increased the plant growth and yield.

The plants can only absorb P in two soluble forms, the monobasic ($H_2PO_4^-$) and the dibasic (HPO_4^{2-}) ions (Glass 1989). Several phosphate-solubilizing microorganisms (PSMs) are now recorded to convert the insoluble form of phosphorus to soluble form through acidification, secretion of organic acids or protons (Richardson et al. 2009) and chelation and exchange reactions (Hameeda et al. 2008). Saprophytic bacteria and fungi are reported for the chelation-mediated mechanisms (Whitelaw 2000) to solubilize phosphate in soil. Release of plant root exudates such as organic ligands can also alter the concentration of P in soil solution (Hinsinger 2001). In certain cases, phosphate solubilization is induced by phosphate starvation (Gyaneshwar et al. 1999). A general sketch of phosphorous solubilization in soil is shown in Fig. 2.3. Inoculation with an efficient P-solubilizing microorganism improves the availability of P from insoluble form of P in soil and enhances use efficiency of phosphatic fertilizer such as super phosphate. There are a number of inoculants which can even degrade rock phosphate and soil fixed P. A number of metabolites are released by these strains which strongly affect the environment and increase nutrient availability for the plants, viz., *B. subtilis*, *B. licheniformis*, *B. megaterium* var. *phosphaticum* and *P. lutea*. Bacterial genera like *Azospirillum*, *Azotobacter*, *Bacillus*, *Beijerinckia*, *Burkholderia*, *Enterobacter*, *Erwinia*, *Flavobacterium*, *Microbacterium*, *Pseudomonas*, *Rhizobium* and *Serratia* are reported as the most significant phosphate-solubilizing bacteria (Mehnaz and Lazarovits 2006). Rhizobacteria can solubilize inorganic P sources and enhance growth and yield of crop plants. Examples of some widely reported P-solubilizing microbial species intimately associated with a large number of agricultural crops like potato, tomato, radish, pulses etc. are *Azotobacter chroococcum* (Kumar and Narula 1999), *Bacillus circulans* and *Cladosporium herbarum* (Singh and Kapoor 1999), *Bradyrhizobium*

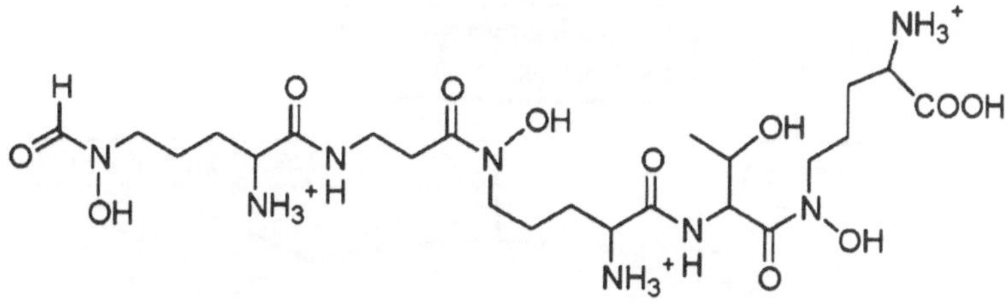

Fig. 2.4 Structure of siderophore

japonicum (Antoun et al. 1998), *Enterobacter agglomerans* (Kim et al. 1998), *Pseudomonas chlororaphis* and *P. putida* (Cattelan et al. 1999) and *Rhizobium leguminosarum* (Chabot et al. 1998). The ability of PGPR to solubilize mineral phosphate, therefore, has been of immense interest to agricultural microbiologists since it can enhance the availability of phosphorus for effective plant growth. PGPR have been recorded to solubilize precipitated phosphates to plants, representing a possible mechanism of plant growth promotion under field conditions (Verma et al. 2001). Synthesis of organic acids by rhizosphere microorganisms could be the possible reason for solubilization of inorganic P sources.

2.1.3.2 Manganese Oxidation

The availability of manganese (Mn) in the rhizosphere is affected by two major factors: redox condition and pH. In oxidized soils manganese is present in its oxidized form, Mn^{4+}, in the low soluble mineral pyrolusite. Some rhizosphere bacteria (*Bacillus*, *Pseudomonas* and *Geobacter*) can reduce oxidized Mn^{4+} to Mn^{2+}, which is the chemical form that is metabolically useful for plants.

The reaction is as follows:

$$MnO_2 + 4H^+ + 2e^- \leftrightarrow Mn^{2+} + 2H_2O. \quad (2.1)$$

In this reaction two points are important; the reduction of Mn requires electrons and protons. Electrons are supplied by the decomposition of carbonaceous compounds, and protons can be supplied by the proton excretion system of root cells (Marschner 1997). Consequently, the activity of Mn reducers is highly favoured in the rhizosphere. Applications of organic matter also can favour the reduction of Mn (Hue et al. 2001). In alkaline soils where Mn usually is insoluble, the rhizosphere effect is beneficial, but in acidic soils with abundance of Mn minerals, excessive reduction of Mn can induce Mn toxicity in sensitive plants (Hue et al. 1998).

Mn plays an important role in the resistance of plants to disease. Mn, as well as Cu, are required for the synthesis of lignin, which increases the resistance of the root tissues to the penetration of pathogens; consequently Mn-deficient plants are more susceptible to the attack of plant pathogens. *Gaeumannomyces graminis*, like many other soilborne pathogenic fungi, is a powerful oxidizer of Mn that impairs the lignification of roots at infection sites (Graham 1999). Effective rhizosphere Mn reducers (e.g. *Pseudomonas* sp.) could have beneficial effects not only on plant nutrition but also on biocontrol of pathogens. In addition, roots and rhizosphere bacteria can produce chelating agents (phenolic compounds, organic acids) that form soluble complex with Mn and other elements avoiding the reprecipitation of Mn.

2.1.4 Production of Siderophores That Chelate Iron and Make It Available to Plant Root

Iron is one of the bulk minerals present in plenteous amount on Earth, yet it is unavailable in the soil for the plants. This is because Fe^{3+} (ferric ion) is the common form of iron found in nature and is meagrely soluble. To overcome this problem, PGPR secrete siderophores. Siderophores are iron-binding proteins of low molecular mass and have a high binding affinity with ferric ion (Fig. 2.4). Soil Fe is present in oxidized form (Fe^{3+}), as a component of insoluble minerals such

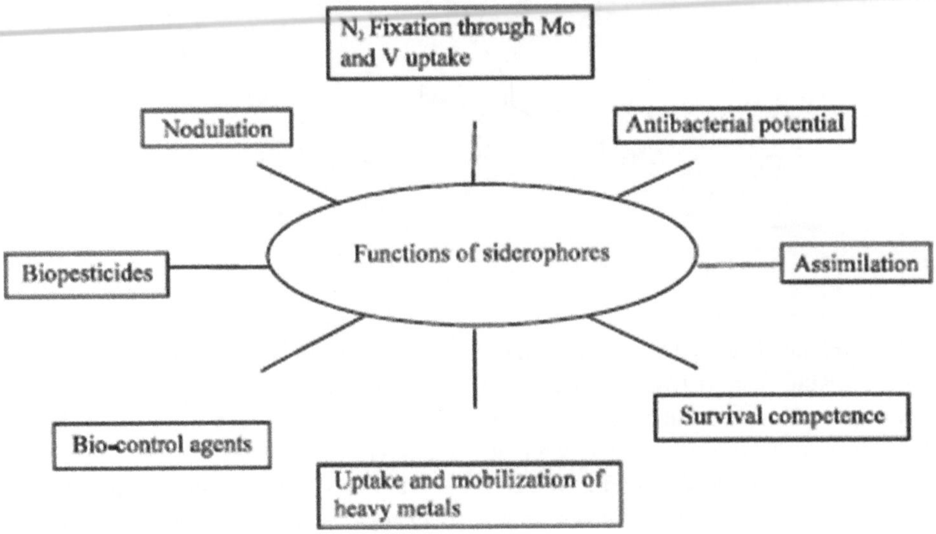

Fig. 2.5 Impact of microbially secreted siderophores on plant growth

as goethite and hematite (Fe_2O_3). Rhizosphere bacteria (*Bacillus*, *Pseudomonas*, *Geobacter*, *Alcaligenes*, *Clostridium* and *Enterobacter*) can reduce Fe^{3+} to Fe^{2+}, the form required by plants. Electrons and protons are available in the rhizosphere and consequently iron is reduced; however it can be reprecipitated. The reactions of reduction are as follows:

$$FeOOH + 3H^+ + e^- \leftrightarrow Fe^{2+} + 2H_2O(1);\ Fe_2O_3 + 6H^+ + 2\ e^- \leftrightarrow 2\ Fe^{2+} + 3H_2O\ (2). \qquad (2.2)$$

Siderophores secreted by PGPR improves plant growth and development by increasing the accessibility of iron in the soil surrounding the roots. The plants utilize siderophores secreted by PGPR for sequestering iron. Cucumber has the ability to use microbial siderophores as the sole source of iron than its own siderophores (phytosiderophores). Microbial siderophores are also reported to increase the chlorophyll content and plant biomass in plants of cucumber.

In addition to transporting iron, siderophores have other functions and effects, including enhancing pathogenicity, acting as intracellular iron storage compounds and suppressing growth of other microorganisms. Siderophores can complex other metals apart from iron, in particular the actinides (Fig. 2.5).

2.1.5 PGPR as Biofertilizer

Biofertilizers are the substances, prepared from living microorganisms which, when applied to the seeds or plant surfaces adjacent to soil, can colonize the rhizosphere or the interior parts of the plants and thereby promotes root growth. The term biofertilizer should not be used interchangeably with green manure, manure, intercrop or organic-supplemented chemical fertilizer. Interestingly some PGPR species have appeared to promote plant growth by acting both as biofertilizer and biopesticide. For instances, strains of *Burkholderia cepacia* have been observed with biocontrol characteristics to *Fusarium* spp., and it can also stimulate growth of maize under iron-poor conditions via siderophore production (Bevivino et al. 1998). *Allorhizobium*, *Azorhizobium*, *Bradyrhizobium*, *Mesorhizobium*, *Rhizobium* and *Sinorhizobium* are reported as the potent PGPR strains for their ability to act as biofertilizers (Vessey 2003). They are an important group of microorganisms used in biofertilizers. Biofertilization accounts for approximately 65 % of the nitrogen supply to crops worldwide.

The relationship between the PGPR and their host can be categorized into two basic levels of complexity: (i) rhizospheric and (ii) endophytic. In rhizospheric relationship, the PGPR can colonize the rhizosphere, the surface of the root or even the superficial intercellular spaces of plant roots (McCully 2001). It is only due to the changes in different physicochemical properties of rhizospheric soil such as soil pH, water potential and partial pressure of O_2 and plant exudation as compared to the bulk soil that in turn can affect the ability of PGPR strains to colonize the rhizosphere (Griffiths et al. 1999). In endophytic relationship, PGPR resides within the apoplastic spaces inside the host plants. There is a direct evidence of existence of endophytes in the apoplastic intercellular spaces of parenchyma tissue (Dong et al. 1997) and xylem vessel (James et al. 2001). Best examples can be cited from legume–rhizobia symbioses in leguminous plants (Vessey 2003). Thus, the means by which the PGPR enhance nutrient status of host plants and thereby act as biofertilizers can be categorized into five distinct areas such as biological N_2 fixation, increasing the availability of nutrients in the rhizosphere, increasing the root surface area, enhancing beneficial symbioses of the host and finally the combinations of all the above modes of action. However, the degree of intimacy between the PGPR and host plant can vary depending on where and how the PGPR colonizes the plant.

Plant growth-promoting rhizobacteria are beneficial soil bacteria that colonize plant roots and enhance plant growth promotion activity by different mechanisms in various ways. There is an ample evidence in the literature indicating that PGPR can be a best alternative to chemical fertilizer for sustainable and eco-friendly agriculture. They will not only provide nutrients to the plants (direct plant growth promotion) and protect plants against the phytopathogens (indirect plant growth promotion) but also increase the soil fertility. Thus, awareness must prevail among the farmers about the negative impact of chemical fertilizers and positive aspects of PGPR as biofertilizer.

2.1.6 Promotion of Mycorrhizal Functioning

Interactions of arbuscular mycorrhizal fungi (AMF) with PGPR have been described with regard to their effect on mycorrhizal development and functioning. PGPR can promote mycorrhizal functioning. Rhizobacteria showing a beneficial effect on mycorrhizae are often referred to as 'mycorrhizae-helper microorganisms'. Bacteria associated to mycorrhizal fungi adhere to fungal spores and hyphal structures and thus spread to the rhizosphere. There is a strong evidence of a vertical transmission of endobacteria through the AM fungus vegetative generation. However, antagonistic effects are often reported in the AM fungi–PGPR interactions. Positive interactions often result in plant growth improvement.

Inoculation with both free-living nitrogen-fixing bacteria such as *Azospirillum brasilense* or *Azotobacter* and AM fungi increases plant productivity. The nitrogen-fixing bacteria stimulate root colonization by AM fungi and increase their number of internal vesicles; they also alter rhizosphere rhizobial populations. It is not clear whether the enhancement of plant growth is due to free-nitrogen fixation or to the production of plant growth-promoting substances.

Some studies considered free-nitrogen fixers like other PGPR species, without reference to nitrogen fixation activity. For instance, in a study using the nitrogen fixer *A. chroococcum* and *P. fluorescens*, the chemotaxis of these two PGPR towards roots of mycorrhizal tomato plants (*Glomus fasciculatum*) was an important step of communication for root colonization. It was found that *G. fasciculatum* alters the characteristics of root exudates which are chemoattractants specific for each PGPR, amino acids for *P. fluorescens* and sugars for *A. chroococcum*. Different combinations between three PGPR species (*A. chroococcum*, *Azospirillum brasilense* and *Burkholderia cepacia*) and two AM fungi (*Glomus clarum* and *G. fasciculatum*) did not show the same trends on root colonization or on the nutritional status of onion and tomato; the highest mycorrhizal colonization was achieved by *Azospirillum brasilense* co-inoculated with each AM species on tomato and by single inoculation with *G. fasciculatum* on onion. Ecology of plant growth-promoting rhizobacteria enhanced root colonization by the AM fungus and improved soil properties such as organic matter content and total N. The use of PGPR and AM fungi has been attempted with the aim of protecting plants against pathogens. The interactions of biocontrol PGPR with AM fungi are often contradictory and probably depend on the tested bacterium, the plant species and the environmental factors.

The possibility of optimizing plant growth by managing interactions between AM fungi, PGPR and the *Rhizobium*–legume symbiosis has been considered as a promising avenue, and synergism resulting from these interactions has been demonstrated earlier. Bacterial culture of fluorescent *Pseudomonas* co-inoculated with *Glomus etunicatum* increased root growth, nodulation and N and P uptake in beans (*Phaseolus vulgaris*).

2.2 Production of ACC Deaminase and Regulation of Ethylene Level in Plants

Although ethylene is essential for normal growth and development in plants, at high concentration it can be harmful as it induces defoliation and other cellular processes that may lead to reduced crop performance. Using their 1-aminocyclopropane-1-carboxylic acid (ACC) deaminase activity, PGPR can divert ACC from the ethylene biosynthesis pathway in the root system of *Arabidopsis thaliana* plant (Fig. 2.6) (Desbrosses et al. 2009). Thus, rhizobacteria assist in diminishing the accumulation of ethylene levels and re-establish a healthy root system needed to cope with environmental stress. The primary mechanism includes the destruction of ethylene via enzyme ACC deaminase. The rhizosphere bacteria such as *Achromobacter*, *Azospirillum*, *Bacillus*, *Enterobacter*, *Pseudomonas* and *Rhizobium* exhibit ACC deaminase activity

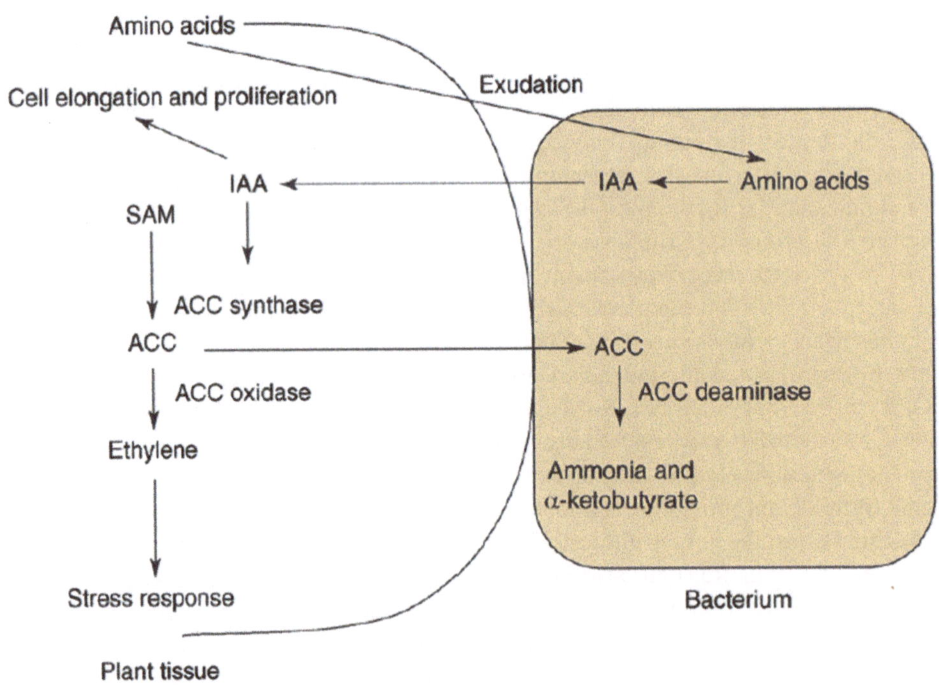

Fig. 2.6 Schematic model to explain how ACC deaminase containing PGPR lowers the ethylene concentration, thereby preventing ethylene-caused inhibition of root elongation (Glick et al. 1999). Key: *IAA* Indole acetic acid, *ACC* 1-aminocyclopropane-1-carboxylic acid, *SAM* S-adenosyl methionine

(Duan et al. 2009). Most of the studies have demonstrated the production of ACC deaminase gene in the plants treated with PGPR under environmental stress. Ghosh et al. (2003) recorded ACC deaminase activity in three *Bacillus* species, namely, *B. circulans* DUC1, *B. firmus* DUC2 and *B. globisporus* DUC3 that stimulated root elongation in *Brassica campestris*. Mayak et al. (2004) observed tomato plants inoculated with the bacterium *Achromobacter piechaudii* under water and saline stress conditions and reported a significant increase in fresh and dry weight of inoculated plants.

PGPR containing ACC deaminase can boost the plant growth particularly under stressed environmental conditions like salinity, drought, waterlogging, temperature, pathogenicity and contaminants in response to a multitude of abiotic and biotic stresses (Saleem et al. 2007). Although efforts have been made to introduce ACC deaminase genes into plants for optimum growth, the genetic modifications for all the plant species are not yet possible due to many handicaps like proprietary rights and international trade agreements on genetically modified (GM) crops and limitations of recombinant DNA technology.

2.3 Production of Volatile Organic Compounds

The discovery of volatile organic compounds (VOCs) produced by rhizobacteria constitute an important mechanism for the elicitation of plant growth by rhizobacteria. Ryu et al. (2003) recorded some PGPR strains namely *Bacillus subtilis* GBO3, *B. amyloliquefaciens* IN937a and *Enterobacter cloacae* JM22 that released a blend of volatile components, particularly, 2,3-butanediol and acetoin, which promoted growth of *Arabidopsis thaliana*, suggesting that synthesis of bioactive VOCs is a strain-specific phenomenon. Acetoin-forming enzymes have been identified earlier (Forlani et al. 1999) in carrot, although their possible functions in plants were not properly established in that period. It has now been established that the VOCs produced by the rhizobacterial strains can act as signalling molecule to mediate plant–microbe interactions as volatiles produced by PGPR colonizing roots are generated at sufficient concentrations to trigger the plant responses (Ryu et al. 2003). Farmer (2001) identified low molecular weight plant volatiles such as terpenes, jasmonates and green leaf components as potent signal molecules for living organisms in different trophic levels. However, to acquire a clear appreciation on the mechanisms of VOCs in signalling plants to register plant defence, more investigations into the volatile components in the plant–rhizobacteria system should follow.

2.4 Rhizosphere Engineering

Rhizosphere microbial populations are tremendously affected by the interactions between the plants and the soil environment. Rhizosphere engineering involves the selection of beneficial microbial populations by plant rhizosphere. For instances, some crop species or cultivars select populations of antibiotic-producing strains that play a major role in soils, naturally suppressive to soilborne fungal pathogens (Ryan et al. 2009). Persistent organic pollutants such as polychlorinated biphenyls (PCBs) are a global problem. Using root-associated microbes in rhizospheric engineering approach, the levels of PCBs can be successfully depleted as these microbes can use plant secondary metabolites such as phenylpropanoids (Narasimhan et al. 2003). Similar technology has been developed by Lugtenberg et al. (2001) during their investigation on the growth of microbes with the ability to metabolize exotic nutrients exuded by plants. One of the earliest successes of this technology was based on the favourable partitioning of the exotic nutrient opines, produced by the transgenic plants (Oger et al. 1997) that led to the improved and competitive growth of the metabolizing strains in comparison with the microbes unable to metabolize opines. Rhizosphere engineering ultimately reduces our reliance on agrochemicals by replacing their functions with beneficial microbes, biodegradable biostimulants or transgenic plants (Ryan et al. 2009). It is now possible to create a nutritional bias that may be especially successful in identifying microbial populations due to the general nutrient-limiting conditions in the rhizosphere.

Molecular microbiological advances have been exploited in order to achieve a complete knowledge of the complex chemical and biological interactions that generally occur in the rhizosphere, ensuring that the strategies to engineer the rhizosphere are safe and eco-friendly to agricultural systems. For example, plants are genetically engineered to modify the rhizosphere pH to release the compounds that could improve nutrient availability, protect plants against biotic and abiotic stresses or encourage the proliferation of beneficial microorganisms. Growth stimulation can be mediated directly through enhanced nutrient acquisition or modulation of phytohormone synthesis, while indirect stimulation involves the induction of plant antagonism (Ryan et al. 2009). Sundheim et al. (1988) observed that an engineered strain of *Pseudomonas* expressing chitinase gene from *Serratia marcescens* more effectively controlled *Fusarium oxysporum* f. sp. *redolens* and *Gaeumannomyces graminis* var. *tritici* in vitro. Recently, experiments targeting on the DAPG-producing PGPR strain *Pseudomonas fluorescens* (phlD?) have demonstrated that plant species can differentially enrich and support different microbial populations (De La Fuente et al. 2006) and genotypes (Landa et al. 2006) in the rhizosphere. Notz et al. (2001) significantly correlated DAPG accumulation by *Pseudomonas fluorescens* CHAO with the expression of DAPG biosynthesis gene phlA and observed that the expression was significantly greater in the rhizosphere of monocots than dicots. Although the exact mechanism is not totally understood, Di Gregorio et al. (2006) noticed a combined application of Triton X-100 and *Sinorhizobium* sp. Pb002 inoculums for the improvement of lead phytoextraction by *Brassica juncea* in EDTA-amended soil.

2.5 PGPR as Biotic Elicitors

Elicitors are chemicals or biofactors of various sources that can trigger physiological and morphological responses and phytoalexin accumulation in plants. It may be abiotic elicitors such as metal ions or inorganic compounds and biotic elicitors, basically derived from fungi, bacteria, viruses, plant cell wall components and chemicals that are released due to antagonistic reaction of plants against phytopathogens or herbivore attack. It has now been observed that the treatment of plants with biotic elicitors can cause an array of defence reactions including the accumulation of a range of plant defensive bioactive molecules such as phytoalexins in the intact plants. Thus, elicitation is being used to induce the expression of genes responsible for the synthesis of antimicrobial metabolites. Rhizosphere microbes are best known to act as biotic elicitors, which can induce the synthesis of secondary products in plants (Sekar and Kandavel 2010). Signal perception is the first committed step towards the biotic elicitor signal transduction pathway in plants. Jasmonic acid and its methyl ester are the signal transducers in a wide range of plant cell cultures that could accumulate rapidly when the suspension cultures of *Rauvolfia canescens* and *Eschscholtzia californica* are treated with a yeast elicitor (Roberts and Shuler 1997). Some of the well-reported PGPR as biotic elicitors have been exemplified in Table 2.4.

Table 2.4 PGPR species as biotic elicitors to elicit plant response

PGPR species	Plant	Metabolite induced	References
Pseudomonas fluorescens	*Catharanthus roseus*	Ajmalicine	Jaleel et al. (2007)
		Serpentine	Jaleel et al. (2009)
P. putida and *P. fluorescens*	*Hyoscyamus niger*	Hyoscyamine and scopolamine	Ghorbanpour et al. (2010)
Bacillus subtilis	*Crocus sativus*	Picrocrocin, crocetin and safranal compounds	Sharaf-Eldin et al. (2008)
B. cereus	*Salvia miltiorrhiza*	Tanshinone	Zhao et al. (2010)

Ajmalicine, serpentine, picrocrocin, crocetin, hyoscyamine and scopolamine, safranal compounds and tanshinone are recorded as the important metabolites produced by PGPR species in eliciting the physiological and morphological responses in crop plants.

2.6 Antagonistic Activity of PGPR

Rhizobacteria can suppress the growth of various phytopathogens in a variety of ways like competing for nutrients and space, limiting available Fe supply through producing siderophores and producing lytic enzymes and antibiosis (Jing et al. 2007). Among PGPR, fluorescent pseudomonads are widely reported for their broad-spectrum antagonistic activity against a number of phytopathogens. Deliveries of microbial antagonists with urban and agricultural wastes are believed to be the most effective means in suppressing root pathogens of avocado and citrus (Sultana et al. 2006). Recently, different PGPR strains of *Rhizobium meliloti* have been reported to produce siderophores (Arora et al. 2001) in iron stress conditions and thereby added an advantage to exclude the pathogen *Macrophomina phaseolina*. Application of *Pseudomonas aureoginosa* in combination with common medicinal plant *Launaea nudicaulis* also holds good promises for effective control of root-infecting fungi (Mansoor et al. 2007).

Competition for nutrients, niche exclusion, induced systemic resistance and production of antifungal metabolites (AFMs) is the probable means responsible for biocontrol activity of PGPR (Bloemberg and Lugtenberg 2001). Most of the PGPR are recorded to produce AFMs, of which phenazines, pyrrolnitrin, 2,4-diacetylphloroglucinol (DAPG), pyoluteorin, viscosinamide and tensin are the frequently detected classes. Among PGPR, *Pseudomonas* is the best-characterized biocontrol agent at the molecular level. *P. fluorescens* strain WCS374 has been recorded to suppress *Fusarium* wilt in radish leading to an average increase of 40 % in yield (Bakker et al. 2007). The individual genes such as phzO and phzH responsible for the presence of functional group on phenazine compound have been detected (Chin-A-Woeng et al. 2001). More recently, information has been generated on the biosynthesis of pyoluteorin in *P. fluorescens* Pf5 and 2, 4-diacetylphloroglucinol in *P. fluorescens* Q2-87 (Kidarsa et al. 2011). Biocontrol activity of *Streptomyces* spp. is reported by Kumar et al. (2009) indicating the tremendous potentiality of PGPR as an alternative in controlling plant diseases in agriculture than that of conventional fungicides. *Azospirillum*, *Azotobacter*, *Bacillus*, *Enterobacter*, *Paenibacillus*, *Pseudomonas* and *Streptomyces* are recorded as the potent genera of rhizobacteria acting against the pathogens like tomato mottle virus, *Rhizoctonia bataticola*, *Myzus persicae*, *Acyrthosiphon kondoi* and *Fusarium oxysporum*. Experiments on the dual effect of PGPR and AM fungi on *Fusarium oxysporum* f. sp. *melongenae* causing brinjal wilt have been made by Kalita et al. (2009). PGPR strains such as *Azotobacter* sp., *Azospirillum* sp. and *P. fluorescens* are recorded as the most promising microbes to suppress the wilt disease of brinjal in vitro. The rhizobaceria when inoculated as mixtures exhibited maximum efficiency in the suppression of diseases along with increase in chlorophyll content, total number of leaves and shoot height, thereby increasing overall crop yield than when inoculated singly. However, the application of these PGPR strains did not affect populations of beneficial indigenous rhizosphere bacteria including the fluorescent pseudomonads and the siderophore-producing bacterial strains.

2.7 Rhizoremediation

The application of PGPR in rhizoremediation technologies is now being considered as effective, since inoculation of PGPR strains could aid remarkable enhancement in plant growth and development on contaminated agroclimatic conditions. Rhizobacteria can directly assist rhizoremediation by producing IAA, biological nitrogen fixation, solubilizing P and secreting siderophores (Denton 2007). PGPR strains, pseudomonads and *Acinetobacter* enhance uptake of Fe, Zn, Mg, Ca, K and P by crop plants (Esitken et al. 2006).

PGPR along with AM fungi are now being utilized in the nutrient-poor agricultural soils to increase the solubility of heavy metals, thereby increasing the chances of success in rhizoremediation. Investigations on the application of PGPR strains in decreasing the bioavailability of toxicity resulting in better growth and development in heavy metal contaminated soils through recycling of nutrients, maintaining soil structure, detoxifying chemicals and controlling pests are also well studied (Denton 2007). Studies on certain rhizobacteria in Ni uptake by *Alyssum murale* indicated that this group of bacteria can release the metal from its non-soluble phase by decreasing the pH of the environment (Zhuang et al. 2007). *Azotobacter* spp. facilitate the mobility of heavy metals in the soil and thus enhance bioremediation of soil from heavy metals, such as cadmium, mercury and lead.

Metal contamination of soil has an important bearing on PGPR functions. Metal homeostasis resistance in bacteria is often maintained by sequestration, active efflux, reduced uptake, detoxification and synthesis of binding protein (Choudhary and Shrivastava 2001). In some cases, a few mechanisms may also co-exist. Strain Psd was able to resist Cd, Al and Zn and thus could be able to survive for carrying out its PGPR functions in soil containing high concentrations of these metal ions. Phosphate solubilization by strain Psd is another important property as nonavailability of phosphate can be limiting for plants. Strain Psd could solubilize mineral source of complex phosphate as well as release phosphate from organic sources via two phosphatase enzymes. Mineral phosphate solubilization in bacteria occurs by production of organic acids, and organic phosphate release is aided by acid and alkaline phosphatases (Rodriguez and Fraga 1999). The complete genome sequence analysis of *P. fluorescens* Pf5 and detailed molecular genetic analysis of *P. fluorescens* CHAO have firmly established the biocontrol capabilities and its regulation (Zuber et al. 2003). Phosphate-solubilizing bacteria are common in the rhizosphere (Nautiyal et al. 2000), and secretion of organic acids and phosphatase is the common method of facilitating the conversion of insoluble forms of P to plant available forms (Kim et al. 1998). The solubilization of P in the rhizosphere is the most common mode of action implicated in PGPR that increases nutrient availability to host plants (Richardson 2001). More importantly, increases in root length and root surface area are sometimes reported (Holguin and Glick 2001).

2.8 Resistance to Water Stress

Drought stress causes limitation to the plant growth and productivity of agricultural crops particularly in arid and semi-arid areas. Inoculation of plants with PGPR enhances the production of IAA, cytokinins, antioxidants and ACC deaminase. PGPR are reported as beneficial to the plants like tomatoes and peppers growing on water-deficient soils for conferring resistance to water stress conditions (Aroca and Ruiz-Lozano 2009). More investigations into the mechanisms by which PGPR elicit tolerance to specific stress factors would improve our knowledge on the use of these rhizobacteria in agriculture to provide induced systemic tolerance to water stress.

2.9 Salt Tolerance

Salinity is a major abiotic stress reducing the yield of a wide variety of crops all over the world. Globally, more than 770,000 km^2 of land is salt affected by secondary salinization: 20 % of irrigated land and about 2 % of dry agricultural land. The restriction of plant growth and productivity due to salinity is especially acute in arid and semiarid regions around the world.

PGPR (*Staphylococcus kloosii* EY37 and *Kocuria erythromyxa* EY43) inoculated onto seeds improves emergence performance, growth and nutrient uptake and could induce tolerance to salt stress in radish (Yildirim et al. 2008). Similar findings were reported in the previous studies showing that application of PGPR (*Azospirillum* spp., *Achromobacter* spp., *Serratia* spp., *Rhizobium* spp., *Aeromonas* spp. and *Bacillus* spp.) may stimulate yield, growth and PNE uptake from soil in different crops such as tomatoes (Mayak et al. 2004) and lettuce (Barassi et al. 2006) under salinity conditions.

References

Amer GA, Utkhede RS (2000) Development of formulations of biological agents for management of root rot of lettuce and cucumber. Can J Microbiol 46:809–816

Antoun H, Beauchamp CJ, Goussard N, Chabot R, Lalande R (1998) Potential of *Rhizobium* and *Bradyrhizobium* species as plant growth promoting rhizobacteria on nonlegumes: effect on radishes (*Raphanus sativus* L.). Plant Soil 204:57–67

Aroca R, Ruiz-Lozano JM (2009) Induction of plant tolerance to semi-arid environments by beneficial soil microorganisms-a review. In: Lichtouse E (ed) Climate change, intercropping, pest control and beneficial microorganisms, sustainable agriculture reviews, vol 2. Springer, Dordrecht, pp 121–135

Arora NK, Kang SC, Maheshwari DK (2001) Isolation of siderophore-producing strains of *Rhizobium meliloti* and their biocontrol potential against *Macrophomina phaseolina* that causes charcoal rot of groundnut. Curr Sci 81:673–677

Bakker PAHM, Pieterse CMJ, van Loon LC (2007) Induced systemic resistance by fluorescent *Pseudomonas* spp. Phytopathology 97:239–243

Barassi CA, Ayrault G, Creus CM, Sueldo RJ, Sobrero MT (2006) Seed inoculation with Azospirillum mitigates NaCl effects on lettuce. Sci Hortic 109:8–14

Bashan Y, de-Bashan LE (2010) How the plant growth-promoting bacterium *Azospirillum* promotes plant growth–a critical assessment. Adv Agron 108:77–136

Bevivino A et al (1998) Characterization of a free-living maize rhizosphere population of *Burkholderia cepacia*: effect of seed treatment on disease suppression and growth promotion of maize. FEMS Microbiol Ecol 27:225–237

Bloemberg GV, Lugtenberg BJJ (2001) Molecular basis of plant growth promotion and biocontrol by rhizobacteria. Curr Opin Plant Biol 4:343–350

Burelle K, Vavrina CS, Rosskopf EN, Shelby RA (2002) Field evaluation of plant growth promoting rhizobacteria amended transplant mixes and soil solarization for tomato and pepper production in Florida. Plant Soil 238:257–266

Cattelan AJ, Hartel PG, Fuhrmann JJ (1999) Screening for plant growth-promoting rhizobacteria to promote early soybean growth. Soil Sci Soc Am J 63:1670–1680

Chabot R, Beauchamp CJ, Kloepper JW, Antoun H (1998) Effect of phosphorus on root colonization and growth promotion of maize by bioluminescent mutants of phosphate-solubilizing *Rhizobium leguminosarum* biovar *phaseoli*. Soil Biol Biochem 30:1615–1618

Chin-A-Woeng TF, Thomas-Oates JE, Lugtenberg BJ, Bloemberg GV (2001) Introduction of the phzH gene of *Pseudomonas chlororaphis* PCL1391 extends the range of biocontrol ability of phenazine-1-carboxylic acid producing *Pseudomonas* spp. strains. Mol Plant Microbe Interact 14:1006–1015

Choudhary R, Shrivastava S (2001) Curr Sci 70:768–781

De La Fuente L, Landa BB, Weller DM (2006) Host crop affects rhizosphere colonization and competitiveness of 2, 4-diacetylphloroglucinol-producing *Pseudomonas fluorescens*. Phytopathology 96:751–762

De Salamone IEG, Hynes RK, Nelson LM (2001) Can J Miccrobiol 47:404–411

Denton B (2007) Advances in phytoremediation of heavy metals using plant growth promoting bacteria and fungi. MMG 445 Basic Biotechnol 3:1–5

Desbrosses G, Contesto C, Varoquaux F, Galland M, Touraine B (2009) PGPR-Arabidopsis interactions is a useful system to study signalling pathways involved in plant developmental control. Plant Signal Behav 4:321–323

Di Gregorio S, Barbafieri M, Lampis S, Sanangelantoni AM, Tassi E, Vallini G (2006) Combined application of Triton X-100 and *Sinorhizobium* sp. Pb002 inoculum for the improvement of lead phytoextraction by *Brassica juncea* in EDTA amended soil. Chemosphere 63:293–299

Dong Z, McCully ME, Canny MJ (1997) Does *Acetobacter diazotrophicus* live and move in the xylem of sugarcane stems? Anatomical and physiological data. Ann Bot 80:147–158

Duan J, Muller KM, Charles TC, Vesely S, Glick BR (2009) 1-Aminocyclopropane-1-carboxylate (ACC) deaminase genes in rhizobia from Southern Saskatchewan. Microb Ecol 57:423–436

Esitken A, Pirlak L, Turan M, Sahin F (2006) Effects of floral and foliar application of plant growth promoting rhizobacteria (PGPR) on yield, growth and nutrition of sweet cherry. Sci Hort 110:324–327

Estrada de los Santos P, Bustillos-Cristales MR, Caballero-Mellado J (2001) *Burkholderia*, a genus rich in plant-associated nitrogen fixers with wide environmental and geographic distribution. Appl Environ Microbiol 67:2790–2798

Farmer EE (2001) Surface-to-air signals. Nature 411:854–856

Forlani GM, Mantelli M, Nielsen E (1999) Biochemical evidence for multiple acetoin-forming enzymes in cultured plant cells. Phytochemistry 50:255–262

Fuentes-Ramirez LE, Bustillos-Cristales R, Tapia-Hernandez A et al (2001) Novel nitrogen-fixing acetic acid bacteria, *Gluconacetobacter johannae* sp. nov. and *Gluconacetobacter azotocaptans* sp. nov., associated with coffee plants. Int J Syst Evol Microbiol 51:1305–1314

Ghorbanpour MNM, Hosseini S, Rezazadeh M, Omidi KK, Etminan A (2010) Hyoscyamine and scopolamine production of black henbane (*Hyoscyamus niger*) infected with *Pseudomonas putida* and *P. fluorescens* strains under water deficit stress. Planta Med 76(12):167

Ghosh S, Penterman JN, Little RD, Chavez R, Glick BR (2003) Three newly isolated plant growth-promoting bacilli facilitate the seedling growth of canola, *Brassica campestris*. Plant Physiol Biochem 41:277–281

Glass ADM (1989) Plant nutrition: an introduction to current concepts. Jones and Bartlett Publishers, Boston, 234 pp

Glass ADM, Britto DT, Kaiser BN et al (2002) The regulation of nitrate and ammonium transport systems in plants. J Exp Bot 53:855–864

Glick BR (1995) The enhancement of plant growth by free-living bacteria. Can J Microbiol 41:109–117

Glick BR, Patten CL, Holguin G, Penrose DM (1999) Biochemical and genetic mechanisms used by plant growth-promoting bacteria. Imperial College Press, London

Glick BR, Pasternak JJ (2003) Plant growth promoting bacteria. In: Glick BR, Pasternak JJ (eds) Molecular biotechnology – principles and applications of recombinant DNA, 3rd edn. ASM Press, Washington DC, pp 436–454

Graham PH (1999) Biological dinitrogen fixation: symbiotic. In: Sylvia D, Fuhrmann J, Hartel P, Zuberer D (eds) Principles and applications of soil microbiology. Prentice Hall, Upper Saddle River, pp 322–368, 550 pp

Griffiths BS, Ritz K, Ebblewhite N, Dobson G (1999) Soil microbial community structure: effects of substrate loading rates. Soil Biol Biochem 31:145–153

Gutierrez-Manero FJ, Ramos B, Probanza A, Mehouachi J, Talon M (2001) The plant growth promoting rhizobacteria *Bacillus pumilus* and *Bacillus licheniformis* produce high amounts of physiologically active gibberellins. Plant Physiol 111:206–211

Gyaneshwar P et al (1999) Involvement of a phosphate starvation inducible glucose dehydrogenase in soil phosphate solubilization by *Enterobacter asburiae*. FEMS Microbiol Lett 171:223–229

Hameeda B, Harini G, Rupela OP, Wani SP, Reddy G (2008) Growth promotion of maize by phosphate-solubilizing bacteria isolated from composts and macrofauna. Microbiol Res 163:234–242

Hinsinger P (2001) Bioavailability of soil inorganic P in the rhizosphere as affected by root induced chemical changes: A review. Plant Soil 237:173–195

Holguin G, Glick BR (2001) Expression of the ACC deaminase gene from *Enterobacter cloacae* UW4 in *Azospirillum brasilense*. Microb Ecol 41:281–288

Hue NV, Silva J, Uehara G, Hamasaki RT, Uchida R, Bunn P (1998) Managing manganese toxicity in former sugarcane soils of Oahu. University of Hawaii, Cooperative Extension Service, Honolulu. 7 pp

Hue NV, Vega S, Silva J (2001) Manganese toxicity in a Hawaiian oxisol affected by soil pH and organic amendments. Soil Sci Soc Am J 65:153–160

Jaleel CA et al (2007) *Pseudomonas fluorescens* enhances biomass yield and ajmalicine production in *Catharanthus roseus* under water deficit stress. Colloids Surf B: Biointerfaces 60:7–11

Jaleel CA, Gopi R, Gomathinayagam M, Panneerselvam R (2009) Traditional and non-traditional plant growth regulators alter phytochemical constituents in *Catharanthus roseus*. Process Biochem 44:205–209

James EK, Olivares FL, de Oliveira ALM, dos Reis FB, da Silva LG, Reis VM (2001) Further observations on the interaction between sugarcane and *Gluconacetobacter diazotrophicus* under laboratory and greenhouse conditions. J Exp Bot 52:747–760

Jing YD, He ZL, Yang XE (2007) Role of soil rhizobacteria in phytoremediation of heavy metal contaminated soils. J Zhejiang Univ (Sci) 8:192–207

Kalita RB, Bhattacharyya PN, Jha DK (2009) Effects of plant growth promoting rhizobacteria and arbuscular mycorrhizal fungi on *Fusarium oxysporum* causing brinjal wilt. JAPS 4:29–35

Karlidag H, Esitken A, Turan M, Sahin F (2007) Effects of root inoculation of plant growth promoting rhizobacteria (PGPR) on yield, growth and nutrient element contents of leaves of apple. Sci Hort 114:16–20

Kavitha K, Nakkeeran S, Chandrasekar G, Fernando WGD, Mathiyazhagan S, Renukadevi P, Krishnamoorthy AS (2003) Role of antifungal antibiotics, siderophores and IAA production in biocontrol of *Pythium aphanidermatum* inciting damping off in tomato by *Pseudomonas chlororaphis* and *Bacillus subtilis*. In: Proceedings of the 6th international workshop on PGPR, Indian Inst of Spice Res, Calicut, pp 493–497

Khan AA, Jilani G, Akhtar MS, Naqvi SMS, Rasheed M (2009) Phosphorus solubilizing bacteria: occurrence, mechanisms and their role in crop production. J Agric Biol Sci 1:48–58

Kidarsa TA, Goebel NC, Zabriskie TM, Loper JE (2011) Phloroglucinol mediates cross-talk between the pyoluteorin and 2, 4-diacetylphloroglucinol biosynthetic pathways in *Pseudomonas fluorescens* Pf-5. Mol Microbiol 81:395–414

Kim KY, Jordan D, McDonald GA (1998) Effect of phosphate solubilizing bacteria and vesicular-arbuscular mycorrhizae on tomato growth and soil microbial activity. Biol Fertil Soils 26:79–87

Kloepper JW, Schroth MN (1981) Development of powder formulation of rhizobacteria for inoculation of potato seed pieces. Phytopathology 71:590–592

Kumar V, Narula N (1999) Solubilization of inorganic phosphates and growth emergence of wheat as affected by *Azotobacter chroococcum* mutants. Biol Fertil Soils 28:301–305

Kumar V, Haseeb A, Sharma A (2009) Integrated management of *Meloidogyne incognita* and *Fusarium solani* disease complex of chilli. Indian Phytopath 62:324–327

Landa BB, Mavrodi OV, Schroeder KL, Allende-Molar R, Weller DM (2006) Enrichment and genotypic diversity of phlD-containing fluorescent *Pseudomonas* spp., in two soils after a century of wheat and flax monoculture. FEMS Microbiol Ecol 55:351–368

Lugtenberg BJJ, Dekkers L, Bloemberg GV (2001) Molecular determinants of rhizosphere colonization by *Pseudomonas*. Ann Rev Phytopathol 38:461–490

Mansoor F, Sultana V, Haque SE (2007) Enhancement of biocontrol potential of *Pseudomonas aeruginosa* and *Paecilomyces lilacinus* against root rot of mungbean by a medicinal plant *Launaea nudicaulis* L. Pak J Bot 39:2113–2119

Mantelin S, Touraine B (2004) Plant growth-promoting bacteria and nitrate availability: impacts on root development and nitrate uptake. J Exp Bot 55:27–34

Marschner H (1997) Mineral nutrition of higher plants. Academic, London, 889 pp

Mayak S, Tirosh T, Glick BR (2004) Plant growth-promoting bacteria confer resistance in tomato plants to salt stress. Plant Physiol Biochem 42:565–572

McCully ME (2001) Niches for bacterial endophytes in crop plants: a plant biologist's view. Aust J Plant Physiol 28:983–990

Mehnaz S, Lazarovits G (2006) Inoculation effects of *Pseudomonas putida*, *Gluconacetobacter azotocaptans*, and *Azospirillum lipoferum* on corn plant growth under greenhouse conditions. Microb Ecol 51:326–335

Minorsky PV (2008) On the inside. Plant Physiol 146:323–324

Mirza MS, Mehnaz S, Normand P et al (2006) Molecular characterization and PCR detection of a nitrogen fixing *Pseudomonas* strain promoting rice growth. Biol Fertil Soils 43:163–170

Nakkeeran S, Kavitha K, Mathiyazhagan S, Fernando WGD, Chandrasekar G, Renukadevi P (2004) Induced systemic resistance and plant growth promotion by *Pseudomonas chlororaphis* strain PA-23 and *Bacillus subtilis* strain CBE4 against rhizome rot of turmeric (*Curcuma longa* L). Can J Plant Pathol 26:417–418

Narasimhan K, Basheer C, Bajic VB, Swarup S (2003) Enhancement of plant-microbe interactions using a rhizosphere metabolomics driven approach and its application in the removal of polychlorinated biphenyls. Plant Physiol 132:146–153

Nautiyal CS, Bhadauria S, Kumar P, Lal H, Mondal R, Verma D (2000) Stress induced phosphate solubilization in bacteria isolated from alkaline soils. FEMS Microbiol Lett 182:291–296

Notz R, Maurhofer M, Schnider-Keel U, Duffy B, Haas D, Defago G (2001) Biotic factors affecting expression of the 2, 4-diacetylphloroglucinol biosynthesis gene phlA in *Pseudomonas fluorescens* biocontrol strain CHAO in the rhizosphere. Phytopathology 91:873–881

Oger P, Petit A, Dessaux Y (1997) Genetically engineered plants producing opines alter their biological environment. Nat Biotechnol 15:369–372

Reinhold-Hurek B, Hurek T, Gillis M et al (1993) *Azoarcus* gen. nov., nitrogen-fixing Proteobacteria associated with roots of Kallar grass (*Leptochloa fusca* (L.) Kunth), and description of two species, *Azoarcus indigens* sp. nov. and *Azoarcus communis* sp. nov. Int J Syst Bacteriol 43:574–584

Richardson AE (2001) Prospects for using soil microorganisms to improve the acquisition of phosphorus by plants. Aust J Plant Physiol 28:897–906

Richardson AE, Barea JM, McNeill AM, Prigent-Combaret C (2009) Acquisition of phosphorus and nitrogen in the rhizosphere and plant growth promotion by microorganisms. Plant Soil 321:305–339

Riggs PJ, Chelius MK, Iniguez AL, Kaeppler SM, Triplett EW (2001) Enhanced maize productivity by inoculation with diazotrophic bacteria. Aust J Plant Physiol 28:829–836

Roberts SC, Shuler ML (1997) Large scale plant cell culture. Curr Opin Biotechnol 8:154–159

Rodriguez H, Fraga R (1999) Phosphate solubilizing bacteria and their role in plant growth promotion. Biotechnol Adv 17:319–339

Ryan PR, Dessaux Y, Thomashow LS, Weller DM (2009) Rhizosphere engineering and management for sustainable agriculture. Plant Soil 321:363–383

Ryu CM, Farag MA et al (2003) Bacterial volatiles promote growth in Arabidopsis. Proc Natl Acad Sci U S A 100:4927–4932

Ryu CM, Farag MA, Hu CH, Reddy MS, Kloepper JW, Pare PW (2004) Bacterial volatiles induce systemic resistance in Arabidopsis. Plant Physiol 134:1017–1026

Saleem M, Arshad M, Hussain S, Bhatti AS (2007) Perspective of plant growth promoting rhizobacteria (PGPR) containing ACC deaminase in stress agriculture. J Ind Microbiol Biotechnol 34:635–648

Salisbury FB (1994) The role of plant hormones. In: Wilkinson RE (ed) Plant–environment interactions. Marcel Dekker, New York, pp 39–81

Sekar S, Kandavel D (2010) Interaction of plant growth promoting rhizobacteria (PGPR) and endophytes with medicinal plants–new avenues for phytochemicals. J Phytol 2:91–100

Sharaf-Eldin M, Elkholy S, Fernandez JA et al (2008) *Bacillus subtilis* FZB24 affects flower quantity and quality of Saffron (*Crocus sativus*). Planta Med 74:1316–1320

Singh S, Kapoor KK (1999) Inoculation with phosphate-solubilizing microorganisms and a vesicular-arbuscular mycorrhizal fungus improves dry matter yield and nutrient uptake by wheat grown in a sandy soil. Biol Fertil Soils 28:139–144

Sultana V, Ara J, Parveen G, Haque SE, Ahmad VU (2006) Role of Crustacean chitin, fungicides and fungal antagonists on the efficacy of *Pseudomonas aeruginosa* in protecting chilli from root rot. Pak J Bot 38:1323–1331

Sundheim L, Poplawsky AR, Ellingboe AH (1988) Molecular cloning of two chitinase genes from *Serratia marcescens* and their expression in *Pseudomonas* species. Physiol Mol Plant Pathol 33:483–491

Vavrina CS (1999) The effect of LS213 (*Bacillu pumilus*) on plant growth promotion and systemic acquired resistance in muskmelon and watermelon transplants and subsequent field performance. In: Proceedings of the international symposium stand establishment, vol 107,p 111

Verma SC, Ladha JK, Tripathi AK (2001) Evaluation of plant growth promoting and colonization ability of endophytic diazotroph from deep water rice. J Biotechnol 91:127–141

Vessey JK (2003) Plant growth promoting rhizobacteria as biofertilizers. Plant Soil 255:571–586

Vidhyasekaran P, Sethuraman K, Rajappan K, Vasumathi K (1997) Powder formulation of *Pseudomonas fluorescens* to control pigeonpea wilt. Biol Control 8:166–171

Vivekananthan R, Ravi M, Ramanathan A, Samiyappan R (2004) Lytic enzymes induced by *Pseudomonas fluorescens* and other biocontrol organisms mediate defence against the anthracnose pathogen in mango. World J Microbiol Biotechnol 20:235–244

Whitelaw MA (2000) Growth promotion of plants inoculated with phosphate-solubilizing fungi. Adv Agron 69:99–151

Yildirim E, Turan M, Donmez MF (2008) Mitigation of salt stress in radish (*Raphanus sativus* L.) by plant growth promoting rhizobacteria. Roumanian Biotechnol Lett 13:3933–3943

Zhao J, Zhou L, Wub J (2010) Promotion of *Salvia miltiorrhiza* hairy root growth and tanshinone production by polysaccharide–protein fractions of plant growth-promoting rhizobacterium *Bacillus cereus*. Process Biochem 45:1517–1522

Zhuang X, Chen J, Shim H, Bai Z (2007) New advances in plant growth-promoting rhizobacteria for bioremediation. Environ Int 33:406–413

Zuber S, Carruthers F, Keel C, Mattart A, Blumer C, Pessi G, Gigot-Bonnefoy C, Schnider-Keel U, Heeb S, Reimmann C, Haas D (2003) Mol Plant Microbe Interact 616–634

3 Pest Management

Plant growth-promoting rhizobacteria (PGPR) are being exploited commercially for plant protection to induce systemic resistance against various pests and diseases. Mixtures of different PGPR strains have resulted in increased efficacy by inducing systemic resistance against several pathogens attacking the same crop. Seed treatment with PGPR causes cell wall structural modifications and biochemical/physiological changes leading to the synthesis of proteins and chemicals involved in plant defence mechanisms. Lipopolysaccharides, siderophores and salicylic acid are the major determinants of PGPR-mediated ISR. The performance of PGPR has been successful against certain pathogens, insect and nematode pests under field conditions.

3.1 Disease Management

Plant diseases are in association with crop plants since agriculture began and were managed through synthetic pesticides to increase food production. But continuous usage of pesticides has resulted in the outbreak of pathogens resistant to fungicides apart from environmental pollution. Introduction of PGPR for increasing plant growth promotion during the 1950s from the Soviet Union and in Western countries (Backman et al. 1997) opened new vistas to use PGPR as an alternative to chemical pesticides for the management of soilborne pathogens (Kloepper 1993). Application of PGPR either as single strain or strain-mixture-based formulations checked disease spread besides increasing growth and yield.

PGPR have diverse applications for the management of plant diseases in agriculture, horticulture and forestry. Rhizosphere colonization is one of the first steps in the pathogenesis of soilborne microorganisms. *Pseudomonas* species are often used as model root-colonizing bacteria (Lugtenberg et al. 2001).

Plant-associated and root-associated bacteria have been extensively examined for their roles in natural and induced suppressiveness of soilborne diseases. Rhizobacteria are a subset of total rhizosphere bacteria which have the capacity, upon reintroduction to seeds or vegetative plant parts (such as potato seed pieces), to colonize the developing root system in the presence of competing soil microflora. Root colonization is typically examined by quantifying bacterial populations on root surfaces; however, some rhizobacteria can also enter roots and establish at least a limited endophytic phase.

PGPR are indigenous to soil and the plant rhizosphere and play a major role in the biocontrol of plant pathogens. They can suppress a broad spectrum of bacterial, fungal and nematode diseases. PGPR can also provide protection against viral diseases (Table 3.1). The use of PGPR has become a common practice in many regions of the world. Although significant control of plant pathogens has been demonstrated by PGPR in laboratory and greenhouse studies, results in the field have been inconsistent. Recent progress in

P.P. Reddy, *Plant Growth Promoting Rhizobacteria for Horticultural Crop Protection*,
DOI 10.1007/978-81-322-1973-6_3, © Springer India 2014

Table 3.1 Management of horticultural crop diseases using PGPR

PGPR	Crop	Disease	Mode of application
Actinoplanes spp.	Beetroot	*Pythium ultimum*	Soil application
Pseudomonas fluorescens	Banana	Panama wilt, *Fusarium oxysporum* f. sp. *cubense*	Sucker treatment
		Bunchy top virus	Foliar spray
	Mulberry	Leaf spot, *Cercospora moricola*	Foliar spray
	Mango	Anthracnose, *Colletotrichum gloeosporioides*	Foliar spray
	Apple	Grey mould, *Botrytis cinerea*	Flower/fruit spray
	Pear	Fire blight, *Erwinia amylovora*	Foliar spray
	Strawberry	Grey mould, *B. cinerea*	Flower/fruit spray
	Potato	Soft rot, *Erwinia carotovora*	Seed/soil treatment
		Bact. wilt, *Ralstonia solanacearum*	Soil treatment
	Tomato	Wilt, *F. oxysporum* f. sp. *lycopersici*	Seed/soil treatment
		Cucumber mosaic virus	Foliar spray
		White rot, *Sclerotinia sclerotiorum*	Soil treatment
		Bact. wilt, *R. solanacearum*	Soil treatment
	Brinjal	Root rot, *Rhizoctonia solani*	Soil treatment
		Blight, *Pythium vexans*	Seed treatment
	Chilli	Fruit rot and dieback, *Colletotrichum capsici*	Seed/seedling treatment, foliar spray
		Wilt, *F. oxysporum*	Soil treatment
		Macrophomina phaseolina	Soil treatment
		Powdery mildew, *Leveillula taurica*	Foliar spray
	Pea	Damping off, *P. ultimum*	Soil treatment
		Wilt, *F. oxysporum* f. sp. *pisi*	Soil treatment
		Root rot, *Aphanomyces euteiches*	Seed treatment
	Pigeon pea	Wilt, *F. oxysporum* f. sp. *udum*	Seed treatment
	Onion	Tip blight, *Alternaria* sp.	Foliar spray
	Cluster bean	Bacterial blight, *Xanthomonas axonopodis* pv. *cyamopsidis*	Foliar spray
	Radish	Wilt, *F. oxysporum* f. sp. *raphani*	Soil treatment
	Beetroot	*Pythium debaryanum*, *P. ultimum*	Seed treatment
	Cucumber	Wilt, *F. oxysporum* f. sp. *cucumerinum*, *R. solani*	Soil treatment
		Damping off, *Pythium aphanidermatum*	Soil treatment
		P. ultimum	Seed treatment
		P. ultimum, *R. solani*	Seed treatment
		Anthracnose, *Colletotrichum orbiculare*	Foliar spray
		Angular leaf spot, *Pseudomonas syringae* pv. *lachrymans*	Foliar spray
		Cucumber mosaic virus	Foliar spray
	Carnation	Wilt, *F. oxysporum* f. sp. *dianthi*	Soil application
	Areca nut	Fruit rot, *Phytophthora arecae*, *C. capsici*	Foliar spray
	Black pepper	Foot rot, *P. capsici*	Foliar spray
		Anthracnose, *C. gloeosporoides*	Foliar spray
	Ginger	Rhizome rot, *P. myriotylum*, *P. aphanidermatum*	Seed treatment
		Bact. wilt, *R. solanacearum*	Soil treatment
	Fenugreek	Root rot, *R. solani*	Seed treatment
	Vanilla	Root rot, *Phytophthora meadii*, *F. oxysporum* f. sp. *vanillae*	Soil treatment

(continued)

Table 3.1 (continued)

PGPR	Crop	Disease	Mode of application
P. syringae	Apple	Blue mold, *Penicillium expansum*; grey mould, *B. cinerea*	Preharvest fruit spray
	Peach	Brown rot, *Monilinia fructicola*	Post-harvest fruit dipping
P. aeruginosa	Tomato	Damping off, *P. aphanidermatum*	Seed treatment
	French bean	*B. cinerea*	Foliar spray
P. putida	Tomato	Wilt, *F. oxysporum* f. sp. *lycopersici*	Seed/soil treatment
	Pea	Damping off, *P. ultimum*	Seed treatment
	Cucumber	Wilt, *F. oxysporum* f. sp. *cucumerinum*, *R. solani*	Soil treatment
		P. ultimum	Seed treatment
		P. aphanidermatum	Soil treatment
	Muskmelon	Wilt, *F. oxysporum* f. sp. *melonis*	Seed treatment
	Radish	*F. oxysporum* f. sp. *raphani*	Soil treatment
	Carnation	Wilt, *F. oxysporum* f. sp. *dianthi*	Root dip treatment
P. corrugata	Cucumber	Damping off, *P. aphanidermatum*	Soil treatment
P. chlororaphis	Tomato	Damping off	Seed treatment
		F. oxysporum f. sp. *radicis-lycopersici*	Soil treatment
P. aureofaciens AB244	Tomato	*P. ultimum*	Soil application
P. aureofaciens 63-28	Cucumber	*P. aphanidermatum*	Soil application
Bacillus subtilis	Apple	Blue mold, *P. expansum*; grey mould, *B. cinerea*	Preharvest fruit spray
		Replant disease	Soil treatment, root dip treatment
		Phytophthora cactorum	Soil treatment
	Peach	Brown rot, *M. fructicola*	Post-harvest fruit line spray
	Potato	Scab, *Streptomyces scabies*	Soil treatment
		Bact. wilt, *R. solanacearum*	Soil treatment
	Tomato	Wilt, *F. oxysporum* f. sp. *lycopersici*	Seed/soil treatment
		Damping off, *P. aphanidermatum*	Seed treatment
		Bacterial spot and late blight	Foliar spray
		Tomato mottle virus	Foliar spray
		Cucumber mosaic virus	Foliar spray
	Brinjal	Collar rot, *S. sclerotiorum*	Soil treatment
	French bean	Root rot, *R. solani*	Seed treatment
	Pigeon pea	Wilt, *F. oxysporum* f. sp. *udum*	Seed treatment
	Cucumber	Angular leaf spot, *P. syringae* pv. *lachrymans*	Foliar spray
		Cucumber mosaic virus	Foliar spray
	Lettuce	Root rot, *P. ultimum*	Soil treatment
	Black pepper	Stem rot, *Sclerotium rolfsii*	Soil treatment
	Turmeric	Rhizome rot	Rhizome treatment
B. pumilus	Strawberry	Grey mould, *B. cinerea*	Flower/fruit spray
	Tomato	Wilt, *F. oxysporum* f. sp. *lycopersici*	Seed/soil treatment
		Tomato mottle virus	Foliar spray
		Cucumber mosaic virus	Foliar spray
	Cucumber	Angular leaf spot, *P. syringae* pv. *lachrymans*	Foliar spray
		Bacterial wilt	Soil treatment
B. coagulans	Mango	Bacterial canker, *Xanthomonas campestris* pv. *mangiferaeindicae*	Foliar spray

(continued)

Table 3.1 (continued)

PGPR	Crop	Disease	Mode of application
B. amyloliquefaciens	Tomato	Tomato mottle virus	Foliar spray
Paenibacillus sp. 300	Cucumber	*F. oxysporum* f. sp. *cucumerinum*	Soil treatment
Burkholderia cepacia	Pea	Root rot, *Aphanomyces euteiches*	Seed treatment
		Damping off, *P. ultimum*	Seed treatment
	French bean	Root rot, *R. solani*	Seed treatment
		Wilt, *Sclerotium rolfsii*	Soil treatment
	Cucumber	*P. ultimum*	Soil treatment
Enterobacter aerogenes	Apple	Collar rot, *Phytophthora cactorum*	Soil treatment
E. cloacae	Cucumber	Damping off, *P. aphanidermatum*	Seed treatment
	Lettuce	Root rot, *P. ultimum*	Soil treatment
Erwinia herbicola	Apple	*Erwinia amylovora*	Foliar spray
Streptomyces griseoviridis	Cauliflower	Blight, *Alternaria brassicicola*	Seed treatment
	Bell pepper	Damping off, *P. aphanidermatum*	Seed/soil treat.
	Narcissus	*F. oxysporum* f. sp. *narcissi*	Bulb treatment
Streptomyces spp.	Cucumber	Wilt, *F. oxysporum* f. sp. *cucumerinum*	Soil treatment
Serratia marcescens	Cucumber	Wilt, *F. oxysporum* f. sp. *cucumerinum*, *R. solani*	Soil treatment
S. plymuthica	Beetroot	Damping off, *P. aphanidermatum*	Soil treatment
	Cucumber	*P. ultimum*	Soil application
S. liquifaciens	Carnation	Wilt, *F. oxysporum* f. sp. *dianthi*	Soil treatment
Bradyrhizobium japonicum	Okra	*R. solani*	Seed treatment
Agrobacterium radiobacter K84, K1026	Peach	Crown gall, *Agrobacterium tumefaciens*	Root dip treatment
Avirulent *Ralstonia solanacearum*	Ginger	Bact. wilt, *R. solanacearum*	Soil treatment

our understanding of their diversity, colonizing ability and mechanism of action, formulation and application should facilitate their development as reliable biocontrol agents against plant pathogens. Some of these rhizobacteria may also be used in integrated pest management programmes. Greater application of PGPR is possible in agriculture for biocontrol of plant pathogens (Siddiqui 2005).

3.1.1 Pseudomonads

A major group of rhizobacteria with potential for biological control is the pseudomonads (Kremer and Kennedy 1996). *Pseudomonas* spp. are ubiquitous bacteria in agricultural soils. Tremendous progress has been made in characterizing the process of root colonization by pseudomonads, the biotic and abiotic factors affecting colonization, bacterial traits and genes contributing to rhizosphere competence and the mechanisms of pathogen suppression (Weller 2007). Pseudomonads possess many traits that make them well suited as biocontrol and growth-promoting agents (Weller 1988). These include the ability to (1) grow rapidly *in vitro* and to be mass produced, (2) rapidly utilize seed and root exudates, (3) colonize and multiply in the rhizosphere and spermosphere environments and in the interior of the plant, (4) produce a wide spectrum of bioactive metabolites (i.e. antibiotics, siderophores, volatiles and growth-promoting substances), (5) compete aggressively with other microorganisms and (6) adapt to environmental stresses. In addition, pseudomonads are responsible for the natural suppressiveness of some soils to soilborne pathogens (Weller et al. 2002).

Alabouvette et al. (1993) showed that in addition to nonpathogenic *Fusarium oxysporum*, *P. fluorescens* and *P. putida* are the main candidates for the biological control of Fusarium wilts. The fluorescent pseudomonads are involved in the natural suppressiveness of some soils to Fusarium wilts, and they have been applied successfully to suppress Fusarium wilts of various plant species (Lemanceau and Alabouvette 1993). For many pseudomonads, production of metabolites such as antibiotics, siderophores and hydrogen cyanide (HCN) is the primary mechanism of biocontrol (Weller and Thomashow 1993). Several evidences indicate that siderophore production when iron is limited is responsible for the antagonism of some strains of *P. aeruginosa* against *Pythium* spp., the causal agents of damping off and root rot of many crops (Charest et al. 2005). The antibiotics produced by bacterial biocontrol agents and their role in microbial interaction were reviewed by Raaijmakers et al. (2002). *P. fluorescens* CHAO isolated and intensively studied by the group of Défago in Switzerland produces several bioactive compounds (antibiotics, siderophores, HCN, indole acetic acid), giving it one of the broadest spectra of potential biocontrol and growth-promoting mechanisms of known PGPR (Weller and Thomashow 1993).

In Canada, *Pseudomonas* species were developed for the biological control of *Pythium* diseases in hydroponic systems for greenhouses (Paulitz and Bélanger 2001). In a spring cucumber crop, *P. corrugata* strain 13 and *P. fluorescens* strain 15 produced 88 % more marketable fruit, while in a fall crop with severe disease pressure due to higher slab temperatures, both strains significantly increased by 600 % marketable fruit. Strain 15 also increased fruit production in treatments not inoculated with pathogen (Paulitz and Bélanger 2001). Several reports show the critical role played by fluorescent *Pseudomonas* spp. in naturally occurring soils that are suppressive to Fusarium wilt (Mazzola 2002).

The major weakness of pseudomonads as biocontrol agents is their inability to produce resting spores (as do many *Bacillus* spp.), which complicates formulation of the bacteria for commercial use. Fluorescent *Pseudomonas* species have been studied for decades for their plant growth-promoting effects through effective suppression of soilborne plant diseases. Among various biocontrol agents, fluorescent pseudomonads, equipped with multiple mechanisms for biocontrol of phytopathogens and plant growth promotion, are being used widely (Banasco et al. 1998) as they produce a wide variety of antibiotics, chitinolytic enzymes, growth-promoting hormones, siderophores, HCN and catalase and can solubilize phosphorous (Banasco et al. 1998). Cold-tolerant fluorescent *Pseudomonas* isolated from Garhwal Himalayas act as potential plant growth-promoting and biocontrol agents in pea (Negi et al. 2005).

All species and strains of *Pseudomonas* are Gram-negative rods. Exceptions to this classification have recently been discovered in *Pseudomonas* biofilms. A significant number of cells can produce exopolysaccharides known as biofilms (Hassett et al. 2002). Secretion of exopolysaccharides such as alginate makes it difficult for Pseudomonads to be phagocytosed by mammalian white blood cells (Ryan et al. 2004). Exopolysaccharide production also contributes to surface-colonizing biofilms which are difficult to remove from preparation surface.

Certain members of the *Pseudomonas* genus have been applied directly to soils as a way of preventing the growth or establishment of crop pathogens. The biocontrol properties of *P. fluorescens* strains CHAO or Pf5 are currently best understood, although it is not clear exactly how the plant growth-promoting properties of *P. fluorescens* are activated. Theories include that the bacteria might induce systemic resistance in the host plant, so it can better resist attack by the pathogens; the bacteria might outcompete other (pathogenic) soil microbes, e.g. by siderophores giving a competitive advantage at scavenging for iron; the bacteria might produce compounds antagonistic to other soil microbes, such as phenazine type antibiotics or HCN. There is an experimental evidence to support all of these theories, in certain conditions (Haas and Defago 2005).

Of the various rhizospheric bacteria, *Pseudomonas* spp. are aggressive colonizers of the rhizosphere of various crop plants and have a broad spectrum of antagonistic activity against plant pathogens (Schroth and Hancock 1982). Among *Pseudomonas* species, *P. aeruginosa*, a plant growth-promoting rhizobacterium, has been found to be an effective biocontrol agent of root pathogens (Burdman et al. 1996). *Septoria tritici* (*Mycosphaerella graminicola*) was suppressed by *P. aeruginosa* strain *leci* (Leavy et al. 1992). Fuhrmann and Wollum (1989) reported that co-inoculation of siderophore-producing pseudomonads with mixtures of the competing *Bradyrhizobium* typically enhanced nodulation by *B. japonicum* strain USDA 110. Many strains of pseudomonads can indirectly protect the plants by inducing systemic resistance against various pests and diseases (Ramamoorthy et al. 2001). The beneficial effect on plant shoot dry mass was more pronounced with HCN producing *Pseudomonas* strain (Goel et al. 2002). The *Pseudomonas* bacteria were inoculated into the rhizosphere and remained spatially separated from the pathogen that was inoculated on the aboveground plant parts, either into the stem (Van Peer et al. 1991) or on the leaf surface (Wei et al. 1991).

Fluorescent pseudomonads were first developed as talc-based formulation for the treatment of potato seed tubers for growth promotion (Kloepper and Schroth 1981). Talc-based formulation of *P. fluorescens* strain Pf1 and Pf2 increased grain yield of pigeon pea besides the control of pigeon pea wilt (Vidhyasekaran et al. 1997). Seed treatment of pigeon pea with peat-based formulation of *B. subtilis* supplemented with 0.5 % chitin or with 0.5 % of sterilized *Aspergillus* mycelium controlled wilt of pigeon pea. It also increased growth promotion even in the presence of inoculum pressure (Manjula and Podile 2001). Chitin supplementation enhances the biocontrol efficacy of formulations. But incorporation of chitin will increase the production cost of biopesticides. Hence, identification of cheap and easily available source of chitin is essential. Seed treatment with wettable powder formulation of *P. putida* strain 30 and 180 suppressed wilt of musk melon to the extent of 63 and 50 % after 90 days of transplanting muskmelon in the field.

Growth of pseudomonads on spoiling foods can generate a 'fruity odour'. *P. aeruginosa* is a highly relevant opportunistic human pathogen. One of the most worrying characteristics of *P. aeruginosa* is its low antibiotic susceptibility.

3.1.2 Bacilli

Bacillus subtilis strain QST713 has natural fungicidal activity and is employed as a biocontrol agent. *B. subtilis* has shown antagonistic activity towards *Fusarium solani in vitro* (Kendrick 1963). Similarly, *Bacillus* species have been tested on a wide variety of plant species for their ability to control diseases (Cook and Baker 1983). *Bacillus* species are able to form endospores that allow them to survive for extended periods of time under adverse environmental conditions. Some members of the group are diazotrophs, and *B. subtilis* was isolated from the rhizosphere of a range of plant species at a concentration as high as 10^7 per gram of rhizosphere soil (Wipat and Harwood 1999).

Bacillus subtilis, due to its broad host range and its ability to form endospores and produce different biologically active compounds with a broad spectrum of activity, is potentially useful biocontrol agent (Nagórska et al. 2007). *B. subtilis* strain RB14 produces the cyclic lipopeptide antibiotics iturin A and surfactin active against several phytopathogens. The antifungal activity of the selected isolates was associated with their ability to produce inhibitory volatile substances and diverse and complex lytic chitinases (Fig. 3.1).

Bacillus megaterium from tea rhizosphere is able to solubilize phosphate and produce IAA, siderophore and antifungal metabolite and thus helps in the plant growth promotion and reduction of disease intensity (Chakraborty et al. 2006).

Many species of *Bacillus* including *B. subtilis*, *B. licheniformis*, *B. pumilus*, *B. amyloliquefaciens*, *B. cereus* and *B. mycoides* are known to suppress growth of several fungal pathogens such as *Rhizoctonia*, *Fusarium*, *Sclerotinia*, *Sclerotium*, *Gaeumannomyces*, *Nectria*, *Pythium*, *Phytophthora* and *Verticillium*. The main property of antagonist bacterial strains is production

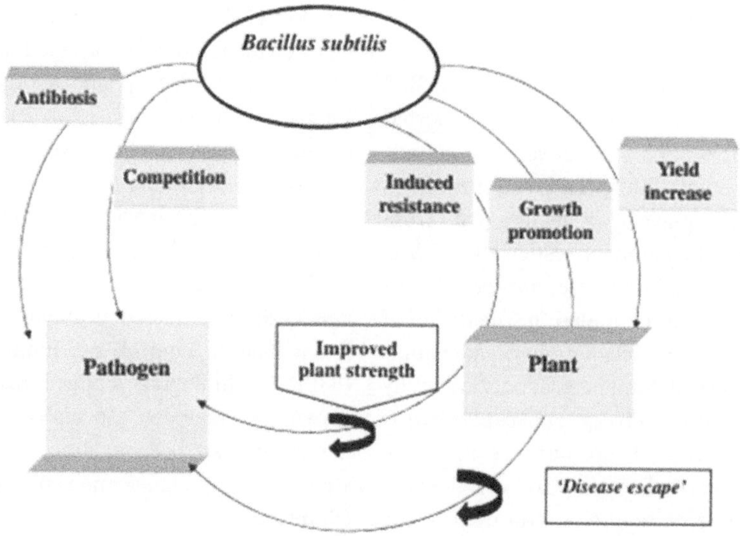

Fig. 3.1 Mode of action of *Bacillus subtilis* strain FZB24 promoting plant growth and disease control (Adapted from Kilian et al. 2000)

of antifungal antibiotics, which seem to play a major role in biological control of plant pathogens and post-harvest spoilage fungi. The main antibiotic producers of this genus are *B. brevis* (gramicidin, tyrothricin), *B. licheniformis* (bacitracin), *B. polymyxa* (polymyxin, colistin), *B. pumilus* (pumulin), *B. subtilis* (polymyxin, difficidin, subtilin, mycobacillin, bacitracin), *B. cereus* (cerexin, zwittermicin), *B. circulans* (circulin) and *B. laterosporus* (laterosporin). Many of these antifungal substances have been characterized and identified as peptide antibiotics. Antifungal peptides (iturins) produced by *Bacillus* species include mycosubtilins, bacillomycins, surfactins, fungistatins and subsporins. *Bacillus* spp. produces also a range of other metabolites, including phytoalexins, chitinases and other cell wall-degrading enzymes, and volatile compounds which elicit plant resistance mechanisms against attack by fungi, bacteria and pathogenic nematodes.

3.1.3 Diazotrophic PGPR

Azotobacter spp. protect several plants from root disease caused by soilborne fungi through HCN and siderophore production. Hydrogen cyanide production by *Azotobacter* isolates accounts for about 60 % of the inhibition on phytopathogens observed in soil. Most of the *Azotobacter* isolates (AZT1, AZT3, AZB4, AZC6, AZC7 and AZC8) are producing HCN as well as siderophore and act as potent antifungal agents. Synergistic interaction of these two with other metabolites may further function as stress factors including local and systemic host resistance that led for the suppression of the root pathogens. Out of ten *Azotobacter* isolates, four isolates (AZT1, AZT3, AZB4 and AZC6) were exhibiting high antifungal activity against *A. flavus*, *Cercospora* and *F. oxysporum*. The antifungal activity of the screened *Azotobacter* isolates was concentration dependent and statistically significant ($p<0.05$).

3.1.4 Rhizobia

Rhizobia have a good potential to be used as biological control agents against some plant pathogens. Strains of *Sinorhizobium meliloti* are antagonistic to *Fusarium oxysporum* (Antoun et al. 1998), and rhizobia antagonistic to *F. solani* f. sp. *phaseoli* isolated from commercial snap bean appeared to have a good potential for controlling Fusarium rot (Buonassisi et al. 1986).

3.1.5 Actinomycetes

Recently, actinobacteria residing in plants, called endophytic actinomycetes, have been reported as new sources for bioactive compounds (Taechowisan et al. 2005) and had beneficial effects to the host plant by protecting plant from pathogens. The metabolites, especially indole acetic acid (IAA) and chitinase, are recorded as responsible for the degradation of different complex and relatively recalcitrant organic compounds present in soil. Similar antagonistic activity of endophytic *Streptomyces griseorubiginosus* against *F. oxysporum* f. sp. *cubense* has been recorded on banana by Cao et al. (2004). Evidences are now available on actinobacteria used in the control of *Rhizoctonia solani* and *Ralstonia solanacearum* in tomato (Sabaratnam and Traquair 2002) and *Colletotrichum musae* in banana (Taechowisan et al. 2003).

3.2 Nematode Management

PGPR destroy nematodes in virtually all soils because of their constant association in the rhizosphere. The PGPR-nematode interactions have been extensively studied with the aim to manage plant-parasitic nematodes. These studies involve the selection of bacteria that can be used as biocontrol agents against nematodes. The non-parasitic rhizobacteria reduce nematode populations by colonizing the rhizosphere of the host plant. A large number of rhizobacteria are known to reduce nematode populations and important genera include *Agrobacterium*, *Alcaligenes*, *Bacillus*, *Clostridium*, *Desulfovibrio*, *Pseudomonas*, *Serratia* and *Streptomyces*. Practical control systems, formulation, mechanisms of nematode suppression and possible uses of bacteria in nematode biocontrol are suggested (Siddiqui and Mahmood 1999). Application of some of these bacteria has given very promising results (Table 3.2).

In the last few years, other bacterial species have shown biocontrol potential against nematodes. Bacteria isolated from the root of nematicidal plants and identified as *Stenotrophomonas maltophilia*, *Bacillus mycoides* and *Pseudomonas* sp. reduced trichodorid nematode density on potato by 56–74 %, respectively. These bacteria were characterized for production of hydrolytic enzymes and HCN, phenol oxidation and antifungal activity (Insunza et al. 2002). *Rhizobium etli* has been reported to have a biocontrol effect against the root-knot nematode, *M. incognita*, and showed the capacity to colonize plant roots and nematode galls (Hallmann et al. 2001).

PGPR also induce systemic resistance against nematode pests. Though attempts to use PGPR for nematode control are limited, the use of PGPR as biological control agents of plant-parasitic nematodes especially for potato cyst nematode has been reported as a successful strategy in the management of these nematodes (Sikora 1992).

3.2.1 Bacilli

3.2.1.1 *Bacillus licheniformis*

Two bacteria *B. licheniformis* and *Pseudomonas mendocina* and two fungi *Acrophialophora fusispora* and *Aspergillus flavus* were evaluated for their efficacy as biocontrol agents individually and in various combinations against *M. incognita* infecting tomato. Individually, *B. licheniformis* gave best control although when all four agents were used, nematode multiplication was reduced by 82.06 %.

3.2.1.2 *B. subtilis*

Multiplication rate of *R. reniformis* was reduced significantly (44.5 %) when soil drenching with *B. subtilis* (isolate Bs_t) was done a week before nematode inoculation. But, it was not so in the case of treatments that received simultaneous inoculations of the bacterium and the nematode. These results indicated that *B. subtilis* could induce systemic resistance in tomato against *R. reniformis* (Niknam and Dhawan 2001a).

3.2.2 Pseudomonads

3.2.2.1 *Pseudomonas fluorescens*

It plays a major role in plant growth promotion, induced systemic resistance, biological control of pathogens, etc. PGPR are known to enhance

Table 3.2 Biological control of plant parasitic nematodes in horticultural crops using PGPR

Bacteria	Crop	Nematode	References
Pseudomonas fluorescens	Citrus (sweet orange, lemon)	*Tylenchulus semipenetrans*	Santhi et al. (1999)
	Citrus (acid lime)	*T. semipenetrans*	Parvatha Reddy et al. (2000)
	Banana	*Radopholus similis* and *Meloidogyne incognita*	Aalten and Gowen (1998)
		R. similis	Aalten and Gowen (1998) and Santhi (2003)
	Grapevine	*M. incognita*	Santhi et al. (1998)
	Tomato	*M. incognita*	Verma et al. (1999) and Santhi and Sivakumar (1995)
		Rotylenchulus reniformis	Niknam and Dhawan (2001b)
	Potato	*M. incognita*	Mani et al. (1998)
	Carrot	*R. reniformis*	Rao and Shylaja (2004)
	Okra	*M. incognita*	Sobita Devi and Dutta (2002)
	Black pepper	*M incognita*	Sosamma and Koshy (1995)
	Turmeric	*M. incognita*	Seenivasan et al. (2001)
P. fluorescens + oil cakes	Citrus (acid lime)	*T. semipenetrans*	Parvatha Reddy et al. (2000)
P. stutzeri	Black pepper	*M. incognita*	Eapen et al. (1997)
	Turmeric	*M. incognita*	Seenivasan et al. (2001)
	Tomato	*M. incognita*	Khan and Tarannum (1999)
P. chitinolytica	Tomato	*M. javanica*	Spiegel et al. (1991)
	Black pepper	*M. incognita*	Sosamma and Koshy (1995)
Agrobacterium radiobacter/*Bacillus sphaericus*	Potato	*Globodera pallida*	Racke and Sikora (1992)
Azotobacter chroococcum	Tomato	*M. incognita*	Chahal and Chahal (2003)
	Brinjal	*M. incognita*	Chahal and Chahal (1988)
		M. javanica	Bansal and Verma (2002)
Bacillus subtilis	Grapevine	*M. incognita*	Mani et al. (1999)
	Tomato	*M. incognita*	Khan and Tarannum (1999)
		R. reniformis	Niknam and Dhawan (2001a)
	Pigeon pea	*Heterodera cajani*	Siddiqui and Mahmood (1995)

plant growth promotion and reduce severity of many nematode diseases.

P. fluorescens produce the antibiotics 2,4-diacetylphloroglucinol (DAPG), phenazine-1-carboxylic acid (PCA) and phenazine (Phe) and antimicrobial metabolites such as pyoluteorin (Plt), hydrogen cyanide (HCN), siderophore pyoverdine (Pvd), salicylic acid (Sal) and pyochelin (Pch) which inhibit a broad spectrum of plant-parasitic nematodes (Boruah and Kumar 2002). There are many evidences substantiating the importance of DAPG in biological control. The production of DAPG is governed by *Phl* gene. The population size of DAPG producers in the rhizosphere is correlated with the disease suppressiveness of the soil and *in situ* antibiotic production. The diverse DAPG-producing *Pseudomonas* spp. have been isolated from the rhizosphere of various crop plants, and their role in promoting plant growth and inhibiting nematode diseases is the subject of ongoing investigations worldwide.

The level of infestation of root-knot nematode *M. incognita* on tomato was reduced with fewer galls and egg masses in the soil following root dipping with *P. fluorescens* strain Pf1 (Santhi and Sivakumar 1995).

3.2.2.2 *P. chitinolytica*

Application of the bacterium *P. chitinolytica* reduced the root-knot nematode infection in tomato crop (Spiegel et al. 1991). *P. aeruginosa*

Table 3.3 PGPR used as bio control agents against insect pests affecting different horticultural crops

PGPR	Crops	Disease/pathogen/insect
Pseudomonas putida 89b-27	Cucumber	Striped cucumber beetle, *Acalymma vittatum*; spotted cucumber beetle, *Diabrotica undecimpunctata howardi*
Bacillus sp.	Cucumber	Cotton aphids
B. licheniformis	Bell pepper	*Myzus persicae*
B. cereus MJ-1	Red pepper	*Myzus persicae*
B. subtilis G803	Bell pepper	*Myzus persicae*
	Tomato	Whitefly
B. amyloliquefaciens	Bell pepper	*Myzus persicae*
Serratia marcescens	Cucumber	Spotted cucumber beetle, *Diabrotica undecimpunctata howardi*
Flavimonas oryzihabitans INR-5	Cucumber	Striped cucumber beetle, *Acalymma vittatum*; spotted cucumber beetle, *Diabrotica undecimpunctata howardi*

78 produce a polar substance, heat labile, sensitive to extreme pH values causing *in vitro* juvenile mortality of the root-knot nematode, *M. javanica* (Ali et al. 2002).

3.2.3 Diazotrophic PGPR

3.2.3.1 *Azotobacter chroococcum*

Chahal and Chahal (1986) reported that *A. chroococcum* caused significant inhibition of hatching of egg masses of *M. incognita* and did not allow the larvae to penetrate into the roots of brinjal to form galls. *A. chroococcum* was also responsible for reduction in soil and root population of root-knot nematodes and root galling on tomato and brinjal. Among different strains of *A. chroococcum* tested, strain 23 proved better than other strains (Chahal and Chahal 1988, 1999). Chahal and Chahal (2003) reported significant reduction in the gall formation and soil and root population of *M. incognita* both in tomato and brinjal by *A. chroococcum.*

3.2.4 Rhizobia

Reitz et al. (2000) showed that the lipopolysaccharides of *Rhizobium etli* G12 induce the systemic resistance to infection by the cyst nematode *Globodera pallida* in potato roots.

3.3 Insect Pest Management

Reports on PGPR-mediated ISR against insects are restricted to very few crops (Table 3.3). The majority of bacterial pathogens of insects and related taxa occur in the families Bacillaceae, Pseudomonadaceae, Enterobacteriaceae, Streptococcaceae and Micrococcaceae. Most of these bacteria are weak pathogens that infect insects subject to environmental stress, but a minority is highly virulent.

3.3.1 Pseudomonads

Generally, fluorescent pseudomonads influence growth and development of insects at all stages of their growth. Induction of systemic resistance by PGPR strains, namely, *P. putida* strain 89b-27, has significantly reduced populations of the striped cucumber beetle, *Acalymma vittatum*, and the spotted cucumber beetle, *Diabrotica undecimpunctata howardi*, on cucumber (Zehnder et al. 1997a, b). As certain fluorescent pseudomonads are effective rhizosphere colonizers and are endophytic in nature in the plant system, attempts have been made to transfer the insecticidal crystal protein from *Bacillus thuringiensis* to *P. fluorescens*. *P. fluorescens*, thus genetically engineered, was effective against a lepidopteron insect pest. Thus, PGPR treatment of crops can be effective for insect pest management and has a great potential for future use.

A few studies have explored the effects of PGPR treatment on plant–insect interactions. Field studies with PGPR-treated cucumber plants revealed PGPR-related resistance to cucurbit wilt but the exact mechanism of resistance remained unknown until a later study (Tuzun and Kloeppper 1995). Eventually, it was revealed that the PGPR-treated cucumber plants produced less cucurbitacin which reduced feeding by cucumber beetles known to transmit the wilt disease (Zehnder et al. 1997a, b).

3.3.2 Bacilli

Evidence that PGPR play an important role in plant–insect interactions was also supported in later studies. While evaluating a PGPR strain of *Bacillus subtilis* known to increase plant tolerance to salinity stress, investigators noted lower whitefly populations on PGPR-treated tomato plants grown under greenhouse conditions (Hanafi et al. 2007). The mechanism of this resistance was not identified, but a later study showed that treatment of tomato plants with PGPR reduced adult whitefly emergence and even restored whitefly population suppression normally achieved by jasmonic acid (JA)-pathway plant responses (Valenzuela-Soto et al. 2010). Although the factors responsible for this resistance are not fully understood, the effects observed further supported the theory that these bacteria may be important for resistance to insect pests.

Because large monoculture systems contain high densities of a single crop type, it would be beneficial for crop protection to reduce oviposition of lepidopteron insect pests whose larvae are the most damaging life stage. It is well documented that herbivorous insects and their natural enemies alike are able to utilize plant volatiles in order to help them locate their respective hosts (Dicke and Van Loon 2000). Considering the established effects of headspace volatiles on the oviposition behaviour of herbivorous insects and the various benefits of PGPR treatment (*Bacillus pumilus* strain INR-7) of plants, it is possible that treating plants with PGPR may reduce the oviposition of insect pests. If they do provide plants with resistance from insect pests, PGPR could be applied in new ways and could be integrated into current pest management programmes. PGPR spores are easily stored, can be applied through sprayers, are appropriate for use by organic growers and are added easily to seed coat formulations. PGPR are already being used in agriculture and understanding the effects they may be having on pest populations, whether positive or negative, is important.

3.3.3 Enterobacteria

Induction of systemic resistance by PGPR *S. marcescens* strain 90-166 has significantly reduced populations of the striped cucumber beetle, *Acalymma vittatum*, and the spotted cucumber beetle, *Diabrotica undecimpunctata howardi*, on cucumber. Its efficacy was better than application of the insecticide esfenvalerate (Zehnder et al. 1997a, b).

3.3.4 *Flavimonas* spp.

Induction of systemic resistance by PGPR *Flavimonas oryzihabitans* strain INR-5 has significantly reduced populations of the striped cucumber beetle, *Acalymma vittatum*, and the spotted cucumber beetle, *Diabrotica undecimpunctata howardi*, on cucumber (Zehnder et al. 1997a, b).

3.4 Integrated Pest Management

3.4.1 Integration and Mixtures of PGPR

A single biocontrol agent may not perform well at all times in all kinds of soil environment to suppress plant pathogens (Raupach and Kloepper 1998). In addition application of single biocontrol agent-based formulation might have resulted in inadequate colonization; inability to tolerate the extremes of soil pH, moisture and temperature; and fluctuations in the production of antimicrobial substances (Weller and Thomashow 1993).

In nature biocontrol results from mixtures of antagonists, rather than from high populations of a single antagonist. Moreover, mixtures of antagonists are considered to account for protection of disease-suppressive soils (Schippers 1992). Consequently, application of a mixture of introduced biocontrol agents would more closely mimic the natural situation and may broaden the spectrum and enhance the efficacy and reliability of biocontrol (Duffy and Weller 1995). Strategies for formulating mixtures of biocontrol agents could be envisioned including mixtures of organisms with differential plant colonization patterns; biocontrol agents that control different pathogens; antagonists with different mechanisms of disease suppression; and taxonomically different organisms and antagonists with different optimum temperature, pH and moisture conditions for plant colonization (Backman et al. 1997). Combination of various mechanisms of biocontrol is useful in achieving the goal without genetic engineering (Janisiewicz 1996).

Inconsistent performance of biocontrol agents was overcome by the combined application of several biocontrol strains that mimic the natural environment (Raupach and Kloepper 1998). Development of cocktail formulation with compatible isolates will improve disease control through synergy in crosstalk between the isolates that lead to increased production of antibiotics at the site of colonization and thereby could suppress the establishment of pathogenic microbes. Advantages of strain mixtures include broad spectrum of action, enhanced efficacy and reliability and also allow the combination of various traits without genetic engineering (Janisiewicz 1996). Application of mixed PGPR strain-based formulations to field might ensure at least one of the mechanisms to operate under variable environment that exists under field conditions (Duffy et al. 1996).

Studies on combinations of biocontrol agents for plant disease control have included mixtures of bacteria and mixtures of fungi and bacteria.

3.4.1.1 Mixtures of PGPR

PGPR strains *Bacillus pumilus* (INR 7), *B. subtilis* (GBO3) and *Curtobacterium flaccumfaciens* (ME1) were tested alone and in combination for biocontrol against *Colletotrichum orbiculare* (causing anthracnose), *Pseudomonas syringae* pv. *lachrymans* (causing angular leaf spot) and *Erwinia tracheiphila* (causing cucurbit wilt disease). Greater suppression and enhanced consistency was observed against multiple cucumber pathogens using strain mixture. This might be due to different mechanisms of action for each PGPR strain (Raupach and Kloepper 1998). Application of a mixture of two chitinolytic bacterial strains, namely, *Paenibacillus* sp. and *Streptomyces* sp., in the ratio of 1:1 or 4:1 was more effective than when they were applied individually for the control of Fusarium wilt of cucumber incited by *F. oxysporum* f. sp. *cucumerinum* (Singh et al. 1999).

Combined application of *Pichia guilliermondii* and *Bacillus mycoides* (B16) reduced the infection of *Botrytis cinerea* by 75 % on fruits in strawberry plants grown commercially under greenhouse conditions. But the individual application of either antagonist resulted in 50 % reduction of strawberry fruit infection. Population of yeast increased when applied as mixture rather than single application (Guetsky et al. 2002).

The potential for using fluorescent *Pseudomonas* and *Rhizobium* in pea production has been shown in field studies where there was a reduction in the number of *Fusarium oxysporum* infected peas grown in infested soils and an improvement of plant growth in terms of shoot height and dry weight. The strains used exhibited antifungal activity and produced siderophores (Kumar et al. 2001).

Plant disease control using combinations of antagonistic rhizobacteria has been presented in Table 3.4.

PGPR formulations comprising of bacterial strain mixtures having the capability to induce

Table 3.4 Efficacy of mixtures of PGPR formulations against plant disease and growth promotion

Formulation	Crop	Results	References
Talc-based formulation of *B. subtilis* and *P. chlororaphis* (PA23)	Tomato	Increased growth promotion and significant reduction of damping off	Kavitha et al. (2003)
Talc-based *B. subtilis* (BSCBE4) and *P. chlororaphis* (PA23)	Turmeric	Significant reduction of rhizome rot and yield increase of rhizomes	Nakkeeran et al. (2004)
Talc-based *B. subtilis* (BSCBE4), *P. chlororaphis* (PA23) and *P. fluorescens* (ENPF1)	*Phyllanthus amarus*	Significant reduction of stem blight caused by *Corynespora cassiicola* under field conditions	Mathiyazhagan et al. (2004)
Phosphobacterium + *Azospirillum* + neem cake	Grapevine	*M. incognita*	Sivakumar and Vadivelu (1999)

chitinase in plant play an important role in hydrolyzing chitin, the structural component in gut linings of insects, and would lead to better control of insect pest (Broadway et al. 1998). In addition certain PGPR strains also activate octadecanoid, shikimate and terpenoid pathways. This in turn alters the volatile production in the host plant leading to the attraction of natural enemies (Bell and Muller 1993). Identification of entomopathogenic PGPR strains that have the capability to colonize phylloplane in a stable manner will be a breakthrough in the management of foliar pests (Otsu et al. 2004). Combined application of entomopathogenic strains with compatible PGPR strains that have the ability to suppress plant diseases has to be developed for broad-spectrum action. On the contrary, certain studies explain that some strain mixtures perform even lower than that of individual strains. So, the basic knowledge on molecular signalling mechanisms between related strains and species has to be understood for the development of a better formulation that could suppress a broad spectrum of pathogens and pests besides plant growth promotion.

3.4.1.2 Mixtures of PGPR and Fungi

Development of mixtures of biocontrol agents should be emphasized, because these may result in better plant colonization, better adapt to the environmental changes that occur throughout the growing season, have a larger number of pathogen-suppressive mechanisms and protect against a broader range of pathogens. Increasing the genetic diversity of biocontrol systems by the mixture of microorganisms may persist longer in the rhizosphere and utilize a wider array of biocontrol mechanisms (e.g. induction of resistance, production of antibiotics and competition for nutrients) under a broader range of environmental conditions (Pierson and Weller 1994). Multiple organisms may enhance the level and consistency of control by providing multiple mechanisms of action, a more stable rhizosphere community and effectiveness over a wide range of environmental conditions. In particular combinations of fungi and bacterial PGPR provide potential green alternative protection at different times or under different conditions and occupy different or complementary niches. Such combinations may overcome inconsistencies in the performance of individual isolates.

Several researchers have observed improved disease control using combinations of multiple compatible biocontrol organisms and have demonstrated enhanced biocontrol of Fusarium wilt by combining certain nonpathogenic strains of *F. oxysporum* with fluorescent strains of *Pseudomonas*. Increased suppression of *Fusarium* wilt of carnation was observed by combining *P. putida* WCS358 with nonpathogenic *Fusarium oxysporum* Fo47 (Lemanceau et al. 1993). The enhanced disease suppression may be due to siderophore-mediated competition for iron by WCS358, which makes the pathogenic *F. oxysporum* strain more sensitive to competition for glucose by the nonpathogenic strain Fo47. Furthermore, strains of nonpathogenic *Verticillium lecanii*, *Acremonium rutilum* or *Fusarium oxysporum* with the fluorescent *Pseudomonas* spp. strains WCS358, WCS374 or WCS417 resulted in significantly better suppression of *Fusarium* wilt of radish compared to the single organism (Leeman et al. 1996).

Limited numbers of compatible and effective mixtures of biocontrol agents are available. The majorities of mixtures have no benefit or detrimental effects on biocontrol activity. Further, a mixture that improves activity under one set of conditions may be antagonistic under another set of conditions. A biocontrol product composed of a mixture of strains has a potential economical constraint. Production and registration of such a product will be more costly than a product composed of single strain.

In few cases combinations of biocontrol agents do not result in improved suppression of disease. Tomato seedlings were treated with the potential biocontrol agents such as nonpathogenic strains of *Fusarium* spp., *Trichoderma* spp., *Gliocladium virens*, *Pseudomonas fluorescens*, *Burkholderia cepacia* and others in the greenhouse and transplanted into pathogen-infested field soil. Combinations of antagonists like multiple *Fusarium* isolates, *Fusarium* with bacteria and *Fusarium* with other fungi also reduced disease, but did not provide better control than the nonpathogenic *Fusarium* (Larkin and Fravel 1998). A *T. harzianum* strain with a strain of *P. fluorescens* was able to suppress root rot of pea caused by *Aphanomyces euteiches* f. sp. *pisi* but did not result in better disease suppression (Dandurand and Knudsen 1993).

Positive and negative interactions of introduced microorganisms and indigenous microflora can influence their performance in the rhizosphere. For example, two groups of microorganisms that occupy the same ecological niche and have the same nutritional requirements are bound to compete for nutrients. Siderophore-mediated competition for iron between the two biocontrol agents *P. putida* WCS358 and *P. fluorescens* WCS374 decreased colonization of radish roots by the latter strain (Raaijimakers et al. 1995). Hubbard et al. (1983) described negative effects of endemic *Pseudomonas* spp. on *T. harzianum*. They suggested that negative effects were caused by effective competition for iron by the *Pseudomonas* spp. because addition of iron to naturally infested soil suppressed growth inhibition of *T. harzianum* and also suppressed *Pythium* seed rot of pea. Negative interaction between two biocontrol agents may also be due to detrimental effects of secondary metabolites produced by one organism on the other (Mew et al. 1994). Thus, an important prerequisite for the desired effectiveness of strains appears to be compatibility of the co-inoculated microorganisms. Numerous biotic and abiotic factors contribute to this inconsistent performance of biocontrol agents (Weller 1988). Inadequate colonization of the rhizosphere, limited tolerance to changes in environmental conditions and fluctuation in the production of antifungal metabolites are among the most important factors (Duffy et al. 1996). Antagonism between the indigenous microbial population and biocontrol agent or mixture of biocontrol agents applied can also influence the performance of a biocontrol agent in the rhizosphere.

These results indicate that specific interactions of biocontrol agents influence disease suppression in combination. It is necessary therefore to further investigate microbial interactions that enhance or detract biocontrol efficacy (Handelsman and Stabb 1996) to understand and predict the performance of mixtures of biocontrol agents.

Plant disease control using combinations of antagonistic rhizobacteria and parasitic fungi has been presented in Table 3.5.

3.4.2 Spectrum of Protection by PGPR

Recently, work on the spectrum of PGPR-mediated ISR against different pathogens in different crop plants has been gaining importance. Seed treatment with *P. fluorescens* strain WCS 417 has protected radish through induction of systemic resistance not only against the fungal

Table 3.5 Efficacy of mixtures of PGPR and parasitic fungi formulations against plant disease

Vegetable crop	Components of disease complex	Components of INM	References
Banana	Panama wilt, *Fusarium oxysporum* f. sp. *cubense*	*T. viride* + *P. fluorescens*-sucker treat., pseudo-stem injection, soil treatment	
Capsicum	*M. incognita* + *R. solanacearum*	*P. fluorescens* + *P. chlamydosporia*/*T. harzianum*	Rao et al. (2002)
	M. incognita	*P. chlamydosporia* + *P. fluorescens*	Naik (2004)
Cabbage	*M. incognita* + *Plasmodiophora brassicae*	*P. fluorescens* + *T. viride* + chitin	Samiyappan (2009)
Okra	*M. incognita*	*P. fluorescens* + *G. fasciculatum* + FYM	Verma (2001)
Pumpkin, cluster bean, chilli, watermelon	*M. javanica* + *M. phaseoli*/ *R. solani*/*F. solani*/*F. oxysporum*	*P. aeruginosa* + *P. lilacinus*	Perveen et al. (1998)
Beetroot	Damping off, *Pythium debaryanum*, *P. ultimum*	*Penicillium* spp. + *P. fluorescens* – seed treatment	
Black pepper	Quick wilt, *Phytophthora* sp.	*T. harzianum* (IISR P-26) + *P. fluorescens* (TNAU Pf1) – soil treatment	

root pathogen *F. oxysporum* f. sp. *raphani* but also against the bacterial leaf pathogen *P. syringae* pv. *tomato* and fungal leaf pathogens *Alternaria brassicicola* and *F. oxysporum* (Hoffland et al. 1996). This implies that the same PGPR strain can induce resistance against multiple pathogens in the same crop. ISR by *P. putida* strain 89 B-27 and *S. marcescens* strain 90-166 against anthracnose of cucumber was established by Wei et al. (1991). Later studies showed that the same PGPR strains induced systemic protection against angular leaf spot caused by *P. syringae* pv. *lachrymans* (Liu et al. 1993a), Fusarium wilt incited by *F. oxysporum* f. sp. *cucumerinum* (Liu et al. 1993b) and cucurbit wilt caused by *Erwinia tracheiphila* (Kloepper et al. 1993). Seed treatment of *S. marcescens* strain 90-166 has shown ISR in cucumber against anthracnose, cucumber mosaic virus, bacterial angular leaf spot and cucurbit wilt diseases (Kloepper et al. 1993; Liu et al. 1995a, b). In addition, the same bacterial strain has also been effective in controlling the striped cucumber beetle, *Acalymma vittatum*, and spotted cucumber beetle, *Diabrotica undecimpunctata howardi* (Zehnder et al. 1997a, b). Parallel experiments have shown that the *P. fluorescens* strain Pf1 induces resistance against different pathogens in different crops, namely, *Rhizoctonia solani* in rice (Nandakumar 1998), *Colletotrichum falcatum* in sugarcane (Viswanathan 1999) and *Pythium aphanidermatum* in tomato (Ramamoorthy et al. 1999). The broad spectrum of PGPR-mediated ISR is more rewarding than narrow spectrum of disease protection. Hence, selecting a suitable strain having potential to induce systemic resistance against multiple pathogens and pests is the most important task in the delivery of microbial agents to the field.

References

Aalten PM, Gowen SR (1998) Entomopathogenic nematodes and fluorescent *Pseudomonas* rhizosphere bacteria inhibiting *Radopholus similis* invasion in banana roots. Brighton Crop Prot Conf Pests Dis 2:675–680

Alabouvette C, Lemanceau P, Steinberg C (1993) Recent advances in the biological control of fusarium wilts. Pestic Sci 37:365–373

Ali NI, Siddiqui IA, Shahid J, Shaukat S, Zaki MJ (2002) Nematicidal activity of some strains of *Pseudomonas* spp. Soil Biol Biochem 34:1051–1058

Antoun H, Beauchamp CJ, Goussard N, Chabot R, Lalande R (1998) Potential of *Rhizobium* and *Bradyrhizobium* species as plant growth promoting rhizobacteria on nonlegumes: effect on radishes (*Raphanus sativus* L.). Plant Soil 204:57–67

Backman PA, Wilson M, Murphy JF (1997) Bacteria for biological control of plant diseases. In: Rechcigl NA,

Rechecigl JE (eds) Environmentally safe approaches to crop disease control. Lewis Publishers, Boca Raton, pp 95–109

Banasco P, Fuente DeLa L, Gaultieri G, Noya F, Arias A (1998) Fluorescent *Pseudomonas* spp. as biocontrol agents against forage legume root pathogenic fungi. Soil Biol Biochem 10(Suppl 10–11):1317–1323

Bansal RK, Verma VK (2002) Antagonistic efficacy of *Azotobacter chroococcum* against *Meloidogyne javanica* infecting brinjal. Indian J Nematol 32:132–134

Bell E, Muller JE (1993) Characterization of an *Arabidopsis lipoxygenase* gene responsive to methyl jasmonate and wounding. Plant Physiol 103:1133–1137

Boruah HP, Kumar BS (2002) Biological activity of secondary metabolites produced by a strain of Pseudomonas fluorescens. Folia Microbiol 47:359–363

Broadway RM, Gongora C, Kain WC, Sanderson JA, Monroy JA, Bennett KC, Warner JB, Hoffman MP (1998) Novel chitinolytic enzymes with biological activity against herbivorous insects. J Chem Ecol 24:985–998

Buonassisi AJ, Copeman RJ, Pepin HS, Eaton GW (1986) Effect of *Rhizobium* spp. on *Fusarium solani* f. sp. *phaseoli*. Can J Phytopathol 8:140–146

Burdman S, Volpin H, Kigel J, Kaplunic Y, Okon Y (1996) Appl Environ Microbiol 62:3030–3033

Cao L, Qiu Z, Dai X, Tan H, Lin Y, Zhou S (2004) Isolation of endophytic actinomycetes from roots and leaves of banana (*Musa acuminata*) plants and their activities against *Fusarium oxysporum* f. sp. *cubense*. World J Microbiol Biotechnol 20:501–504

Chahal PPK, Chahal VPS (1986) Effect of *Azotobacter chroococcum* on the hatching of egg masses and eggs of *Meloidogyne incognita*. Plant Soil 95:289–292

Chahal PPK, Chahal VPS (1988) Biological control of root-knot nematode on brinjal (*Solanum melongena* L.) with *Azotobacter chroococcum*. In: Maqbool MA, Golden AM, Gaffar A, Krusberg LQ (eds) Advances in plant nematology. National Nematological Research Centre, University of Karachi, Karachi, pp 257–263

Chahal VPS, Chahal PPK (1999) Final technical report of PL – 480 project on microbial control of plant parasitic nematodes submitted to USDA

Chahal VPS, Chahal PPK (2003) *Bacillus thuringiensis* for the control of *Meloidogyne incognita*. In: Trivedi PC (ed) Nematode management in plants. Scientific Publishers (India), Jodhpur, pp 251–257

Chakraborty U, Chakraborty B, Basnet M (2006) Plant growth promotion and induction of resistance in *Camellia sinensis* by *Bacillus megaterium*. J Basic Microbiol 46(Suppl 3):186–195

Charest MH, Beauchamp CJ, Antoun H (2005) Effects of the humic substances of de-inking paper sludge on the antagonism between two compost bacteria and *Pythium ultimum*. FEMS Microbiol Ecol 52:219–227

Cook RJ, Baker KF (1983) The nature and practice of biological control of plant pathogens. American Phytopathological Society, St. Paul

Dandurand LM, Knudsen GR (1993) Influence of *Pseudomonas fluorescens* on hyphal growth and biocontrol efficacy of *Trichoderma harzianum* in the spermosphere and rhizosphere of pea. Phytopathology 83:265–270

Dicke M, Van Loon JJA (2000) Multitrophic effects of herbivore-induced plant volatiles in an evolutionary context. Entomol Exp Appl 97:237–249

Duffy BK, Weller DM (1995) Use of *Gaeumannomyces graminis* var. *graminis* alone and in combination with fluorescent *Pseudomonas* spp. to suppress take-all of wheat. Plant Dis 79:907–911

Duffy BK, Simon A, Weller DM (1996) Combination of *Trichoderma koningii* with fluorescent pseudomonads for control of take-all on wheat. Phytopathology 86:188–194

Eapen SJ, Ramana KV, Sarma YR (1997) Evaluation of *Pseudomonas fluorescens* isolates for control of *Meloidogyne incognita* in black pepper (*Piper nigrum* L). In: Edison S, Ramana KV, Sasikumar B, Babu KN, Eapen SJ (eds) Biotechnology of spices, medicinal & aromatic plants. Indian Institute of Spices Research, Calicut, pp 129–133

Fuhrmann J, Wollum AG (1989) Biol Fertil Soils 7:108–112

Goel AK, Sindhu SS, Dadarwal KR (2002) Biol Fertil Soils 36:391–396

Guetsky R, Elad Y, Shtienberg D, Dinoor A (2002) Establishment, survival and activity of the biocontrol agents *Pichia guilermondii* and *Bacillus mycoides* applied as a mixture on strawberry plants. Biocontrol Sci Technol 12:705–714

Haas D, Defago G (2005) Nat Rev Microbiol 3:307–319

Hallmann J, Quadt-Hallmann A, Miller WG, Sikora RA, Lindow SE (2001) Endophytic colonization of plants by the biocontrol agent *Rhizobium etli* G12 in relation to *Meloidogyne incognita* infection. Phytopathology 91:415–422

Hanafi A, Traore M, Schnitzler WH, Woitke M (2007) Induced resistance of tomato to whiteflies and Pythium with the PGPR Bacillus subtilis in a soilless crop grown under greenhouse conditions. In: Hanafi A, Schnitzler WH (eds) Proceedings of the VIIIth IS on protected cultivation in mild winter climates, pp 315–322. Acta Hort. 747

Handelsman J, Stabb EV (1996) Biocontrol of soilborne plant pathogens. Plant Cell 8:1855–1869

Hassett D, Cuppoletti J, Trapnell B, Lymar S, Rowe J, Yoon S, Hilliard G, Parvatiyar K, Kamani M, Wozniak D, Hwang S, McDermott T, Ochsner U (2002) Adv Drug Deliv Rev 54:1425–1443

Hoffland E, Hakulinem J, Van Pelt JA (1996) Comparison of systemic resistance induced by avirulent and non-pathogenic *Pseudomonas* species. Phytopathology 86:757–762

Hubbard JP, Harmand GE, Hadar Y (1983) Effect of soil-borne *Pseudomonas* spp. on the biological control agent, *Trichoderma hamatum*, on pea seeds. Phytopathology 73:655–659

Insunza V, Alstrom S, Eriksson KB (2002) Root bacteria from nematicidal plants and their biocontrol potential against trichodorid nematodes in potato. Plant Soil 241:271–278

Janisiewicz WJ (1996) Ecological diversity, niche overlap and coexistence of antagonists used in developing mixtures for biocontrol of postharvest diseases of apples. Phytopathology 86:473–479

Kavitha K, Nakkeeran S, Chandrasekar G, Fernando WGD, Mathiyazhagan S, Renukadevi P, Krishnamoorthy AS (2003) Role of antifungal antibiotics, siderophores and IAA production in biocontrol of *Pythium aphanidermatum* inciting damping off in tomato by *Pseudomonas chlororaphis* and *Bacillus subtilis*. Proceedings of the 6th international workshop on PGPR. Indian Institute of Spice Research, Calicut, pp 493–497

Kendrick WB (1963) Mycopath Appl 19:241–245

Khan MR, Tarannum Z (1999) Effects of field application of various micro-organisms on *Meloidogyne incognita* on tomato. Nematol Medit 27:233–238

Kilian M, Steiner U, Krebs B, Junge H, Schmiedeknecht G, Hain R (2000) FZB24 *Bacillus subtilis* – mode of action of a microbial agent enhancing plant vitality. Pflanzenschutz-Nachrichten Bayer 1:72–93, 1/00

Kloepper JW (1993) Plant growth promoting rhizobacteria as biological control agents. In: Metting FB Jr (ed) Soil microbial ecology- applications in agricultural and environmental management. Marcel Dekker, New York, pp 255–274

Kloepper JW, Schroth MN (1981) Development of powder formulation of rhizobacteria for inoculation of potato seed pieces. Phytopathology 71:590–592

Kloepper JW, Tuzun S, Liu L, Wei G (1993) Plant growth-promoting rhizobacteria as inducers of systemic disease resistance. In: Lumsden RD, Waughn JL (eds) Pest management: biologically based technologies. American Chemical Society Books, Washington, DC, pp 156–165

Kremer RJ, Kennedy AC (1996) Rhizobacteria as biocontrol agents of weeds. Weed Technol 10(Suppl 3): 601–609

Kumar BSD, Berggren I, Martensson AM (2001) Potential for improving pea production by co-inoculation with fluorescent *Pseudomonas* and *Rhizobium*. Plant Soil 229:25–34

Larkin RP, Fravel DR (1998) Efficacy of various fungal and bacterial biocontrol organisms for control of Fusarium wilt of tomato. Plant Dis 82:1022–1028

Leavy E, Eyal Z, Chet I, Hochman A (1992) Physiol Mol Plant Pathol 40:163–171

Leeman M, Den Ouden FM, van Pelt JA, Dirks FPM, Steiji H (1996) Iron availability affects induction of systemic resistance to Fusarium wilt of radish by *Pseudomonas fluorescens*. Phytopathology 86:149–155

Lemanceau P, Alabouvette C (1993) Suppression of Fusarium wilts by fluorescent *Pseudomonas*: mechanisms and applications. Biocontrol Sci Technol 3:219–234

Lemanceau P, Bakker PAHM, de Kogel WJ, Alabouvette C, Schippers B (1993) Antagonistic effect of non-pathogenic *Fusarium oxysporum* Fo47 and pseudobactin 358 upon pathogenic *Fusarium oxysporum* f. sp. *dianthi*. Appl Environ Microbiol 59:74–82

Liu L, Kloepper JW, Tuzun S (1993a) Induction of systemic resistance against cucumber bacterial angular leaf spot caused by *Pseudomonas syringae* pv. *lachrymans* with two plant growth promoting rhizobacteria (Abstr.). Phytopathology 82:1340

Liu L, Kloepper JW, Tuzun S (1993b) Induction of systemic resistance to *Fusarium oxysporum* by PGPR strains (Abstr.). In: Proceedings of the sixth international congress of plant pathology, vol 24

Liu L, Kloepper JW, Tuzun S (1995a) Induction of systemic resistance in cucumber against Fusarium wilt by plant growth promoting rhizobacteria. Phytopathology 85:695–698

Liu L, Kloepper JW, Tuzun S (1995b) Induction of systemic resistance in cucumber against bacterial leaf spot by plant growth promoting rhizhobacteria. Phytopathology 85:843–847

Lugtenberg BJJ, Dekkers L, Bloemberg GV (2001) Molecular determinants of rhizosphere colonization by *Pseudomonas*. Ann Rev Phytopathol 38:461–490

Mani MP, Rajeswari S, Sivakumar CV (1998) Management of the potato cyst nematodes, *Globodera* spp. through plant rhizosphere bacterium *Pseudomonas fluorescens* Migula. J Biol Control 12:131–134

Mani MP, Sivakumar CV, Ramakrishnan S (1999) Status of *Pasteuria penetrans* in root-knot nematode infested vineyards of Tamil Nadu. Indian J Nematol 29:104

Manjula K, Podile AR (2001) Chitin supplemented formulations improve biocontrol and plant growth promoting efficiency of *Bacillus subtilis* AF1. Can J Microbiol 47:618–625

Mathiyazhagan S, Kavitha K, Nakkeeran S, Chandrasekar G, Manian K, Renukadevi P, Krishnamoorthy AS, Fernando WGD (2004) PGPR mediated management of stem blight of *Phyllanthus amarus* (Schum and Thonn) caused by *Corynespora cassiicola* (Berk and Curt) Wei. Arch Phytopathol Plant Prot 33:183–199

Mazzola M (2002) Mechanisms of natural soil suppressiveness to soilborne diseases. Antonie Van Leeuwenhoek 81:557–564

Mew TW, Rosales AM, Maningas TW (1994) Biological control of rhizoctonia sheath blight and blast of rice. In: Ryder MH, Stephens PM, Bowen GD (eds) Improving plant productivity with rhizosphere bacteria. Proceedings of the third international workshop on plant growth promoting rhizobacteria, Adelaide, South Australia

Nagórska K, Bikowski M, Obuchowski M (2007) Multicellular behaviour and production of a wide variety of toxic substances support usage of *Bacillus subtilis* as a powerful biocontrol agent. Acta Biochim Pol 54(Suppl 3):495–508

Naik D (2004) Biotechnological approaches for the management of wilt disease complex in capsicum (Capsicum annum *L.*) and Egg Plant (Solanum melongena) with special emphasis on biological control. Ph. D. thesis, Kuvempu University, Shimoga

Nakkeeran S, Kavitha K, Mathiyazhagan S, Fernando WGD, Chandrasekar G, Renukadevi P (2004) Induced systemic resistance and plant growth promotion by *Pseudomonas chlororaphis* strain PA-23 and *Bacillus subtilis* strain CBE4 against rhizome rot of turmeric (*Curcuma longa* L). Can J Plant Pathol 26:417–418

Nandakumar R (1998) Induction of systemic resistance in rice with fluorescent pseudomonads for the management of sheath blight disease. M.Sc. (Agri.). thesis, TNAU, Coimbatore, India, 105 pp

Negi YK, Garg SK, Kumar J (2005) Cold-tolerant fluorescent *Pseudomonas* isolates from Garhwal Himalayas as potential plant growth promoting and biocontrol agents in pea. Curr Sci 89(Suppl 12):25

Niknam GR, Dhawan SC (2001a) Induction of systemic resistance by Bacillus subtilis isolate Bs1 against Rotylenchulus reniformis in tomato. National Congress on Centenary of Nematology in India – appraisal & future plans. Indian Agricultural Research Institute, New Delhi, pp 143–144

Niknam GR, Dhawan SC (2001b) Effect of seed bacterization, soil drench and bare root-dip application methods of *Pseudomonas fluorescens* isolate Pf1 on the suppression of *Rotylenchulus reniformis* infecting tomato. National Congress on Centenary of Nematology in India – appraisal & future plans. Indian Agricultural Research Institute, New Delhi, p 144

Otsu Y, Matsuda Y, Mori H, Ueki H, Nakajima T, Fujiwara K, Matsumoto M, Azuma N, Kakutani K, Nonomura T, Sakuratani Y, Shinogi T, Tosa Y, Mayama S, Toyoda H (2004) Stable phylloplane colonization by entomopathogenic bacterium *Pseudomonas fluorescens* KPM-018P and biological control of phytophagous ladybird beetles *Epilacna vigintioctopunctata* (Coleoptera: Coccinellidae). Biocontrol Sci Technol 14:427–439

Parvatha Reddy P, Nagesh M, Rao MS, Rama N (2000) Management of *Tylenchulus semipenetrans* by integration of *Pseudomonas fluorescens* with oil cakes. In: Proceedings of the international symposium on citriculture, Nagpur, India, pp 830–833

Paulitz TC, Belanger RB (2001) Biological control in greenhouse systems. Ann Rev Phytopathol 39:103–133

Perveen S, Ehteshamul-Haque S, Ghaffar A (1998) Efficacy of *Pseudomonas aeruginosa* and *Paecilomyces lilacinus* in the control of root rot-root knot disease complex on some vegetables. Nematol Medit 26:209–212

Pierson EA, Weller DM (1994) Use of mixtures of fluorescent pseudomonads to suppress take-all and improve the growth of wheat. Phytopathology 84:940–947

Raaijimakers MJ, Leeman M, Mark MP, Schot V (1995) Dose response relationships in biological control of *Fusarium* wilt of radish by *Pseudomonas* sp. Phytopathology 85:1075–1081

Raaijmakers JM, Vlami M, de Souza JT (2002) Antibiotic production by bacterial biocontrol agents. Antonie Van Leeuwenhoek 81:537–547

Racke J, Sikora RA (1992) Isolation, formulation and antagonistic activity of rhizosphere bacteria toward the potato cyst nematode. Globodera pallida Soil Biol Biochem 24:521–526

Ramamoorthy V, Raguchander T, Samiyappan R (1999) Isolation, characterization and screening fluorescent pseudomonads for managing damping-off disease of major vegetable crops. Symposium on plant disease management for sustainable agriculture. Indian Phytopathological Society, Southern Zone held at C.P.C.R.I. Kayankulam, Kerala, India, p 26

Ramamoorthy V, Viswanathan R, Raguchander T, Prakasam V, Samiyappan R (2001) Induction of systemic resistance by plant growth promoting rhizobacteria in crop plants against pests and diseases. Crop Prot 20:1–11

Rao MS, Shylaja M (2004) Role of *Pseudomonas fluorescens* (Migula) in induction of systemic resistance (ISR) and managing *Rotylenchulus reniformis* (Linford and Oliveira) on carrot (*Daucus carota* L.). Pest Manag Hort Ecosyst 10:87–93

Rao MS, Naik D, Shylaja M, Parvatha Reddy P (2002) Prospects for the management of nematode disease complex in capsicum using biological control agents. In: Proceedings of international conference on vegetables, Bangalore, pp 347–351

Raupach GS, Kloepper JW (1998) Mixtures of plant growth promoting rhizobacteria enhance biological control of multiple cucumber pathogens. Phytopathology 88:1158–1164

Reitz M, Rudolph K, Schröder I, Hoffmann-Hergarten S, Hallmann J, Sikora RA (2000) Lipopolysaccharides of *Rhizobium etli* strain G12 act in potato roots as an inducing agent of systemic resistance to infection by the cyst nematode *Globodera pallida*. Appl Environ Microbiol 66:3515–3518

Ryan KJ, Ray CG, Sherris (2004) Medical microbiology (4th ed). McGraw Hill

Sabaratnam S, Traquair JA (2002) Formulation of a *Streptomyces* biocontrol agent for the suppression of *Rhizoctonia* damping-off in tomato transplants. Biol Control 23:245–253

Samiyappan R (2009) Biological control of fungal nematode complex diseases by plant growth promoting rhizobacteria (PGPR). In: Rajendran G, Ramakrishnan S, Subramanian S, Jonathan EI, Sivakumar M, Kumar S (eds) Biological control of plant parasitic nematodes. Agrotech Publishing Academy, Udaipur, pp 69–77

Santhi A (2003) Management of Radopholus similis, Helicotylenchus multicinctus and Pratylenchus coffeae in Banana. Ph. D. thesis, Tamil Nadu Agricultural University, Coimbatore

Santhi A, Sivakumar CV (1995) Biocontrol potential of *Pseudomonas fluorescens* (Migula) against root-knot nematode, *Meloidogyne incognita* (Kofoid and White, 1919) Chitwood, 1949 on tomato. J Biol Control 9:113–115

Santhi A, Rajeswari S, Sivakumar CV (1998) Soil application of *Pseudomonas fluorescens* (Migula) for the control of root-knot nematode (*Meloidogyne incognita*) on grapevine (*Vitis vinifera* Linn.). In: Mehta UK (ed) Nematology – challenges and opportunities in

21st century. Sugarcane Breeding Institute, Coimbatore, pp 203–206

Santhi A, Sundarababu R, Sivakumar CV (1999) Field evaluation of rhizobacterium, Pseudomonas fluorescens for the management of the citrus nematode, Tylenchulus semipenetrans. Proceedings of the national symposium on rational approaches in nematode mangement for sustainable agriculture. Indian Agricultural Research Institute, New Delhi, pp 38–42

Schippers B (1992) Prospects for management of natural suppressiveness to control soilborne pathogens. In: Tiamos EC, Panavizas GC, Cook RJ (eds) Biological control of plant diseases, progress and challenges for the future. NATO ASI series A: life sciences, vol 230. Plenum Press, New York, pp 21–34

Schroth MN, Hancock JG (1982) Science 216:1376–1381

Seenivasan N, Parameswaran S, Sridar P, Gopalakrishnan C, Gnanamurthy P (2001) Application of bioagents and neem cake as soil application for the management of root-knot nematode in turmeric. National Congress on Centenary of Nematology in India – appraisal & future plans. Indian Agricultural Research Institute, New Delhi, p 164

Siddiqui ZA (2005) PGPR: prospective biocontrol agents of plant pathogens. In: Siddiqui ZA (ed) PGPR: biocontrol and biofertilization. Springer, Dordrecht, pp 111–142

Siddiqui ZA, Mahmood I (1995) Biological control of *Heterodera cajani* and *Fusarium udum* by *Bacillus subtilis*, *Bradyrhizobium japonicum* and *Glomus fasciculatum* on pigeonpea. Fund Appl Nematol 18:559–566

Siddiqui ZA, Mahmood I (1999) Role of bacteria in the management of plant parasitic nematodes: a review. Bioresour Technol 69:167–179

Sikora RA (1992) Management of the antagonistic potential in agricultural ecosystems for the biological control of plant-parasitic nematodes. Ann Rev Phytopathol 30:245–270

Singh PP, Shin YC, Park CS, Chung YR (1999) Biological control of *Fusarium* wilt of cucumber by chitinolytic bacteria. Phytopathology 89:92–99

Sivakumar M, Vadivelu S (1999) Management of *Meloidogyne incognita* on grapevine using biocontrol agents, botanicals and biofertilizers. Pest Manag Hort Ecosyst 5:127–131

Sobita Devi L, Dutta U (2002) Effect of *Pseudomonas fluorescens* on root-knot (*Meloidogyne incognita*) on okra plant. Indian J Nematol 32:215

Sosamma VK, Koshy PK (1995) Effect of *Pasteuria penetrans* and *Paecilomyces lilacinus* on population build up of root-knot nematode, *Meloidogyne incognita* on black pepper. National symposium on nematode problems of India – an appraisal of the nematode mangement with eco-friendly approaches and biocomponents. Indian Agricultural Research Institute, New Delhi, p 47

Spiegel Y, Cohn E, Galper S, Sharon E, Chet I (1991) Evaluation of a newly isolated bacterium, *Pseudomonas chitinolytica* sp. nov., for controlling the root-knot nematode *Meloidogyne javanica*. Biocontrol Sci Technol 1:115–125

Taechowisan T, Peberdy JF, Lumyong S (2003) Isolation of endophytic actinomycetes from selected plants and their antifungal activity. World J Microbiol Biotechnol 19:381–385

Taechowisan T, Lu C, Shen Y, Lumyong S (2005) Secondary metabolites from endophytic *Streptomyces aureofaciens* CMUAc130 and their antifungal activity. Microbiology 151:1691–1965

Tuzun S, Kloeppper J (1995) Practical application and implementation of induced resistance. In: Hammerschmidt R, Kuć J (eds) Induced resistance to disease in plants. Kluwer Academic Publishers, Dordrecht, pp 152–168

Valenzuela-Soto JH, Estrada-Herna'ndez MG, Ibarra-Laclette E, De'lanoFrier JP (2010) Inoculation of tomato plants (*Solanum lycopersicum*) with growth promoting *Bacillus subtilis* retards whitefly *Bemisia tabaci* development. Planta 231:397–410

Van Peer R, Niemann GJ, Schippers B (1991) Induced resistance and phytoalexin accumulation in biological control of Fusarium wilt of carnation by *Pseudomonas* sp. WCS417r. Phytopathology 81:728–734

Verma AC (2001) N.D. University of agriculture and technology, Kumarganj, Faizabad, Uttar Pradesh. In: Dhawan SC et al. (eds) Indian nematology-progress and perspectives, Division of Nematology, Indian Agricultural Research Institute, New Delhi, pp 121–125

Verma KK, Gupta DC, Paruthi IJ (1999) Preliminary trial on the efficacy of *Pseudomonas fluorescens* as seed treatment against *Meloidogyne incognita* in tomato. Proceedings of the national symposium on rational approaches in nematode management for sustainable agriculture. Indian Agricultural Research Institute, New Delhi, pp 79–81

Vidhyasekaran P, Sethuraman K, Rajappan K, Vasumathi K (1997) Powder formulation of *Pseudomonas fluorescens* to control pigeonpea wilt. Biol Control 8:166–171

Viswanathan R (1999) Induction of systemic resistance against red rot disease in sugarcane by plant growth promoting rhizobacteria. Ph.D. thesis, Tamil Nadu Agricultural University, Coimbatore, India, 175 pp

Wei G, Kloepper JW, Tuzun S (1991) Induction of systemic resistance of cucumber to *Colletotrichum orbiculare* by select strains of plant growth promoting rhizobacteria. Phytopathology 81:1508–1512

Weller DM (1988) Biological control of soil borne plant pathogens in the rhizosphere with bacteria. Ann Rev Phytopathol 26:379–407

Weller DM (2007) *Pseudomonas* biocontrol agents of soilborne pathogens: looking back over 30 years. Phytopathology 97:250–256

Weller DM, Thomashow LS (1993) Use of rhizobacteria for biocontrol. Curr Opin Biotechnol 4:306–311

Weller DM, Raaijmakers JM, McSpadden Gardener BB, Thomashow LS (2002) Microbial populations responsible for specific soil suppressiveness to plant pathogens. Ann Rev Phytopathol 40:309–348

Wipat A, Harwood CR (1999) FEMS Microb Ecol 28:1–9

Zehnder G, Kloepper J, Tuzun S, Yao CB, Wei G, Chambliss O, Shelby R (1997a) Insect feeding on cucumber mediated by rhizobacteria-induced plant resistance. Entomol Exp Appl 83:81–85

Zehnder G, Kloepper JW, Yao C, Wei G (1997b) Induction of systemic resistance in cucumber against cucumber beetles (Coleoptera: Chrysomelidae) by plant growth promoting rhizobacteria. J Econ Entomol 90:196–391

4 Mechanisms of Biocontrol

Plant pathogens such as fungi, bacteria, viruses, nematodes, etc., which cause various diseases in crop plants, are controlled by PGPR. The widely recognized mechanisms of biocontrol mediated by PGPR to a broad spectrum of pathogens are as follows:

- Competitive rhizosphere colonization
- Production of antibiotics and bacteriocins
- Production of siderophores that chelate iron, making it unavailable for growth of pathogens
- Production of lytic enzymes that can lyse cell wall of pathogenic fungi
- Production of antifungal metabolites such as hydrogen cyanide which suppress growth of fungal pathogens
- Detoxification and degradation of toxin produced by pathogens
- Induction of systemic resistance
- Production of biochemical compounds associated with host defence
- Regulation of plant ethylene levels through the ACC deaminase enzyme

PGPR may use more than one of these mechanisms as experimental evidence suggests that biocontrol of plant pathogens is the net result of multiple mechanisms that may be activated simultaneously.

Biocontrol may also be improved by genetically engineered PGPR to overexpress one or more of these traits so that strains with several different anti-pathogen traits can act synergistically (Glick and Bashan 1997).

4.1 Competitive Rhizosphere Colonization

Despite their potential as low-input practical agents of crop protection, application of PGPR has been hampered by inconsistent performance in field tests; this is usually attributed to their poor rhizosphere competence. Rhizosphere competence of biocontrol agents comprises effective root colonization combined with the ability to survive and proliferate along growing plant roots over a considerable time period, in the presence of the indigenous microflora (Lugtenberg and Dekkers 1999). Given the importance of rhizosphere competence as a prerequisite for effective biological control, understanding root–microbe communication (Bais et al. 2004), as affected by genetic (Okubara et al. 2004) and environmental (Pettersson and Baath 2004) determinants in spatial (Bais et al. 2004) and temporal (Chatterton et al. 2004) contexts, will significantly contribute to improve the efficacy of these biocontrol agents.

4.1.1 Competition for Root Niches and Bacterial Determinants

The root surface and the surrounding rhizosphere are significant carbon sinks (Rovira 1965). Photosynthate allocation to this zone can be as high as 40 % (Degenhardt et al. 2003). Thus, along root surfaces there are various suitable nutrient-rich niches attracting a great diversity of

P.P. Reddy, *Plant Growth Promoting Rhizobacteria for Horticultural Crop Protection*,
DOI 10.1007/978-81-322-1973-6_4, © Springer India 2014

microorganisms, including phytopathogens. Competition for these nutrients and niches is a fundamental mechanism by which PGPR protects plants from phytopathogens (Duffy 2001). PGPR reach root surfaces by active motility facilitated by flagella and are guided by chemotactic responses (Nelson 2004). Known chemical attractants present in root exudates include organic acids, amino acids, and specific sugars (Welbaum et al. 2004). Some exudates can also be effective as antimicrobial agents and thus give ecological niche advantage to organisms that have adequate enzymatic machinery to detoxify them (Bais et al. 2004). The quantity and composition of chemoattractants and antimicrobials exuded by plant roots are under genetic and environmental control (Bais et al. 2004). This implies that PGPR competence highly depends either on their abilities to take advantage of a specific environment or on their abilities to adapt to changing conditions. As an example, *Azospirillum* chemotaxis is induced by sugars, amino acids and organic acids, but the degree of chemotactic response to each of these compounds differs among strains (Reinhold et al. 1985).

Bacterial lipopolysaccharides (LPS), in particular the O-antigen chain, can also contribute to root colonization (Dekkers et al. 1998). However, the importance of LPS in this colonization might be strain dependent since the O-antigenic side chain of *Pseudomonas fluorescens* WCS374 does not contribute to potato root adhesion (De Weger et al. 1989), whereas the O-antigen chain of *P. fluorescens* PCL1205 is involved in tomato root colonization (Dekkers et al. 1998). Furthermore, the O-antigenic aspect of LPS does not contribute to rhizoplane colonization of tomato by the plant beneficial endophytic bacterium *P. fluorescens* WCS417r, but, interestingly, this bacterial determinant was involved in endophytic colonization of roots (Duijff et al. 1997).

It has also been recently demonstrated that the high bacterial growth rate and ability to synthesize vitamin B_1 and exude NADH dehydrogenases contribute to plant colonization by PGPR (Dekkers et al. 1998). Another determinant of root colonization ability by bacteria is type IV pili which play a role in plant colonization by endophytic bacteria such as *Azoarcus* sp. (Steenhoudt and Vanderleyden 2000).

4.1.2 Root Colonization and Site-Specific Recombinase

Bacterial traits required for effective root colonization are subject to phase variation, a regulatory process for DNA rearrangements orchestrated by site-specific recombinase (Van der Broek et al. 2003). In certain PGPR, efficient root colonization is linked to their ability to secrete a site-specific recombinase (Dekkers et al. 1998). Transfer of the site-specific recombinase gene from a rhizosphere-competent *P. fluorescens* into a rhizosphere-incompetent *Pseudomonas* strain enhanced its ability to colonize root tips (Dekkers et al. 2000).

4.1.3 Utilization of Root Exudates and Root Mucilage by PGPR

Since root exudates are the primary source of nutrients for rhizosphere microorganisms (Van Overbeek and Van Elsas 1995), rhizosphere competence implies that PGPR are well adapted to their utilization (Lugtenberg et al. 1999). Despite the fact that sugars have often been reported as the major carbon source in exudates, the ability to use specific sugars does not play a major role in tomato root colonization (Lugtenberg et al. 1999). Similarly, although amino acids are present in root exudates, the bioavailability of amino acids alone is considered insufficient to support root tip colonization by auxotrophic mutants of *P. fluorescens* WCS365 (Simons et al. 1997). In contrast, Simons et al. (1997) reported that amino acid synthesis is required for root colonization by *P. fluorescens* WCS365, indicating that amino acid prototrophy is involved in rhizosphere competence. In addition, PGPR regulate the rate of uptake of polyamines such as putrescine, spermine and spermidine, since their high titer could retard bacterial growth and reduce their ability to competitively colonize roots (Kuiper et al. 2001). Root mucilage also offers a utilizable carbon source for PGPR (Knee et al. 2001) to use for the competitive colonization.

4.2 Production of Antibiotics and Bacteriocins

4.2.1 Antibiotics

Fluorescent pseudomonads and *Bacillus* species play an active role in the suppression of pathogenic microorganisms. These bacterial antagonists enforce suppression of plant pathogens by the secretion of extracellular metabolites that are inhibitory at low concentration. Antibiotics are generally considered to be organic compounds of low molecular weight produced by microbes. Antibiosis plays an active role in the biocontrol of plant disease and it often acts in concert with competition and parasitism. Antibiosis has been postulated to play an important role in disease suppression by rhizobacteria. The primary mechanism of biocontrol by fluorescent pseudomonads involves the production of antifungal antibiotics such as 2,4-diacetylphloroglucinol, phenazine-1-carboxylic acid, phenazine-1-carboxamide, pyoluteorin, pyrrolnitrin, oomycin A, viscosinamide, butyrolactones, kanosamine, zwittermycin A, aerugine, rhamnolipids, cepaciamide A and ecomycins; antibacterial antibiotics pseudomonic acid and azomycin; antitumor antibiotics FR901463 and cepafungins; and antiviral antibiotic karalicin (Table 4.1). These antibiotics are known to possess antiviral, antimicrobial, insect and mammalian antifeedant, antihelminthic, phytotoxic, antioxidant, cytotoxic, antitumour and plant growth-promoting activities.

In addition to direct anti-pathogenic action, antibiotics also serve as determinants in triggering induced systemic resistance (ISR) in the plant system and contribute to disease suppression by conferring a competitive advantage to biocontrol agents. Synergism between antibiotics and ISR may further increase host resistance to plant pathogens.

The major antibiotics that play a vital role in the suppression of plant pathogens are grouped into non-volatile and volatile antibiotics.

Table 4.1 Antibiotics produced by rhizobacteria

PGPR	Antibiotics
Pseudomonas spp.	*Antifungal antibiotics*
	Phenazines
	Phenazine-1-carboxylic acid
	Phenazine-1-carboxamide
	Pyrrolnitrin
	Pyoluteorin
	2,4-diacetylphloroglucinol
	Rhamnolipids
	Oomycin A
	Cepaciamide A
	Ecomycins
	DDR
	Viscosinamide
	Sulphonamide
	Pyocyanin
	Butyrolactones
	N-butylbenzene
	Antibacterial antibiotics
	Pseudomonic acid
	Azomycin
	Antitumour antibiotics
	FR901463
	Cepafungins
	Antiviral antibiotic
	Karalicin
Bacillus spp.	Kanosamine
	Zwittermycin A
	Iturin A (cyclopeptide)
	Bacillomycin
	Plipastatins A and B

4.2.1.1 Non-volatile Antibiotics

- Polyketides (2,4-diacetylphloroglucinol, pyoluteorin, mupirocin)
- Heterocyclic nitrogenous compounds (phenazine derivatives)
- Phenylpyrrole (pyrrolnitrin)
- Cyclic lipopeptides
- Lipopeptides (iturin, bacillomycin, plipastatin, surfactin)
- Aminopolyols (zwittermycin A)

4.2.1.2 Volatile Antibiotics

- Hydrogen cyanide
- Aldehydes
- Alcohols
- Ketones
- Sulfides

Table 4.2 Antibiotics of PGPR in the management of soilborne diseases

Antibiotics	PGPR	Pathogen	Crop	References
Phenazine	*Pseudomonas* sp.	*F. oxysporum*	Tomato	Chin-A-Woeng et al. (1998)
Aerugine	*P. fluorescens*	*Phytophthora*	Pepper	Lee et al. (2003)
		Colletotrichum orbiculare	Cucumber	Lee et al. (2003)
Pyrrolnitrin	*Burkholderia cepacia*	*Fusarium sambucinum*	Potato	Burkhead et al. (1994)
	P. fluorescens	*Rhizoctonia solani*	Cucumber	Hammer et al. (1997)
	P. cepacia	*F. sambucinum*	Potato	Burkhead et al. (1994)
Pantocin A, B	*P. agglomerans*	*Erwinia herbicola*	Apple	Wright et al. (2001)

Root-associated fluorescent pseudomonads produce and excrete secondary metabolites which are inhibitory to plant pathogenic organisms including fungi, bacteria and nematodes. Among these metabolites the polyketide compound, DAPG, has received particular attention because of its broad spectrum antifungal, antibacterial and antihelminthic activity. Phenazines (PHZ) are N-containing heterocyclic pigments synthesized by species of *Pseudomonas, Streptomyces, Burkholderia* and *Brevibacterium.* Pyrrolnitrin (PRN) is a broad spectrum antifungal metabolite produced by many fluorescent and nonfluorescent strains of the genus *Pseudomonas*. A phenyl pyrrol derivative of PRN has been developed as an agricultural fungicide. Pyrrolnitrin persists actively in the soil for at least 30 days. Pyoluteorin (PLT) is an aromatic polyketide antibiotic consisting of a resorcinol ring derived through polyketide biosynthesis. PLT is produced by several *Pseudomonas* spp. including strains that suppress plant diseases caused by phytopathogenic fungi (Maurhofer et al. 1994). PLT mainly inhibits the oomycetous fungi including *Pythium ultimum* against which it is strongly active when applied to seeds. PLT-producing pseudomonads decrease the severity of *Pythium* damping off. *P fluorescens* strain CHAO and its antibiotic overproducing derivative CHAO/PME 3424 repeatedly reduced *M. incognita* galling in tomato, brinjal, mung and soybean in early growth stage. A strong negative correlation existed between rhizobacteria colonization and nematode invasion.

Antibiotics, such as polymyxin, circulin and colistin, produced by the majority of *Bacillus* spp. are active against Gram-positive and Gram-negative bacteria as well as many pathogenic fungi (Table 4.2). The *B. cereus* UW85 strain, which suppresses oomycete pathogens and produces the antibiotics zwittermycin A (aminopolyol) and kanosamine (aminoglycoside), contributes to the biocontrol of alfalfa damping off. Regarding bacteria as biocontrol agents to act as a biological solution, some researchers have highlighted the use of sporulating Gram-positive species such as *Bacillus* and *Paenibacillus* spp., which can confer higher population stability during formulation and storage of inoculant products.

4.2.2 Bacteriocins

Other molecules used in microbial defence systems are bacteriocins. According to a review by Riley and Wertz (2002), bacteriocins differ from traditional antibiotics in one critical way: they commonly have a relatively narrow killing spectrum and are only toxic to bacteria closely related to the producing strain. Almost all bacteria may make at least one bacteriocin, and many bacteriocins isolated from Gram-negative bacteria appear to have been created by recombination between existing bacteriocins (Riley 1993). The colicins, proteins produced by some strains of *Escherichia coli* that are lethal for related strains, are the most representative bacteriocins produced by Gram-negative bacteria. Like colicin, a name derived from *E. coli*, other bacteriocins have been thus defined and named, such as pyocins from *P. pyogenes* strains, cloacins from *Enterobacter cloacae*, marcescins from *Serratia marcescens* and megacins from *B. megaterium* (Cascales et al. 2007). Interestingly, bacteriocins from *Bacillus* spp. are increasingly becoming more important due to their sometimes broader spectra of inhibition (as compared with most lactic bacterial bacteriocins), which may include Gram-negative bacteria, yeasts or fungi, in addition to Gram-positive species, some of which are known to be pathogenic to humans and/or animals (Abriouel et al. 2011).

4.3 Production of Siderophores That Chelate Iron, Making It Unavailable for Growth of Pathogens

Siderophores are low molecular weight extracellular compounds with a high affinity for ferric iron that are secreted by microorganisms to take up iron from the environment, and their mode of action in suppression of disease was thought to be solely based on competition for iron with the pathogen. Fluorescent pseudomonads are characterized by the production of yellow-green pigments termed pyoverdines which fluoresce under UV light and function as siderophores. The siderophores produced by fluorescent pseudomonads promoted plant growth. The siderophores of fluorescent pseudomonads were later reported to be implicated in the suppression of plant pathogens. Competition for iron between pathogens and siderophores of fluorescent pseudomonads has been implicated in the biocontrol of wilt diseases caused by *Fusarium* and *Pythium*. Pyoverdines chelate iron in the rhizosphere and deprive pathogens of iron which is required for their growth and pathogenesis. Rhizobacteria produce various types of siderophores (pseudobactin and ferrooxamine B) that chelate the scarcely available iron and thereby prevent pathogens from acquiring iron.

The involvement of siderophore production in disease suppression by *Pseudomonas* strain WCS 358 was studied on carnation and radish using *F. oxysporum* f. sp. *dianthi* and *F. oxysporum* f. sp. *raphani*, respectively. Siderophore production by a plant growth-promoting fluorescent *Pseudomonas* sp. RBT 13 was found effective against several fungal and bacterial pathogens. Several strains of siderophore-producing *P. fluorescens* have been shown to inhibit *F. oxysporum* f. sp. *cubense*, *F. oxysporum* f. sp. *vasinfectum* and *Rhizoctonia solani*.

Pseudomonas culture and purified siderophores showed good antifungal activity against the plant deleterious fungi, namely, *Aspergillus niger*, *A. flavus*, *A. oryzae*, *F. oxysporum* and *Sclerotium rolfsii*. Though siderophores are part of primary metabolism (iron is an essential element), on occasions they also behave as antibiotics which are commonly considered to be secondary metabolites. The siderophores exerted maximum impact on *F. oxysporum* than on *Alternaria* sp. and *Colletotrichum capsici*.

The fluorescent pseudomonads had the property to form ferric siderophore complex which prevents the availability of iron to the microorganisms. Ultimately this led to iron starvation and prevented the survival of the microorganisms including nematodes. *Pseudomonas aeruginosa* strain IE-6 and its streptomycin-resistant strain IE-6+ markedly suppressed nematode population densities in root and subsequent root-knot development. The iron concentration in soil was lowered by the addition of an iron chelator.

4.4 Production of Lytic Enzymes That Can Lyse Cell Wall of Pathogenic Fungi

A variety of microorganisms also exhibit hyperparasitic activity, attacking pathogens by excreting cell wall hydrolases (Chernin and Chet 2002). Chitinase produced by *Serratia plymuthica* C48 inhibited spore germination and germ-tube elongation in *Botrytis cinerea* (Frankowski et al. 2001). The ability to produce extracellular chitinases is considered crucial for *S. marcescens* to act as antagonist against *Sclerotium rolfsii* (Ordentlich et al. 1988), and for *Paenibacillus* sp. strain 300 and *Streptomyces* sp. strain 385 to suppress *F. oxysporum* f. sp. *cucumerinum*. It has been also demonstrated that extracellular chitinase and laminarinase synthesized by *Pseudomonas stutzeri* digest and lyse mycelia of *F. solani* (Lim et al. 1991). Although, chitinolytic activity appears less essential for PGPR such as *S. plymuthica* IC14 when used to suppress *Sclerotinia sclerotiorum* and *B. cinerea*, synthesis of proteases and other biocontrol traits is involved (Kamensky et al. 2003). The β-1,3-glucanase synthesized by *Paenibacillus* sp. strain 300 and *Streptomyces* sp. strain 385 lyse fungal cell walls of *F. oxysporum* f. sp. *cucumerinum* (Singh et al. 1999). *B. cepacia* synthesizes β-1,3-glucanase that destroys the integrity of *R. solani*, *S. rolfsii* and *Pythium ultimum* cell walls (Fridlender et al. 1993). Similar to siderophores and antibiotics, regulation of lytic enzyme production (proteases and chitinases in particular) involves the GacA/GacS (Corbell and Loper 1995) or GrrA/GrrS regulatory systems (Ovadis et al. 2004) and colony phase variation (Lugtenberg et al. 2001).

4.5 Production of Antifungal Metabolites Such as HCN Which Suppress Growth of Fungal Pathogens

The cyanide ion is exhaled as HCN and metabolized to a lesser degree into other compounds. HCN first inhibits the electron transport, and the energy supply to the cell is disrupted leading to the death of the organisms. It inhibits proper functioning of enzymes and natural receptor's reversible mechanism of inhibition and it is also known to inhibit the action of cytochrome oxidase. HCN is produced by many rhizobacteria and is postulated to play a role in biological control of pathogens. *P. fluorescens* HCN inhibited the mycelial growth of *Pythium* in vitro. The cyanide-producing strain CHAO stimulated root hair formation, indicating that the strain induced and altered plant physiological activities. Four of the six PGPR strains that induced systemic resistance in cucumber against *Colletotrichum orbiculare* produced HCN. Fluorescent *Pseudomonas* strain RRS1 isolated from tuberose produced HCN and the strain improved seed germination and root length. Low oxygen concentrations are a prerequisite for the activity of the transcription factor *ANR* which positively regulates HCN biosynthesis.

HCN from *P. fluorescens* strain CHAO not repressed by fusaric acid played a significant role in disease suppression of *F. oxysporum* f. sp. *radicis-lycopersici* in tomato. HCN is a broad-spectrum antimicrobial compound involved in the biological control of root disease by many plant-associated fluorescent pseudomonads. Further, the enzyme HCN synthase is encoded by three biosynthetic genes (*henA*, *henB* and *henC*).

4.6 Detoxification and Degradation of Toxin Produced by Pathogens

Another mechanism of biological control is the detoxification of pathogen virulence factors. For example, certain biocontrol agents are able to detoxify albicidin toxin produced by *Xanthomonas albilineans* (Dong et al. 2000). The detoxification mechanisms include the production of a protein that reversibly binds the toxin in both *Klebsiella oxytoca* (Walker et al. 1988) and *Alcaligenes denitrificans* (Basnayake and Birch 1995), as well as an irreversible detoxification of albicidin mediated by an esterase that occurs in *Pantoea dispersa* (Zhang and Birch 1997). Several different microorganisms, including strains of *Burkholderia cepacia* and *Ralstonia solanacearum*, can also hydrolyze fusaric acid, a phytotoxin produced by various *Fusarium* species (Toyoda et al. 1988). More often though, pathogen toxins display a broad-spectrum activity and can suppress growth of microbial competitors, or detoxify antibiotics produced by some biocontrol microorganisms, as a self-defence mechanism against biocontrol agents (Schouten et al. 2004).

Although biocontrol activity of microorganisms involving synthesis of allelochemicals has been studied extensively with free-living rhizobacteria, similar mechanisms apply to endophytic bacteria (Lodewyckx et al. 2002), since they can also synthesize metabolites with antagonistic activity towards plant pathogens (Chen et al. 1995). For example, Castillo et al. (2002) demonstrated that munumbicins, antibiotics produced by the endophytic bacterium *Streptomyces* sp. strain NRRL 30562 isolated from *Kennedia nigriscans*, can inhibit in vitro growth of phytopathogenic fungi *P. ultimum* and *F. oxysporum*. Subsequently, it has been reported that certain endophytic bacteria isolated from field-grown potato plants can reduce the in vitro growth of *Streptomyces scabies* and *Xanthomonas campestris* through production of siderophore and antibiotic compounds (Sessitsch et al. 2004). Interestingly, the ability to inhibit pathogen growth by endophytic bacteria, isolated from potato tubers, decreases as the bacteria colonize the host plant's interior, suggesting that bacterial adaptation to this habitat occurs within their host and may be tissue type and tissue site specific (Sturz et al. 1999). Aino et al. (1997) have also reported that the endophytic *P. fluorescens* strain FPT 9601 can synthesize DAPG and deposit DAPG crystals around and in the roots of tomato, thus demonstrating that endophyte can produce antibiotics in planta.

4.7 Induction of Systemic Resistance (ISR)

Induced systemic resistance (ISR) is defined as an enhancement of the plant's defensive capacity against a broad spectrum of pathogens and pests that is mediated by rhizobacteria. PGPR bring about ISR through fortifying the physical and mechanical strength of the cell wall as well as changing the physiological and biochemical reaction of the host leading to the synthesis of defence chemicals against the challenge pathogen. Induction of ISR is through jasmonic acid (JA) and ethylene (ET) signalling pathways.

4.7.1 Structural and Ultrastructural Cell Wall Modifications in the Host Plants

PGPR can induce structural changes in the host and these changes were characterized by a considerable enlargement of the callose-enriched wall appositions deposited onto the inner surface of cell wall in the epidermis and outer cortex (Benhamou et al. 1998), callose deposition (M'Piga et al. 1997) and lignification (Kloepper 1993). A strain of *P. fluorescens* functions as an activator of plant disease resistance by inducing callose synthesis in tomato (M'Piga et al. 1997). Bean roots bacterized with a saprophytic fluorescent pseudomonad had higher lignin content than control (Anderson and Guerra 1985).

Treatment of PGPR significantly reduced the germination of sporangia and zoospores of *Phytophthora infestans* on the leaf surface of tomato than the leaves of the non-induced control. *Serratia plymuthica* strain R1GC4 sensitizes susceptible cucumber plants to react more rapidly and efficiently against *Pythium ultimum* attack through the formation of physical barriers at sites of fungal entry (Benhamou et al. 2000). *P. fluorescens* induced accumulation of lignin in pea roots (Benhamou et al. 1996a, b). *Bacillus pumilus* SE34 showed a rapid colonization of all tissues including the vascular stele in tomato and induced resistance against *F. oxysporum* (Benhamou et al. 1998). The main facets of the altered host metabolism concerned the induction of a structural response at sites of fungal entry and the abnormal accumulation of electron-dense substances in the colonized areas.

4.7.2 Biochemical/Physiological Changes in the Host Plants

Application of PGPR results in the biochemical or physiological changes in the plants. Normally ISR by PGPR is associated with the accumulation of PR proteins (pathogenesis-related proteins) and synthesis of phytoalexin and other secondary metabolites.

4.7.2.1 PR Proteins (Pathogenesis-Related Proteins)

Colonization of bean root by fluorescent bacteria is correlated with induction of PR proteins and systemic resistance against *Botrytis cinerea* (Zdor and Anderson 1992). In pea, seed treatment with *P. fluorescens* strain 63-28 has produced hydrolytic enzymes such as chitinases and β-1,3-glucanases. These host lytic enzymes accumulate at the site of penetration of the fungus *F. oxysporum* f. sp. *pisi* resulting in the degradation of fungal cell wall (Benhamou et al. 1996b). Inoculation of tomato plants with the same strain has similarly induced the production of plant chitinases when challenged with the wilt pathogen *F. oxysporum* f. sp. *radicis-lycopersici* (M'Piga et al. 1997). ISR has been correlated with a twofold increase in activity of pathogenesis-related peroxidase and chitinase proteins. *P. fluorescens* Pf1 isolate also induced the accumulation PR proteins in response to infection by *C. capsici* in pepper (Ramamoorthy and Samiyappan 2001). The levels of a PR protein increased in bean leaves following seed treatment with PGPR strains (Hynes and Lazarovits 1989)

4.7.2.2 Phytoalexins and Phenolic Compounds

Accumulation of phytoalexin in response to *Pseudomonas* sp. strain WCS 417r treatment in carnation results in the protection of carnation from wilt disease incidence (Van Peer et al. 1991). Increased peroxidase activity and an increase in the level of mRNAs encoding for phenylalanine ammonia lyase (PAL) and chalcone synthase (which lead to synthesis of phytoalexin) are recorded in the early stages of interaction between bean roots and various bacterial endophytes (Zdor and Anderson 1992).

PGPR are known to produce oxidative stress enzymes in response to pathogen attack. Rhizosphere colonization by *Pseudomonas aeruginosa* 7NSK2 activated PAL in bean roots and increased the salicylic acid levels in leaves (De Meyer et al. 1999). Increased activity of PAL was observed in *P. fluorescens*-treated tomato and pepper plants in response to infection by *F. oxysporum* f. sp. *lycopersici* and *Colletotrichum capsici* (Ramamoorthy and Samiyappan 2001). In bean, rhizosphere colonization of various bacteria induced peroxidase (PO) activity (Zdor and Anderson 1992). The higher PO activity was noticed in cucumber roots treated with *Pseudomonas corrugata* and inoculated with *Pythium aphanidermatum* (Chen et al. 2000). Earlier and increased activities of phenylalanine ammonia lyase (PAL), peroxidase (PO) and polyphenol oxidase (PPO) were observed in *P. fluorescens* Pf1-pretreated tomato and hot pepper plants challenged with *P. aphanidermatum*. Phenolic compounds are toxic to pathogens in nature and may increase the mechanical strength of the host cell wall. Accumulation of phenolics by prior application of *P. fluorescens* in pea has been reported against *P. ultimum* and *F. oxysporum* f. sp. *pisi* (Benhamou et al. 1996a). Similarly, *Serratia plymuthica* induced the accumulation of phenolics in cucumber roots following infection by *P. ultimum* (Benhamou et al. 2000). Moreover, *P. fluorescens* Pf1 isolate also induced the accumulation of phenolic substances in response to infection by *F. oxysporum* f. sp. *lycopersici* in tomato (Ramamoorthy et al. 2001).

4.7.3 PGPR Determinants in ISR

ISR requires jasmonic acid (JA) and ethylene (ET) signalling pathways (Van Loon et al. 1998). These accumulating signalling molecules coordinate the defence responses and, when applied exogenously, are sufficient to induce resistance. In *Arabidopsis*, WCS417r-mediated ISR required components of the JA and ET response pathways. Infected plants increased their levels of JA and ET as a sign of active defence (Mauch et al. 1994). These signalling molecules coordinate the activation of a large set of defence responses and, when applied exogenously, can induce resistance themselves (Pieterse et al. 1998). The dependency of ISR on JA and ET is based on enhanced sensitivity to these hormones rather than on an increase in their production (Pieterse et al. 2001). The *Arabidopsis* JA response mutant *jar1* and the

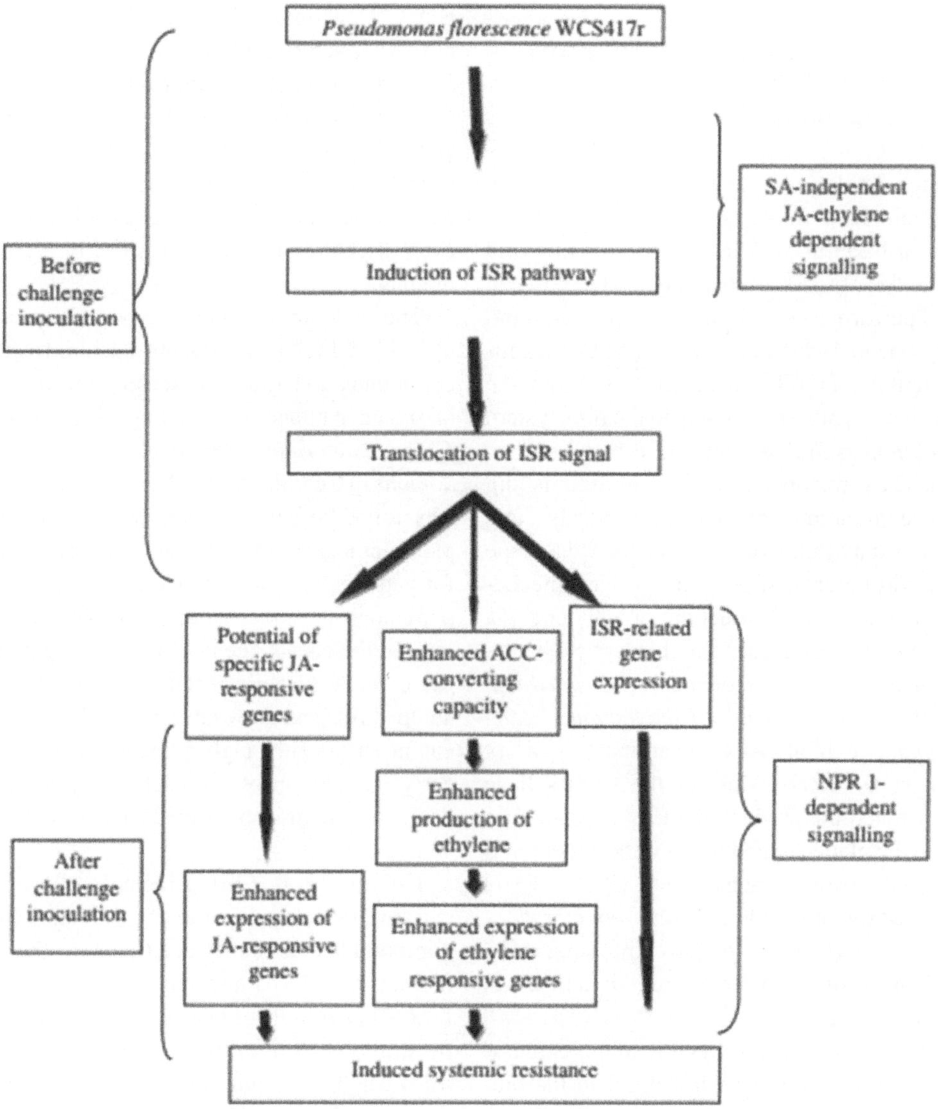

Fig. 4.1 Possible involvement of jasmonic acid and ethylene in *Pseudomonas fluorescens* WCS417r-mediated induced systemic resistance in *Arabidopsis* (Pieterse et al. 2001)

ET response mutant *etr1* were required in the development of ISR (Fig. 4.1).

Pieterse et al. (2001) studied the rhizobacterial strain *P. fluorescens* to enhance the defensive capacity in plants against broad-spectrum foliar pathogens. Based on their experiments, they concluded that *P. fluorescens* strain WCS417r could elicit systemic disease resistance in plants through a variety of signal translocation pathways like SA-independent JA–ethylene-dependent signalling, ISR-related gene expression, NPR 1-dependent signalling, etc.

Methyl jasmonate (MeJA) and the ethylene precursor 1-aminocyclopropane-1-carboxylate (ACC) also promote resistance against *P. syringae* pv. *tomato*. *P. fluorescens* strain WCS417r-mediated ISR follows a signalling pathway in which components from the JA and ethylene response pathways are successively engaged to trigger a defence reaction that is regulated by NPR1 (Pieterse et al. 1998).

4.8 Regulation of Plant Ethylene Levels Through ACC Deaminase Enzyme

Although ethylene is essential for normal growth and development in plants, at high concentration it can be harmful as it induces defoliation and other cellular processes that may lead to reduced crop performance. Using their 1-amino cyclopropane-1-carboxylic acid (ACC) deaminase activity, PGPR can divert ACC from the ethylene biosynthesis pathway in the root system of *Arabidopsis thaliana* plant (Desbrosses et al. 2009). Thus, rhizobacteria assist in diminishing the accumulation of ethylene levels and re-establish a healthy root system needed to cope with environmental stress. The primary mechanism includes the destruction of ethylene via enzyme ACC deaminase. The rhizosphere bacteria such as *Achromobacter, Azospirillum, Bacillus, Enterobacter, Pseudomonas* and *Rhizobium* exhibit ACC deaminase activity (Duan et al. 2009). Most of the studies have demonstrated the production of ACC deaminase gene in the plants treated with PGPR under environmental stress. Grichko and Glick (2001) inoculated tomato seeds with *Enterobacter cloacae* and *Pseudomonas putida* expressing ACC deaminase activity and registered an increase in plant resistance.

Genetic modification of PGPR strains expressing ACC deaminase gene is helpful in the biological control of various plant diseases. PGPR containing ACC deaminase can boost the plant growth particularly under stressed environmental conditions like salinity, drought, waterlogging, temperature, pathogenicity and contaminants in response to a multitude of abiotic and biotic stresses (Saleem et al. 2007).

4.9 Quorum-Sensing Disruption

Recently, it has been discovered that certain PGPR quench pathogen quorum-sensing capacity by degrading autoinducer signals, thereby blocking expression of numerous virulence genes (Newton and Fray 2004). Since most, if not all, bacterial plant pathogens rely upon autoinducer-mediated quorum sensing to turn on gene cascades for their key virulence factors (e.g. cell-degrading enzymes and phytotoxins) (von Bodman et al. 2003), this approach holds tremendous potential for alleviating disease, even after the onset of infection, in a curative manner.

The volatile organic compounds produced by certain PGPR can disrupt bacterial cell–cell communication (quorum sensing) in a number of plant pathogens including *Agrobacterium, Chromobacterium, Pectobacterium* and *Pseudomonas* (Chernin et al. 2011). Quorum sensing was found to exist in many bacteria including plant pathogens. In pathogens, processes critical for pathogenicity were found to be triggered in a population-dependent manner. In *Erwinia carotovora*, the causal agent of soft rot in potato, for example, the bacteria produce enzymes to macerate the host (and draw nutrients) only when present in sufficiently high numbers. This prevents early recognition of the bacteria by the plant's defence systems which could have in turn prevented disease.

Chernin et al. (2011) recently showed that certain PGPR are capable of producing volatile compounds that can reduce (quench) the signal molecules of pathogens. Specifically, they found that volatile organic compounds produced by strains of *P. fluorescens* and *Serratia plymuthica* can disrupt quorum sensing in a number of plant pathogens including *Agrobacterium, Chromobacterium, Pectobacterium* and *Pseudomonas*. Since PGPR are used as agricultural inputs in many crops, 'quorum-quenching' could serve as a new disease management strategy.

The discovery of microbial QS signalling led to identification of numerous enzymatic and nonenzymatic signal interference mechanisms that could quench microbial QS signalling (Zhang and Dong 2004) and inhibition of biofilm formation (Ren et al. 2001). QS activation is mediated by a small autoinducer (AI) molecule, responsible for cell–cell communication and the coordinated action in many bacteria, including PGPR. Commonly reported autoinducer signals are N-acyl homoserine lactones (AHLs) (von Bodman et al. 2003), although half a dozen of other molecules, including diketopiperazines in several Gram-negative bacteria (Holden et al. 1999) and c-butyrolactone in *Streptomyces* (Yamada and Nihira 1998), have also been implicated in density-dependent signalling.

The potential of this tool was further broadened when Gadoury et al. (2012) suggested a similar quorum-sensing mechanism to exist in the grape powdery mildew pathogen *Erysiphe necator*. Here, the authors indicate that production of spores (conidia) is triggered only after the fungus has grown some amount of mycelium and suggested the role of quorum sensing in triggering this condition. This is the first suggestion of quorum sensing in fungal plant pathogens.

Although, the use of quorum-sensing disruption techniques as a tool for disease management is still in its infancy, the potential applications of this idea are truly fascinating.

References

Abriouel H, Franz CM, Ben Omar N, Gálvez A (2011) Diversity and applications of *Bacillus* bacteriocins. FEMS Microbiol Rev 35:201–232

Aino M, Maekawa Y, Mayama S, Kato H (1997) Biocontrol of bacterial wilt of tomato by producing seedlings colonized with endophytic antagonistic pseudomonads. In: Ogoshi A, Kobayashi K, Homma Y, Kodama F, Kondo N, Akino S (eds) Plant growth promoting rhizobacteria: present status and future prospects. Nakanishi Printing, Sapporo, pp 120–123

Anderson AJ, Guerra D (1985) Responses of bean to root colonization with *Pseudomonas putida* in hydroponic system. Phytopathology 75:992–995

Bais HP, Park SW, Weir TL, Callaway RM, Vivanco JM (2004) How plants communicate using the underground information superhighway. Trends Plant Sci 9:26–32

Basnayake WVS, Birch RG (1995) A gene from *Alcaligenes denitrificans* that confers albicidin resistance by reversible antibiotic binding. Microbiology 141:551-560

Benhamou N, Belanger RR, Paulitz T (1996a) Ultrastructural and cytochemical aspects of the interaction between *Pseudomonas fluorescens* and Ri T-DNA transformed pea roots: host response to colonization by *Pythium ultimum* Trow. Planta 199:105–117

Benhamou N, Kloepper JW, Quadt-Hallmann A, Tuzun S (1996b) Induction of defense-related ultrastructural modifications in pea root tissues inoculated with endophytic bacteria. Plant Physiol 112:919–929

Benhamou N, Kloepper JW, Tuzun S (1998) Induction of resistance against Fusarium wilt of tomato by combination of chitosan with an endophytic bacterial strain: ultrastructural and cytochemistry of the host response. Planta 204:153–168

Benhamou N, Gange S, Quere D, Le Dehbi L (2000) Bacterial-mediated induced resistance in cucumber: beneficial effect of the endophytic bacterium *Serratia plymuthica* on the protection against infection by *Pythium ultimum*. Phytopathology 90:45–46

Burkhead KD, Schisler DA, Slininger PJ (1994) Pyrrolnitrin production by biological control agent *Pseudomonas cepacia* B37w in culture and in colonized wounds of potatoes. Appl Environ Microbiol 60:2031–2039

Cascales E, Buchanan SK, Duché D, Kleanthous C, Lloubès R, Postle K, Riley M, Slatin S, Cavard D (2007) Colicin biology. Microbiol Mol Biol Rev 71:158–229

Castillo UF, Strobel GA, Ford EJ, Hess WM, Porter H, Jensen JB, Albert H, Robison R, Condron MAM, Teplow DB, Steevens D, Yaver D (2002) Munumbicins, wide-spectrum antibiotics produced by *Streptomyces* NRRL 30562, endophytic on *Kennedia nigriscans*. Microbiology 148:2675–2685

Chatterton S, Sutton JC, Boland GJ (2004) Timing *Pseudomonas chlororaphis* applications to control *Pythium aphanidermatum*, *Pythium dissotocum*, and root rot in hydroponic peppers. Biol Control 30:360–373

Chen C, Bauske EM, Musson G, Rodriguez-Kabana R, Kloepper JW (1995) Biological control of *Fusarium* wilt on cotton by use of endophytic bacteria. Biol Control 5:83–91

Chen J, Abawi GS, Zuckerman BM (2000) Efficacy of *Bacillus thuringiensis, Paecilomyces marquandii* and *Streptomyces costaricanus* with and without organic amendments against *Meloidogyne hapla* infecting lettuce. J Nematol 32:70–77

Chernin L, Chet I (2002) Microbial enzymes in biocontrol of plant pathogens and pests. In: Burns RG, Dick RP (eds) Enzymes in the environment: activity, ecology, and applications. Marcel Dekker, New York, pp 171–225

Chernin L, Toklikishvili N, Ovadis M, Kim S, Ben-Ari J, Khmel I, Vainstein A (2011) Quorum-sensing quenching by rhizobacterial volatiles. Environ Microbiol Rep 3:698–704

Chin-A-Woeng TFC, Bloemberg GV, van der Bij AJ, van der Drift KMGM, Schripsema J, Kroon B, Scheffer RJ, Keel C, Bakker PAHM, De Bruijn FJ, Thomas-Oates JE, Lugtenberg BJJ (1998) Biocontrol by phenazine-1-carboxamide producing *Pseudomonas chlororaphis* PCL1391 of tomato root rot caused by *Fusarium oxysporum* f. sp. *radicis-lycopersici*. Mol Plant Microbe Interact 10:79–86

Corbell N, Loper JE (1995) A global regulator of secondary metabolite production in *Pseudomonas fluorescens* Pf-5. J Bacteriol 177:6230–6236

De Meyer G, Capieau C, Audenaert K, Buchala A, Metraux JP, Hofte M (1999) Nanogram amounts of salicylic acid produced by the rhizobacterium *Pseudomonas aeruginosa* 7NSK2 activate the systemic acquired resistance pathway in bean. Mol Plant Microbe Interact 12:450–458

De Weger LA, Bakker PAHM, Schippers B, van Loosdrecht MCM, Lugtenberg B (1989) *Pseudomonas* spp. with mutational changes in the O-antigenic side chain of their lipopolysaccharides are affected in their ability to colonize potato roots. In: Lugtenberg BJJ (ed) Signal molecules in plant-microbe interactions. Springer, Berlin, pp 197–202

Degenhardt J, Gershenzon J, Baldwin IT, Kessler A (2003) Attracting friends to feast on foes: engineering terpene emission to make crop plants more attractive to herbivore enemies. Curr Opin Biotechnol 14:169–176

Dekkers LC, Phoelich CC, van der Fits L, Lugtenberg BJJ (1998) A site-specific recombinase is required for competitive root colonization by *Pseudomonas fluorescens* WCS365. Proc Natl Acad Sci U S A 95:7051–7056

Dekkers LC, Mulders IH, Phoelich CC, Chin-A-Woeng TFC, Wijfjes AH, Lugtenberg BJJ (2000) The *sss* colonization gene of the tomato-*Fusarium oxysporum* f. sp. *radicis-lycopersici* biocontrol strain *Pseudomonas fluorescens* WCS365 can improve root colonization of other wild-type *Pseudomonas* spp. bacteria. Mol Plant Microbe Interact 13:1177–1183

Desbrosses G, Contesto C, Varoquaux F, Galland M, Touraine B (2009) PGPR-Arabidopsis interactions is a useful system to study signalling pathways involved in plant developmental control. Plant Signal Behav 4:321–323

Dong YH, Xu JL, Li XZ, Zhang LH (2000) AiiA, an enzyme that inactivates the acylhomoserine lactone quorum-sensing signal and attenuates the virulence of *Erwinia carotovora*. Proc Natl Acad Sci U S A 97:3526–3531

Duan J, Muller KM, Charles TC, Vesely S, Glick BR (2009) 1-Aminocyclopropane-1-carboxylate (ACC) deaminase genes in rhizobia from Southern Saskatchewan. Microb Ecol 57:423–436

Duffy BK (2001) Competition. In: Maloy OC, Murray TD (eds) Encyclopedia of plant pathology. Wiley, New York, pp 243–244

Duijff BJ, Gianinazzi-Pearson V, Lemanceau P (1997) Involvement of the outer membrane lipopolysaccharides in the endophytic colonization of tomato roots by biocontrol *Pseudomonas fluorescens* strain WCS417r. New Phytol 135:325–334

Frankowski J, Lorito M, Scala F, Schmidt R, Berg G, Bahl H (2001) Purification and properties of two chitinolytic enzymes of *Serratia plymuthica* HRO-C48. Arch Microbiol 176:421–426

Fridlender M, Inbar J, Chet I (1993) Biological control of soilborne plant pathogens by a β-1, 3-glucanase-producing *Pseudomonas cepacia*. Soil Biol Biochem 25:1211–1221

Gadoury DM, Wakefield LM, Cadle-Davidson L, Dry IB, Seem RC (2012) Effects of prior vegetative growth, inoculum density, light, and mating on conidiation of Erysiphe necator. Phytopathology 102:65–72

Glick BR, Bashan Y (1997) Genetic manipulation of plant growth-promoting bacteria to enhance biocontrol of phytopathogens. Biotechnol Adv 15:353–376

Grichko VP, Glick BR (2001) Amelioration of flooding stress by ACC deaminase-containing plant growth-promoting bacteria. Plant Physiol Biochem 39:11–17

Hammer PE, Hill DS, Lam ST, van Pee KH, Ligon JM (1997) Four genes from *Pseudomonas fluorescens* that encode the biosynthesis of pyrrolnitrin. Appl Environ Microbiol 63:2147–2154

Holden MTG et al (1999) Quorum-sensing cross-talk: isolation and chemical characterization of cyclic dipeptides from *Pseudomonas aeruginosa* and other gram-negative bacteria. Mol Microbiol 33: 1254–1266

Hynes RK, Lazarovits G (1989) Effect of seed treatment with plant growth promoting rhizobacteria on the protein profiles of intercellular fluids from bean and tomato leaves. Can J Plant Pathol 11:191

Kamensky M, Ovadis M, Chet I, Chernin L (2003) Soil-borne strain IC14 of *Serratia plymuthica* with multiple mechanisms of antifungal activity provides biocontrol of *Botrytis cinerea* and *Sclerotinia sclerotiorum* diseases. Soil Biol Biochem 35:323–331

Kloepper JW (1993) Plant growth promoting rhizobacteria as biological control agents. In: Metting FB Jr (ed) Soil microbial ecology- applications in agricultural and environmental management. Marcel Dekker, New York, pp 255–274

Knee EM, Gong FC, Gao M, Teplitski M, Jones AR, Foxworthy A, Mort AJ, Bauer WD (2001) Root mucilage from pea and its utilization by rhizosphere bacteria as a sole carbon source. Mol Plant Microbe Interact 14:775–784

Kuiper I, Bloemberg GV, Noreen S, Thomas-Oates JE, Lugtenberg BJJ (2001) Increased uptake of putrescine in the rhizosphere inhibits competitive root colonization by *Pseudomonas fluorescens* strain WCS365. Mol Plant Microbe Interact 14: 1096–1104

Lee JY, Moon SS, Hwang BK (2003) Isolation and antifungal and antioomycete activities of aerugine produced by *Pseudomonas fluorescens* strain MM-B16. Appl Environ Microbiol 69:2023–2031

Lim HS, Kim YS, Kim SD (1991) *Pseudomonas stutzeri* YPL-1 genetic transformation and antifungal mechanism against *Fusarium solani*, an agent of plant root rot. Appl Environ Microbiol 57:510–516

Lodewyckx C, Vangronsveld J, Porteous F, Moore ERB, Taghavi S, Mezgeay M, van der Lelie D (2002) Endophytic bacteria and their potential applications. Crit Rev Plant Sci 21:583–606

Lugtenberg BJJ, Dekkers LC (1999) What make *Pseudomonas* bacteria rhizosphere competent? Environ Microbiol 1:9–13

Lugtenberg BJ, Kravchenko LV, Simons M (1999) Tomato seed and root exudate sugars: composition, utilization by *Pseudomonas* biocontrol strains, and role in rhizosphere colonization. Environ Microbiol 1:439–446

Lugtenberg BJJ, Dekkers L, Bloemberg GV (2001) Molecular determinants of rhizosphere colonization by *Pseudomonas*. Ann Rev Phytopathol 38:461–490

Mauch F, Hadwiger LA, Boller T (1994) Ethylene: symptom, not signal for the induction of chitinase and β-1, 3-glucanase in pea pods by pathogens and elicitors. Plant Physiol 76:607–611

Maurhofer M, Hase C, Meuwly P, Metraux JP, Defago G (1994) Induction of systemic resistance of tobacco to tobacco necrosis virus by the root colonizing *Pseudomonas fluorescens* strain CHAO: influence of the *gag*A gene and of pyoverdine production. Phytopathology 84:678–684

M'Piga P, Belanger RR, Paulitz TC, Benhamou N (1997) Increased resistance to *Fusarium oxysporum* f. sp. *radicis-lycopersici* in tomato plants treated with the endophytic bacterium *Pseudomonas fluorescens* strain 63–28. Physiol Mol Plant Pathol 50:301–320

Nelson EB (2004) Microbial dynamics and interactions in the spermosphere. Ann Rev Phytopathol 42:271–309

Newton JA, Fray RG (2004) Integration of environmental and host-derived signals with quorum sensing during plant-microbe interactions. Cell Microbiol 6:213–224

Okubara PA, Kornoely JP, Landa BB (2004) Rhizosphere colonization of hexaploid wheat by *Pseudomonas fluorescens* strains Q8rl-96 and Q2-87 is cultivar-variable and associated with changes in gross root morphology. Biol Control 30:392–403

Ordentlich A, Elad Y, Chet I (1988) The role of chitinase of *Serratia marcescens* in biocontrol of *Sclerotium rolfsii*. Phytopathology 78:84–88

Ovadis M, Liu X, Gavriel S, Ismailov Z, Chet I, Chernin L (2004) The global regulator genes from biocontrol strain *Serratia plymuthica* IC1270: cloning, sequencing, and functional studies. J Bacteriol 186:4986–4993

Pettersson M, Baath E (2004) Effects of the properties of the bacterial community on pH adaptation during recolonization of a humus soil. Soil Biol Biochem 36:1383–1388

Pieterse CMJ, Van Wees SCM, Van Pelt JA, Knoester M, Laan R, Gerrits H, Weisbeek PJ, Van Loon LC (1998) A novel signaling pathway controlling induced systemic resistance in *Arabidopsis*. Plant Cell 10:1571–1580

Pieterse CMJ, Ton J, Van Loon LC (2001) Cross-talk between plant defence signalling pathways: boost or burden? AgBiotechNet 3:ABN068

Ramamoorthy V, Samiyappan R (2001) Induction of defense-related genes in *Pseudomonas fluorescens* treated chilli plants in response to infection by *Colletotrichum capsici*. J Mycol Plant Pathol 31:146–155

Ramamoorthy V, Viswanathan R, Raguchander T, Prakasam V, Samiyappan R (2001) Induction of systemic resistance by plant growth promoting rhizobacteria in crop plants against pests and diseases. Crop Prot 20:1–11

Reinhold B, Hurek T, Fendrik I (1985) Strain-specific chemotaxis of *Azospirillum* spp. J Bacteriol 162:190–195

Ren D, Sims JJ, Wood TK (2001) Inhibition of biofilm formation and swarming of *Escherichia coli* by (5Z)-4-bromo-5-(bromomethylene)-3-butyl-2(5H)-furanone. Environ Microbiol 3:731–736

Riley M (1993) Molecular mechanisms of colicin evolution. Mol Biol Evol 10:1380–1395

Riley MA, Wertz JE (2002) Bacteriocins: evolution, ecology, and application. Ann Rev Microbiol 56:117–137

Rovira AD (1965) Interactions between plant roots and soil microorganisms. Ann Rev Microbiol 19:241–266

Saleem M, Arshad M, Hussain S, Bhatti AS (2007) Perspective of plant growth promoting rhizobacteria (PGPR) containing ACC deaminase in stress agriculture. J Ind Microbiol Biotechnol 34:635–648

Schouten A, van der Berg G, Edel-Hermann V, Steinberg C, Gautheron N, Alabouvette C, de Vos CH, Lemanceau P, Raaijmakers JM (2004) Defense responses of *Fusarium oxysporum* to 2, 4-diacetylphloroglucinol, a broad-spectrum antibiotic produced by *Pseudomonas fluorescens*. Mol Plant Microbe Interact 17:1201–1211

Sessitsch A, Reiter B, Berg G (2004) Endophytic bacterial communities of field-grown potato plants and their plant growth-promoting and antagonistic abilities. Can J Microbiol 50:239–249

Simons M, Permentier HP, de Weger LA, Wijffelman CA, Lugtenberg BJJ (1997) Amino acid synthesis is necessary for tomato root colonization by *Pseudomonas fluorescens* strain WCS365. Mol Plant Microbe Interact 10:102–106

Singh PP, Shin YC, Park CS, Chung YR (1999) Biological control of *Fusarium* wilt of cucumber by chitinolytic bacteria. Phytopathology 89:92–99

Steenhoudt O, Vanderleyden J (2000) *Azospirillum*, a free-living nitrogen-fixing bacterium closely associated with grasses: genetic, biochemical and ecological aspects. FEMS Microbiol Rev 24:487–506

Sturz AV, Christie BR, Matheson BG, Arsenault WJ, Buchanan NA (1999) Endophytic bacterial communities in the periderm of potato tubers and their potential to improve resistance to soil-borne plant pathogens. Plant Pathol 48:360–369

Toyoda H, Hashimoto H, Utsumi R, Kobayashi H, Ouchi S (1988) Detoxification of fusaric acid by a fusaric acid-resistant mutant of *Pseudomonas solanacearum* and its application to biological control of Fusarium wilt of tomato. Phytopathology 78:1307–1311

Van der Broek D, Chin-A-Woeng TFC, Eijkemans K, Mulders IHM, Bloemberg GV, Lugtenberg BJJ (2003) Biocontrol traits of *Pseudomonas* spp. are regulated by phase variation. Mol Plant Microbe Interact 16:1003–1012

Van Overbeek LS, Van Elsas JD (1995) Root exudates-induced promoter activity in *Pseudomonas fluorescens* mutants in the wheat rhizosphere. Appl Environ Microbiol 61:890–898

Van Peer R, Niemann GJ, Schippers B (1991) Induced resistance and phytoalexin accumulation in biological control of Fusarium wilt of carnation by *Pseudomonas* sp. WCS417r. Phytopathology 81:728–734

von Bodman SB, Bauer WD, Coplin DL (2003) Quorum sensing in plant-pathogenic bacteria. Ann Rev Phytopathol 41:455–482

Van Loon LC, Bakker PAHM, Pieterse CMJ (1998) Systemic resistance induced by rhizosphere bacteria. Annu Rev Phytopathol 36:453–483

Walker MJ, Birch RG, Pemberton JM (1988) Cloning and characterization of an albicidin resistance gene from *Klebsiella oxytoca*. Mol Microbiol 2:443–454

Welbaum G, Sturz AV, Dong Z, Nowak J (2004) Fertilizing soil microorganisms to improve productivity of agroecosystems. Crit Rev Plant Sci 23:175–193

Wright SAI, Zumoff CH, Schneider L, Beer SV (2001) *Pantoea agglomerans* strain EH318 produces two antibiotics that inhibit *Erwinia amylovora* in vitro. Appl Environ Microbiol 67:284–292

Yamada Y, Nihira T (1998) Microbial hormones and microbial chemical ecology. In: Barton DHR, Nakanishi K (eds) Comprehensive natural products chemistry, vol 8. Elsevier Sciences, Amsterdam, pp 377–413

Zdor RE, Anderson AJ (1992) Influence of root colonizing bacteria on the defense responses of bean. Plant Soil 140:99–107

Zhang L, Birch RG (1997) The gene for albicidin detoxification from *Pantoea dispersa* encodes an esterase and attenuates pathogenicity of *Xanthomonas albilineans* to sugarcane. Proc Natl Acad Sci U S A 94:9984–9989

Zhang LH, Dong YH (2004) Quorum sensing and signal interference: diverse implications. Mol Microbiol 53:1563–1571

5 Mass Production, Formulation, Delivery and Commercialization

Biocontrol technology could become as a successful component of plant protection only when it is commercialized. The prerequisites for commercialization include mass production and formulation development, fermentation methods, formulation viability, toxicology, proper method of delivery, industrial linkages and quality control.

5.1 Mass Production

The first major concern in commercial production systems involves the achievement of adequate growth of the biocontrol agent. In many cases biomass production of the antagonist is difficult due to the specific requirement of nutritional and environmental conditions for the growth of organism. Mass production is achieved through liquid and semisolid and solid fermentation techniques.

5.1.1 Fermentation

5.1.1.1 Liquid Fermentation

This fermentation system has been adopted for the mass multiplication of bacterial biocontrol agents. For mass multiplication the selected medium should be inexpensive and readily available with appropriate nutrient balance. King's B broth or nutrient broths are used for the mass production of *Pseudomonas* and *Bacillus* spp., through liquid fermentation technology (Nakkeeran et al. 2004).

5.1.1.2 Solid Fermentation

In nature a wide range of organic substrates could be used for the solid-state fermentation for mass multiplication. Solid fermentation media consisting of inert carriers with food bases were used for mass production of biocontrol agents. The media with relatively low microbial content would be suited for solid-state fermentation and for the amendment of biocontrol agents. Solid substrates include straws, wheat bran, sawdust, moistened bagasse, sorghum grains, paddy chaff, decomposed coir pith, farmyard manure and other substrates rich in cellulose for inoculum production. Siddiqui and Mahmood (1996) stated that bacteria had great potential to manage plant-parasitic nematodes. But the practicality of the same could be done by incorporating the antagonistic bacteria to organic manures, followed by incubation at 35 °C for 5–10 days coupled with frequent mixing under sterile environment along with water so as to maintain the organic manure under moist conditions, which aid in the proliferation of the bacteria. The enriched organic manure with biocidal value could be used for the management of pests and plant growth promotion.

Commercial production of disease-suppressive strains of bacteria such as *P. fluorescens* and *B. subtilis* as biocontrol agents in post-harvest diseases requires low cost and high biomass production while maintaining their biocontrol efficacy. Yeast extract as a nitrogen source supports rapid growth and higher cell yields in all of the strains as compared to urea. Yeast extract contains amino acids

P.P. Reddy, *Plant Growth Promoting Rhizobacteria for Horticultural Crop Protection*,
DOI 10.1007/978-81-322-1973-6_5, © Springer India 2014

and peptides, water-soluble vitamins and carbohydrates, which make it an excellent substrate for many microorganisms. Dharani-Aiyer (2004) showed that yeast extract was the best organic nitrogen source for antagonistic bacteria. Nohata and Kurane (1997) considered yeast extract to be too expensive for an industrial process, so it should preferably be replaced by a cheaper industrial product having similar growth characteristics, which should be determined in a economic and technological study. Molasses showed good yield efficacy in both strains (*P. fluorescens* and *B. subtilis*), which may account for the high biomass obtained, because the combination of yeast extract and commercial sucrose also gave high final growth of bacteria. However, a combination of molasses and urea decreased bacterial growth. Luna et al. (2002) showed that a molasses-based medium may be used for production of bacterial biocontrol agents (BAs). Apparently a 1:1 C/N ratio produces optimal growth for bacteria of two different genera (*Pseudomonas* and *Bacillus*) and this may hold true for other genera. The pH is another important parameter for bacterial growth. As a general principle, bacterial growth decreases at more acidic pH values. These observations are in accordance with the results of Costa et al. (2001). Antagonistic bacteria such as *P. fluorescens* and *B. subtilis* can be produced in different media, using various N and C sources while maintaining the efficacy of BAs. By-products such as molasses can serve as an economic culture medium. Future research will concentrate on optimizing growth conditions and possible incorporation of other nutrients into formulations to obtain an even higher biomass.

5.1.2 Mass Multiplication of *Bacillus subtilis*

For mass multiplication of the bacterium, nutrient broth is commonly used. The composition of nutrient broth medium is as follows:

Peptone	–	5 g
Beet extract	–	3 g
Sodium chloride	–	3 g
Distilled water	–	1,000 ml

The medium is prepared and sterilized at 121 °C for 20 min. One loopful of *B. subtilis* is inoculated and incubated for 24 h. From this, talc-based preparation can be made. The talc-based formulation can be used for application against coconut leaf rot and coconut-based mixed crops.

The medium consisting of cotton meal added in hydrolyzed protein provided the highest yield of *B. subtilis* cells (2.44×10^9 cells/mL), after the culture had been incubated for 3 days. This liquid substrate also provided conditions for the bacterium to produce thermostable metabolites, in sufficient amounts to inhibit the plant pathogen's mycelial growth. The production of *B. subtilis* under the solid fermentation system performed better on the brewer's rice substrate; the number of bacterial cells decreased as the substrate concentration increased. In general, the liquid medium yielded a higher amount of *B. subtilis* than the solid medium (Wagner et al. 2005).

The maximum growth of *B. subtilis* strain KO was 0.608 g/100 ml in molasses broth medium supplemented with gelatin 3 % w/v (molasses 100 ml, gelatin 30 g, distilled water 1,000 ml).

The maximum growth of *B. subtilis* strain KO was 0.780 g/100 ml in molasses broth medium supplemented with ammonium phosphate 0.1 % (w/v) (molasses 100 ml, gelatin 30 g, ammonium phosphate 1 g, distilled water 1,000 ml).

5.1.3 Mass Multiplication of *Pseudomonas fluorescens*

A loopful of *P. fluorescens* should be transferred to 100 ml of King's B broth in a 250 ml conical flask under aseptic conditions. The flask can be incubated at room temperature for 48 h. The population of *P. fluorescens* in the broth can be assessed (a minimum of 9×10^9 cfu/ml is necessary).

Recipe for King's B medium (Kings et al. 1954)

Protease peptone	–	20 g
K_2HPO_4	–	1.5 g
$MgSO_4$ $7H_2O$	–	1.5 g
Glycerol	–	20 ml
Agar	–	15 g

(continued)

Distilled water	–	1,000 ml
pH	–	7.2
Cycloheximide	–	100 ppm
Ampicillin	–	50 ppm
Chloramphenicol	–	12.50 ppm

(King's B broth is prepared without the agar and antibiotics)

The pH of the substrate (talc powder) can be adjusted to 7.2 by addition of 150 g calcium carbonate/kg of talc powder and sterilized at 1.1 kg/cm^2 for 30 min for two successive days. One kilogram of the sterilized substrate (talc powder) and 5 g of carboxy methyl cellulose should be placed in a polythene bag or any sterile container under aseptic conditions, 400 ml of *P. fluorescens* suspension to be added and mixed thoroughly. The formulation can be stored in polythene bags for 1 month.

The product should have the following quality parameters:

- Fresh product should contain 2.5×10^8 cfu/g.
- Three months after storage at room temperature, the population should be $8–9 \times 10^7$ cfu/g.
- Storage period is 3–4 months.
- Minimum population load for use is 1×10^8 cfu/g.
- The product should be packed in white polythene bags.
- Moisture content of the final product should not be more than 20 %.
- Population in broth should be $9 \pm 2 \times 10^8$ cfu/ml.

Pseudomonas fluorescens was inoculated in modified King's B liquid medium which includes the amendment of sucrose for increased population and shelf life of *P. fluorescens* in talc formulation.

The following are the composition of modified King's B liquid medium used for growth of *Pseudomonas fluorescens* VP5.

Recipe for modified King's B medium

Proteose peptone # 3	–	20 g/L
K_2HPO_4	–	1.5 g/L
$MgSO_4\,7H_2O$	–	1.5 g/L
Sucrose	–	100 g/L
Colloidal chitin	–	2.5 %

Vermicompost and Farmyard Manure (FYM)-Based Formulation for Mass Multiplication: The talc formulation was mixed with farmyard manure and vermicompost in the ratio of 1:100 and incubated for 1 week. About half kilogram of *P. fluorescens* is added to each plant. In the following table the growth and population density of root-colonizing *P. fluorescens* VP5 strain was calculated by using antibiotic plasmid resistance method. This bioformulation was applied in tea soil every 6 months by using different combinations of treatments (Table 5.1).

5.2 Development of Formulations

PGPR formulation development is essential for easy application, storage, commercialization and field use.

5.2.1 Features of an Ideal Formulation

- Increased shelf life.
- Not phytotoxic to crop plants.
- Dissolve well in water and should release bacteria.
- Tolerate adverse environmental conditions.
- Cost effective and should give reliable control of plant diseases.

Table 5.1 Establishment of *Pseudomonas fluorescens* in tea rhizosphere

	Population density of *P. fluorescens* at 6-month interval			
Treatment details	First 6 months	Second 6 months	Third 6 months	Fourth 6 months
Vermicompost + talc (1:100)	2.1×10^3	5.3×10^3	1.6×10^4	3.7×10^4
Farmyard manure + talc (1:100)	8.0×10^3	3.8×10^4	6.3×10^4	1.4×10^5
Vermicompost + farmyard manure + talc (1:50:50)	2.6×10^4	7.4×10^4	1.9×10^5	2.3×10^5

- Compatible with other agrochemicals.
- Carriers must be cheap and readily available for formulation development (Jeyarajan and Nakkeeran 2000).

5.2.2 Carriers

Carriers should support the survival of bacteria for a considerable length of time. Carriers may be either organic or non-organic. They should be economical and easily available.

5.2.2.1 Organic/Non-organic Carriers

The organic/non-organic carriers used for formulation development include peat, turf, talc, lignite, kaolinite, pyrophyllite, zeolite, montmorillonite, alginate, press mud, sawdust and vermiculite. Carriers increase the survival rate of bacteria by protecting it from desiccation and death of cells (Heijnen et al. 1993). The shelf life of bacteria varies depending upon bacterial genera, carriers and their particle size. The survival rate of *P. fluorescens* (2-79RN10, W4F393) in montmorillonite, zeolite and vermiculite with smaller particle size increased more than in kaolinite, pyrophyllite and talc with bigger particle size. The carriers with smaller particle size have increased surface area, which increase resistance to desiccation of bacteria by the increased coverage of bacterial cells (Dandurand et al. 1994).

5.2.3 Formulations

Formulations of fluorescent pseudomonads were developed through liquid fermentation technology. The fermentor biomass was mixed with different carrier materials (talc/peat/kaolinite/lignite/vermiculite) and stickers (Vidhyasekaran and Muthamilan 1995). Krishnamurthy and Gnanamanickam (1998) developed talc-based formulation of *P. fluorescens* in which methyl cellulose and talc were mixed at 1:4 ratio and blended with equal volume of bacterial suspension at a concentration of 10^{10} cfu/ml. Nandakumar et al. (2001) developed talc-based strain mixture formulation of fluorescent pseudomonads. It was prepared by mixing equal volume of individual strains and blended with talc as per Vidhyasekaran and Muthamilan (1995). Talc-based strain mixtures were effective against diseases and increased plant yield under field conditions than the application of individual strains. Talc- and peat-based formulations of *P. chlororaphis* and *B. subtilis* were prepared and used for the management of turmeric rhizome rot (Nakkeeran et al. 2004).

One school of thought explains that CMC is added as a sticker at 1:4 ratio to talc. Though it is effective in disease management, it would lead to the increase in the production cost, which would prevent the growers to adopt the technology. Moreover another school of thought explains that CMC and talc should be used at 1:100 ratio. Hence, the feasibility of the technique and shelf life of the product has to be evaluated to make the technology as a viable component in disease management so as to promote organic farming.

5.2.3.1 Talc Formulation

Talc is a natural mineral referred as steatite or soapstone that is composed of various minerals in combination with chloride and carbonate. Chemically it is referred as magnesium silicate [$Mg_3Si_4O_3(OH)_2$] and available as powder form from industries suited for a wide range of applications. It has very low moisture equilibrium, relative hydrophobicity, chemical inertness and reduced moisture absorption and prevents the formation of hydrate bridges that enable longer storage periods. Owing to the inert nature of talc and easy availability as raw material from soapstone industries, it is used as a carrier for formulation development. Kloepper and Schroth (1981) demonstrated the potentiality of talc to be used as a carrier for formulating rhizobacteria. The fluorescent pseudomonads did not decline in talc mixture with 20 % xanthan gum after storage for 2 months at 4 °C. *P. fluorescens* isolate Pf1 survived up to 240 days in storage. The initial population of Pf1 in talc-based formulation was 37.5×10^7 cfu/g and declined to 1.3×10^7 cfu/g after 8 months of storage (Vidhyasekaran and Muthamilan 1995). Amendment of sucrose (0.72M) in King's B medium increased the population and shelf life of *P. fluorescens* (P7NF, TL3) in talc-based formulation up to 12 months (Caesar and Burr 1991). *P. putida* strain 30 and 180 survived up to 6 months in talc-based formulations. The population load at the end of sixth month was 10^8 cfu/g of the product (Bora et al. 2004).

5.2.3.2 Peat Formulation

Peat (turf) is a carbonized vegetable tissue formed in wet conditions by decomposition of various plants and mosses. It is formed by the slow decay of successive layers of aquatic and semi-aquatic plants, e.g. sedges, reeds, rushes and mosses. Peat soils are used as carrier materials to formulate PGPR. Though peat carriers are cheap to use, it harbours lot of contaminants. The quality of peat is variable and not readily available worldwide. Sterilization of peat through heat releases toxic substances to the bacteria and thereby reduces bacterial viability (Bashan 1998). Peat-based formulation of *Azospirillum brasilense* had a shelf life up to 4 months. The population load after 4 months of storage was 10^7 cfu/g of the product (Bashan 1998). This population was sufficient for successful plant inoculation. Vidhyasekaran and Muthamilan (1995) reported that the shelf life of *P. fluorescens* in peat-based formulation was maintained up to 8 months (2.8×10^6 cfu/g). The shelf life of *P. chlororaphis* (PA23) and *B. subtilis* (CBE4) in peat carriers was retained for more than 6 months (Nakkeeran et al. 2004).

5.2.3.3 Press Mud Formulation

Press mud is a by-product of sugar industries. It was composted using vermin-composting technique and later used as a carrier for *Azospirillum* spp. This carrier maximizes the survival of *Azospirillum* spp. than lignite, which is predominantly used as a carrier material in India (Muthukumarasamy et al. 1999).

5.2.3.4 Vermiculite Formulation

Vermiculite is a light mica-like mineral used to improve aeration and moisture retention. It is widely used as potting mixture and used as a carrier for the development of formulations for harbouring microbial agents. Vermiculite-based formulation of *P. fluorescens* (Pf1) retained shelf life for a period of 8 months. The viable load of bacteria in the formulation was 1×10^6 cfu/g (Vidhyasekaran and Muthamilan 1995). The shelf life of *Azospirillum* in vermiculite-based formulation was retained up to 10 months. The viable cells after 44 weeks of storage were 1.3×10^7 cfu/g (Saleh et al. 2001).

5.2.3.5 Microencapsulation

Microcapsules of rhizobacteria consist of a cross-linked polymer deposited around a liquid phase, where bacteria are dispersed. Microparticles are characterized based on the distribution of particle size, morphology and bacterial load. The process of microencapsulation involves mixing of gelatin polyphosphate polymer pair (81:19 w/w) at acidic pH with rhizobacteria suspended in oil (Charpentier et al. 1999). Though rhizobacteria have been formulated through microencapsulation method, its shelf life declines at a faster rate, since polymers serve as a barrier for oxygen. This was later improved by developing microcapsules by spray drying. The release of *P. fluorescens* and *P. putida* from the microencapsulated pellets occurred after 15 min immersion in aqueous buffer. It showed that water served as triggering material for the bacterial release (Charpentier et al. 1999).

Though microencapsulation aids in formulating bacteria, still the technology has to be well refined for early release of bacterial cells and for the establishment in the infection court to counterattack the establishment of pathogens. Most of the experiments on microencapsulation have been restricted only to lab. The technology should be standardized for the industrial application so that the technical feasibility could be assessed to popularize the same for field use.

5.2.4 Adjuvants, Spreaders and Stickers

In general, the performance of PGPR formulations in controlling plant diseases is inconsistent. Since disease suppression is the outcome of interactions between biocontrol agents, pathogen, plant and environment, any fluctuations in growing seasons, environmental conditions and high inoculum pressure alter the efficacy of biocontrol formulations. Integrating the usage of formulations with other management strategies that aims at increasing the productivity of the crop could enhance the efficacy of formulations (Larkin et al. 1998). Performance of biocontrol agents in the formulations can be increased by the incorporation of water-soluble adjuvants, oils, stickers and emulsions. It increases the efficacy of biocontrol agents by supplying nutrients and by protecting the microbes from desiccation and death (Ibrahim et al. 1999). Incorporation of carboxy methyl cellulose (CMC) in formulations serves as stickers in uniform seed coating of microbes. Though adjuvants and stickers increase the efficacy of bioproducts, it has its own demerits. Adjuvants/stickers in the formulations will be diluted when exposed to rain or heavy dew. It would alter the efficacy of formulations by reducing the establishment or colonization of PGPR onto the infection court. Sometimes spray application of emulsions or oil-based formulations may be toxic to plants. Hence, a thorough knowledge on the usage of adjuvants and stickers is essential for increasing the efficacy of formulations.

5.3 Application Methods/Modes of Delivery

Rhizobacteria are delivered through seed, soil, foliage and rhizomes or through combination of several methods of delivery.

5.3.1 Seed Treatment

Seed treatment with cell suspensions of PGPR was effective against several diseases. Delivering of *Serratia marcescens* strain 90-166 as seed dip before planting and soil application of 100 ml of the same at the rate of 10^8 cfu/ml to the sterilized soil-less planting mix after seeding reduced bacterial wilt of cucumber and controlled cucumber beetles besides increasing the fruit weight (Zehnder et al. 2001). Transfer of technology for commercial use could be possible if PGPR strains are made available as a product. After realization of the same, several carriers were used for formulation development. Treatment of cucumber seeds with strain mixtures comprising of *Bacillus pumilus* INR7, *B. subtilis* GBO3 and *Curtobacterium flaccumfaciens* ME1 with a mean bacterial density of 5×10^9 cfu/seed reduced the intensity of angular leaf spot and anthracnose equivalent to the synthetic elicitor Actigard and is better than seed treatment with individual strains (Raupach and Kloepper 1998). Treatment of pigeon pea seeds with talc-based formulation of *P. fluorescens* (Pf1) effectively controlled *Fusarium* wilt of pigeon pea under greenhouse and field conditions (Vidhyasekaran et al. 1997). Seed treatment of lettuce with either vermiculite- or kaolin-based carrier of *B. subtilis* (BACT-0) significantly reduced root rot caused by *P. aphanidermatum*, and it also increased the fresh weight of lettuce under greenhouse conditions. Seed treatment with vermiculite-based *P. putida* reduced Fusarium root rot of cucumber and increased the yield and growth of cucumber (Amer and Utkhede 2000). Treatment of tomato seeds with powder formulation of PGPR (*B. subtilis*, *B. pumilus*) reduced symptom severity of TMV and increased the fruit yield (Murphy et al. 2000).

5.3.2 Bio-priming

A successful antagonist should colonize rhizosphere during seed germination (Weller 1983). Priming with PGPR increases germination and improves seedling establishment. It initiates the physiological process of germination but prevents the emergence of plumule and radical. Initiation of physiological process helps in the establishment and proliferation of PGPR on the spermosphere (Taylor and Harman 1990). Biopriming of seeds with bacterial antagonists increases the population load of antagonist to a tune of tenfold on the seeds and thus protected rhizosphere from the ingress of plant pathogens (Callan et al. 1990). Drum priming is a commercial seed treatment method followed to treat seeds with biopesticides. Drum priming of carrot and parsnip seeds with *P. fluorescens* Pf CHAO proliferated well on the seeds and could be explored for realistic scaleup of PGPR (Wright et al. 2003).

5.3.3 Seedling Dip

PGPR are delivered through various means for the management of crop diseases based on the survival nature of the pathogen. In several crops, pathogens gain entry into plants either through seed, root or foliage. Dipping of strawberry roots for 15 min in bacterial suspension of *P. putida* (2×10^9 cfu/ml) isolated from strawberry rhizosphere reduced *Verticillium* wilt of strawberry by 11 % compared to untreated control (Berg et al. 2001). Dipping of *Phyllanthus amarus* seedlings in talc-based formulation of *B. subtilis* (BSCBE4) or *P. chlororaphis* (PA23) for 30 min prior to transplanting reduced stem blight of *P. amarus* (Mathiyazhagan et al. 2004).

5.3.4 Soil Application

Soil being the repertoire of both beneficial and pathogenic microbes, delivering of PGPR strains to soil will increase the population dynamics of augmented bacterial antagonists and thereby would suppress the establishment of pathogenic microbes onto the infection court. Vidhyasekaran and Muthamilan (1995) stated that soil application of peat-based formulation of *P. fluorescens* (Pf1) at the rate of 2.5 kg of formulation mixed with 25 kg of well-decomposed farmyard manure in combination with seed treatment increased rhizosphere colonization of Pf1 and suppressed chickpea wilt caused by *F. oxysporum* f. sp. *ciceris*.

5.3.5 Foliar Spray

The efficacy of biocontrol agents for foliar diseases is greatly affected by fluctuation of microclimate. Phyllosphere is subjected to diurnal and nocturnal, cyclic and noncyclic variation in temperature, relative humidity, dew, rain, wind and radiation. Hence, water potential of phylloplane microbes will be varying constantly. It will also vary between leaves or the periphery of the canopy and on sheltered leaves. Higher relative humidity could be observed in the shaded, dense region of the plant than that of peripheral leaves. Dew formation is greater in centre and periphery. The concentration of nutrients like amino acids, organic acids and sugars exuded through stomata, lenticels, hydathodes and wounds varies highly. It affects the efficacy and survival of antagonist in phylloplane (Andrews 1992).

Application of *B. subtilis* to bean leaves decreased the incidence of bean rust (*Uromyces phaseoli*) by 75 % equivalent to weekly treatments with the fungicide mancozeb (Baker et al. 1983). Pre-harvest foliar application of talc-based fluorescent pseudomonad strain FP7 supplemented with chitin at fortnightly intervals (5 g/l; spray volume 20 L/tree) on mango trees from preflowering to fruit maturity stage induced flowering to the maximum extent, reduced the latent infection caused by *C. gloeosporioides* besides increasing the fruit yield and quality (Vivekananthan et al. 2004). Though seed treatment and foliar application of *P. fluorescens* reduce the severity of rust and leaf spot under field conditions, it is not technically feasible due to increased dosage and economy realized from the crop. Hence, dosage and frequency of application has to be standardized based on the crop value, which could be a reliable and practical approach.

5.3.6 Fruit Spray

Pseudomonas syringae (10 % wettable powder) in the modified packing line was sprayed at the rate of 10 g/l over apple fruit to control blue and grey mould of apple. The population of antagonist increased in the wounds more than tenfold during 3 months in storage (Janisiewicz and Jeffers 1997). Research on the exploration of PGPR has to go a long way to explore its usage to manage post-harvest diseases.

5.3.7 Hive Insert

Honeybees and bumblebees serve as vectors for the dispersal of biocontrol agents for the management of diseases of flowering and fruit crops (Kevan et al. 2003). An innovative method of application of biocontrol agent right in the infection court at the exact time of susceptibility was developed. A dispenser is attached to the hive and loaded with powder formulation of the PGPR or with other desired biocontrol agent. When the foragers exit the hive, the antagonist get dusted on to bee and delivered to the desired crop while attempting to suck the nectar. *Erwinia amylovora* causing fire blight of apple infects through flower and develops extensively on stigma. Colonization by antagonist at the critical juncture is necessary to prevent flower infection. Since flowers do not open simultaneously, the biocontrol agent *P. fluorescens* has to be applied to flowers repeatedly to protect the stigma. Nectar-seeking insects like *Apis mellifera* can be used to deliver *P. fluorescens* to stigma. Bees deposit the bacteria on the flowers soon after opening due to their foraging habits. Honeybees have also been used for the management of grey mould of strawberry and raspberry (Peng et al. 1992; Kovach et al. 2000).

5.3.8 Sucker Treatment

Plant growth-promoting rhizobacteria also play a vital role in the management of soilborne diseases of vegetatively propagated crops. The delivery of PGPR varies depending upon the crop. In banana, rhizobacteria are delivered through sucker treatment or rhizome treatment. Banana suckers were dipped in talc-based *P. fluorescens* suspension (500 g of the product in 50 l of water) for 10 min after paring and pralinage. Subsequently it was followed by capsule application (50 mg of *P. fluorescens* per capsule) on the third and fifth month after planting. It resulted in 80.6 % reduction in panama wilt of banana compared to control (Raghuchander et al. 2001).

5.3.9 Multiple Delivery Systems

Plant pathogens establish host–parasite relationships by entering through infection court such as spermosphere, rhizosphere and phyllosphere. Hence, protection of sites vulnerable for the entry and infection of pathogens would offer a better means for disease management. Seed treatment of pigeon pea with talc-based formulation of fluorescent pseudomonads at the rate of 4 g/kg of seed followed by soil application at the rate of 2.5 kg/ha at 0, 30 and 60 days after sowing controlled pigeon pea wilt incidence under field conditions. The additional soil application of talc-based formulation improved disease control and increased yield compared to seed treatment alone (Vidhyasekaran et al. 1997). The increased efficacy of strain mixtures through combined application might be due to increase in the population of fluorescent pseudomonads in both rhizosphere and phyllosphere. Delivering of rhizobacteria through combined application of different delivery systems will increase the population load of rhizobacteria and thereby might suppress the pathogenic propagules.

5.4 Commercialization of PGPR

5.4.1 Stages of Commercialization

According to Sabitha Doraisamy et al. (2001) different stages in the process of commercialization include isolation of antagonist strains, screening, pot tests and field efficacy, mass production and formulation development, fermentation methods, formulation viability, toxicology, industrial linkages and quality control.

5.4.1.1 Isolation of Antagonist

Isolation of an effective strain plays a prime role in disease management. It is done from the pathogen-suppressive soils either by dilution plate technique or by baiting the soil with fungal structures like sclerotia of pathogen. Consortium of biocontrol agents could be established by isolating the location-specific and crop-specific isolates. It could be used for the development of mixtures of biocontrol agents suited for different ecological niche.

5.4.1.2 Screening of Antagonist

All the strains isolated from different cropping systems have to be ascertained for their virulence and broad spectrum of action against different pathogens causing serious economic threat to cultivation. Selection of an effective strain decides the viability of the technology. Hence, a proper yardstick should be developed to screen the antagonistic potentiality of the biocontrol agents. In vitro screening of the antagonist through dual culture technique alone could not be an effective method for strain selection. To be an effective antagonist, it should possess a high level of competitive saprophytic ability and antibiosis and should have the ability to secrete increased level of cell wall lytic enzymes (chitinases, glucanases and proteases), antibiotics and plant growth promotion. Hence, the yardstick should be developed, comprising of the above-mentioned components. Each component should be given weightage depending upon their role in disease management. This type of rigorous and meticulous screening will lead to identification of an effective biocontrol strain suited for commercialization.

Twenty rhizobacterial isolates from strawberry rhizosphere were evaluated for their antifungal action against *Verticillium dahliae*. The selection of best antagonistic bacterial isolate was done by screening for the antifungal action against different soilborne pathogens apart from the target pathogen. In addition it was also tested for the antifungal mechanism of the rhizobacteria for the production of lytic enzymes (chitinases, glucanases and proteases) and plant growth promotion. Collectively all these parameters were combined based on Bonitur Scale (28 points). The strain that had the highest score was selected for testing its efficacy under greenhouse conditions. Among 20 strains, *P. putida* E2 had the maximum Bonitur Scale of 28 points and was highly effective in suppressing *Verticillium* wilt of strawberry under greenhouse conditions. It was found to perform better than the commercial product Rhizovit (Berg et al. 2001). This clearly explains that selection of bacterial antagonist plays a major role in the commercialization of bacteria for disease management. Initial mistake committed in strain selection will lead to complete failure of the technology.

5.4.1.3 Pot Test and Field Efficacy

The plant, pathogen and antagonists are coexposed to controlled environmental conditions. Exposure of the host to the heavy inoculum pressure of the pathogen along with the antagonist will provide ecological data on the performance of the antagonist under controlled conditions. Promising antagonists from controlled environment are tested for their efficacy under field conditions along with the standard recommended fungicides. Since the variation in the environment under field condition influences the performance of biocontrol agent, trials on the field efficacy should be conducted in at least 15–20 locations under different environmental conditions to promote the best candidate for mass multiplication and formulation development (Jeyarajan and Nakkeeran 2000).

5.4.1.4 Mass Production and Formulation Development

The first major concern in commercial production systems involves the achievement of adequate growth of the biocontrol agent. In many cases biomass production of the antagonist is difficult due to the specific requirements of nutritional and environmental conditions for the growth of organism. Mass production is achieved through liquid and semisolid and solid fermentation techniques.

5.4.1.5 Formulation Viability

Shelf life of the formulations decides the commercialization of biocontrol agents. Formulations should support the viable nature of the product for the increased period of storage. Biocontrol product should have the minimum shelf life of 8–12 months for industrialization. Carrier material should not affect the viable nature of the biocontrol agent. Commercialization of the bioproducts is mainly hampered due to the poor shelf life. Hence, research should be concentrated to increase the shelf life of the formulation by developing superior strains that support the increased shelf life, or the organic formulations that support the maximum shelf life with low level of contaminants must be standardized for making biocontrol as a commercial venture.

5.4.1.6 Toxicology

Safety and environmental considerations could not be taken for granted and it is crucial that biopesticides are regulated in an appropriate way to conform with international standards. The regulatory environment is generally favourable for the biopesticides than the chemical pesticides. However, the cost of carrying out the toxicological study for registration is still prohibitive. Toxicology includes information of antagonist on the safety to human, plants, animals and soil microflora. Cost incurred for the toxicological studies is high. These studies have to be done separately for each and every biocontrol organism. The huge investment on the toxicological studies warrants for the linkages between stakeholders and research organizations (Jeyarajan and Nakkeeran 2000; Sabitha Doraisamy et al. 2001).

5.4.1.7 Industrial Linkages

Industrialization of biocontrol agents requires linkage between corporate and academic bodies. The success and commercialization of a scientific innovation depends on the availability of the technology to the end users. It depends on the linkages between the scientific organization and industries. Biocontrol technology could become as a successful component of crop protection only when it is commercialized.

Research institutes carry out the initial discovery of an effective organism, genetic manipulation of organisms to develop superior strains and studies on mechanisms, field efficacy and protocols for the development of formulations. But to take this technology to the entire country depends on the partnership between the stakeholders and institutes. Corporate resources are required for the large-scale production, toxicology, wide-scale field testing, registration and marketing. Entrepreneurship may be defined as the exchange of intellectual property for research grants, and a royalty stream, with the establishment of university–industry partnership for the benefit of both. The first requirement for the entrepreneurship requires a patent application on the strain and the related technology, especially on the efficacy data, identity of the organism, toxicological data and delivery system. Ideally the process of entrepreneurship will result in an academic corporate research team working towards a common goal.

5.4.1.8 Quality Control

This is very much required to retain the confidence of the farmers on the efficacy of biocontrol agents. Being the living organisms, their population in a product influences the shelf life. The population load of the antagonists decides the minimum level of requirement for bringing the effective biological control of the plant diseases. Depending on the type of the antagonist and formulation, the moisture content and population load varies. The other contaminating organisms should be also under the permissible limits.

5.4.2 Strategies to Promote Commercialization

Commercialization of biocontrol could be promoted by

- Popularization of biocontrol agents
- Industrial linkages

5.4.2.1 Popularization of Biocontrol Agents

Motivating the growers through:

- Publicity
- Field demonstrations
- Farmers days
- Biovillage adoption
- Conducting periodical trainings for commercial producers and farmers to increase/improve the supply

5.4.2.2 Industrial Linkages

- Technical support should be made available to entrepreneurs on quality control and registration.
- Regular monitoring is essential to maintain the quality.

- Constant research support should be extended to standardize the dosage, storage, and delivery systems. There should be positive policy support from the government in using more of biocontrol agents in crop protection.

Research inventions from China, Russia and several other Western countries have now proved the potential use of PGPR towards plant disease management. Among several PGPR strains, *Bacillus*-based products gain momentum for commercialization, because *Bacillus* spp. produce endospores tolerant to extremes of abiotic environments such as temperature, pH, pesticides and fertilizers (Backman et al. 1997). The first commercial product of *Bacillus subtilis* was developed in 1985 in the USA. About 60–75 % of vegetables crops raised in USA are now treated with commercial product of *B. subtilis*, which become effective against soilborne pathogens such as *Fusarium* and *Rhizoctonia* (Nakkeeran et al. 2005). In China, PGPR have been successfully applied over two decades in an area of 20 million hectares of different crop plants for commercial development. Owing to the potentiality of *Bacillus* spp., more than 20 different commercial products of *Bacillus* origin are sold in China to mitigate soilborne diseases (Backman et al. 1997). Besides *Bacillus* spp., certain other PGPR strains belonging to the genera such as *Agrobacterium, Azospirillum, Burkholderia, Pseudomonas* and *Streptomyces* are also used for the production of several commercial products, which are generally being applied against several target pathogens like *Botrytis cinerea, Penicillium* spp., *Mucor piriformis, Geotrichum candidum, Erwinia amylovora*, russet-inducing bacteria, *Fusarium* sp., *Rhizoctonia* sp., *Pythium* sp., *Phytophthora* sp. and *P. tolaasii* (Nakkeeran et al. 2005). Since PGPR have their own potentiality in controlling plant diseases and pest management, these commercial products such as Diegall, Galltrol-A, Zea-Nit, Epic, Quantum 4000, Victus, Mycostop, etc. have, therefore, been registered for their practical use in farming community. Besides, the potentiality of PGPR inoculants in beneficial improvement of agricultural plants in developing countries can never be ignored. In India, more than 40 stakeholders from different provinces have registered themselves for the mass production of PGPR with Central Insecticides Board (CSI), Faridabad, Haryana, through collaboration with Tamil Nadu Agricultural University, Coimbatore, India.

Some of the important PGPR strains along with their commercial products as formulated by Chet and Chernin (2002) and Glick et al. (1999) are represented in Table 5.2.

Table 5.2 Commercial products of PGPR in plant disease management

Product	Target pathogens/diseases	Crops recommended	Manufacturer
Agrobacterium radiobacter – Agrogall 30	Crown gall (*A. tumefaciens*)	Ornamental, fruit and nut plants	Probical Bioestimulantes Foliares Profer, Chile
A. radiobacter – Diegall, K1026	Crown gall (*A. tumefaciens*)	Ornamental and nut plants, pear, blueberry, grape and other plants	Bio-Care Technol. Pvt. Ltd., New South Wales, Australia
A. radiobacter – Galltrol-A	Crown gall (*A. tumefaciens*)	Ornamental, fruit and nut plants	AgBiochem Inc., Orinda, California, USA
A. radiobacter – Nogall	Crown gall (*A. tumefaciens*)	Ornamental, fruit and nut plants	Becker Underwood Pty Ltd., Australia
Bacillus amyloliquefaciens – RhizoVital 42 LI/Rhizo Vital 42TB	Soilborne pathogens	Potato, strawberry, tomato, cucumber, ornamental plants	ABiTEP GmbH, Germany
B. licheniformis SB3086 – Green Releaf	Leaf spot, blight	Ornamental plants, greenhouse and nursery sites	Novozymes A/S, Denmark

(continued)

Table 5.2 (continued)

Product	Target pathogens/diseases	Crops recommended	Manufacturer
B. pumilus QST 2808 – Sonata ASO	Fungal pests such as moulds, mildews, blights, rusts, oak death syndrome	Used in nurseries, landscapes, greenhouse crops	AgraQuest Inc., Davis, USA
B. subtilis – Avogreen	*Colletotrichum gloeosporioides, Cercospora* sp.	Avocado	Ocean Agriculture, South Africa
B. subtilis – Biosafe	Foliar blight	Bean	Lab. Biocontrole Farroupilha, Brazil
B. subtilis – Biosubtilin	*Fusarium, Verticillium, Pythium, Cercospora, Colletotrichum, Alternaria, Ascochyta, Macrophomina, Myrothecium, Ramularia, Xanthomonas, Erysiphe polygoni*	Ornamental plants and vegetable crops	Biotech International Ltd., India
B. subtilis – Cease	Soilborne pathogens (*Rhizoctonia, Pythium, Fusarium, Phytophthora*) and foliar pathogens (*Botrytis, Erwinia, Xanthomonas*)	Several crops	BioWorks Inc., USA
B. subtilis – Ecoshot	Grey mould (*B. cinerea*)	Grape, citrus, legumes, vegetables and others	Kumiai Chemical Industry, Japan
B. subtilis – FZB24 WG, LI and TB	Root rot and wilts (*Alternaria, B. cinerea, Curvularia* spp., *Corynebacterium michiganense, E. carotovora, Fusarium* spp., *Phoma chrysanthemi, Phomopsis sclerotioides, Pyrenochaeta lycopersici, P. ultimum, R. solani, S. sclerotiorum, Verticillium* spp.)	Several crops	ABiTEP GmbH, Germany
B. subtilis – Subtilex/Pro-Mix	Root rot and seed treatments (*Fusarium, Rhizoctonia solani, Aspergillus, Pythium* and *Alternaria*)	Ornamental plants and other crops	Premier Horticulture Inc., Canada
B. subtilis FZB24 – Rhizo-Plus, Rhizo-Plus Konz	*R. solani, Fusarium* spp., *Sclerotinia, Verticillium*	Greenhouses grown crops, ornamentals, shrubs	KFZB Biotechnik GMBH, Berlin, Germany
B. subtilis GBO3 – Campanion	*Rhizoctonia, Pythium, Fusarium, Phytophthora*	Horticultural crops	Growth Products, USA
B. subtilis GBO3 – Kodiak, Kodiak HB, Kodiak AT, Epic, Concentrate, Quantum 4000	*Rhizoctonia solani, Fusarium* spp., *Alternaria* spp., *Aspergillus* spp.	Legumes	Gustafson Inc., Dallas, Texas, USA
B. subtilis MB1600- HiStick N/T, Subtilex	*Fusarium* spp., *Rhizoctonia* spp., *Pythium* spp., *Aspergillus* spp.	Ornamentals, vegetable crops dry/snap beans	Becker Underwood, Ames, IA, USA
B. subtilis QST713 – Rhapsody. Aqueous suspension formulation	Powdery mildew, sour rot, downy mildew, early leaf spot, early blight, late blight, bacterial spot, and walnut blight	Cherries, cucurbits, grapes, leafy vegetables, peppers, potatoes, tomatoes, walnuts	AgraQuest Inc., Davis, USA
B. subtilis QST713 – Serenade. Available as wettable powder	Powdery mildew, downy mildew, *Cercospora* leaf spot, early blight, late blight, brown rot, fire blight	Cucurbits, grapes, vegetables, pome fruits, stone fruits	AgraQuest Inc., Davis, Calif., USA

(continued)

Table 5.2 (continued)

Product	Target pathogens/diseases	Crops recommended	Manufacturer
B. subtilis var. *amyloliquefaciens* FZB24 – Taegro, Tae-Technical	*Rhizoctonia, Fusarium*	Only in greenhouses, ornamentals, shrubs	Earth Biosciences Inc, USA
B. subtilis + *B. amyloliquefaciens* – Bio Yield	Broad-spectrum action against greenhouse pathogens	Tomato, cucumber, pepper, tobacco	Gustafson Inc., Dallas, USA
B. subtilis GBO3 + chemical pesticides – System 3	Seedling pathogens	Beans, pea	Helena Chemical Co., Memphis, USA
B. velezensis – Botrybel	*Botrytis cinerea*	Tomato, lettuce, pepper, grape, strawberry and vegetables	Agricaldes, Spain
Burkholderia cepacia – Botrycid	Soilborne pathogens (*Rhizoctonia, Thielaviopsis, Verticillium, Fusarium* and *Pythium*), disease caused by *Botrytis, Mycosphaerella, Erwinia, Xanthomonas, Agrobacterium* and *Ralstonia solanacearum*	Several crops	SAFER AgroBiologicos, Colombia
B. cepacia – Deny, Intercept, Blue Circle	*Rhizoctonia solani, Pythium, Fusarium*, lesion, spiral, lance, and sting nematodes	Beans, peas, vegetable crops	Stine Microbial Products, Shawnee, KS, USA
Pseudomonas chlororaphis 63-28 – AtEze	*Pythium* spp., *Rhizoctonia solani, Fusarium oxysporum*	Ornamentals, vegetables	Eco Soil Systems Inc., San Diego, CA, USA
P. fluorescens – Conquer	*P. tolaasii*	Mushrooms	Mauri Foods, Kittanning, Philadelphia, USA
P. fluorescens – Victus	*P. tolaasii*	Mushrooms	Sylvan Spawn, Kittanning, PA, USA
P. fluorescens A 506 – BlightBan A506	*Erwinia amylovora* and russet – inducing bacteria	Almond, apple, apricot, blueberry, cherry, pear, peach, strawberry, potato, tomato	Plant Health Technologies, Fresno, Calif., USA
P. syringae ESC-100 – Bio-Save 10, 11, 100, 110, 1,000	*Botrytis cinerea, Penicillium* spp., *Mucor piriformis, Geotrichum candidum*	Pome fruit (Bio-Save 100), citrus (Bio-Save 1,000)	Eco Science Corp., Orlando, FL, USA
Streptomyces lydicus – Actinovate SP	Damping off and root rot (*Pythium, Rhizoctonia, Fusarium, Phytophthora, Verticillium*), powdery mildew (*Erysiphe, Oidium, Podosphaera, Sphaerotheca*), grey mould (*B. cinerea*), downy mildew (*Pseudoperonospora, Peronospora*) and *Alternaria* blight (*Alternaria* spp.)	Ornamental plants, vegetables	Natural Industries Inc., USA
S. griseoviridis K61 – Mycostop	Root rot, damping off and wilt caused by *Fusarium, Alternaria brassicola, Phomopsis, Botrytis, Pythium, Phytophthora* and *Rhizoctonia*	Several crops	Kemira Agro Oy, Helsinki, Finland

References

Amer GA, Utkhede RS (2000) Development of formulations of biological agents for management of root rot of lettuce and cucumber. Can J Microbiol 46:809–816

Andrews JH (1992) Biological control in the phyllosphere. Ann Rev Phytopathol 30:603–635

Backman PA, Wilson M, Murphy JF (1997) Bacteria for biological control of plant diseases. In: Rechcigl NA, Rechecigl JE (eds) Environmentally safe approaches to crop disease control. Lewis Publishers, Boca Raton, pp 95–109

Baker SC, Stavely JR, Thomas CA, Sasser M, Mac Fall SJ (1983) Inhibitory effect of *Bacillus subtilis* on *Uromyces phaseoli* and on development of rust pustules on bean leaves. Phytopathology 73:1148–1152

Bashan Y (1998) Inoculants of plant growth-promoting bacteria for use in agriculture. Biotechnol Adv 16:729–770

Berg G, Fritze A, Roskot N, Smalla K (2001) Evaluation of potential biocontrol rhizobacteria from different host plants of *Verticillium dahliae Kleb*. J Appl Microbiol 91:963–971

Bora T, Ozaktan H, Gore E, Aslan E (2004) Biological control of *Fusarium oxysporum* f. sp. *melonis* by wettable powder formulations of the two strains of *Pseudomonas putida*. J Phytopathol 152:471–475

Caesar AJ, Burr TJ (1991) Effect of conditioning, betaine, and sucrose on survival of rhizobacteria in powder formulations. Appl Environ Microbiol 57:168–172

Callan NW, Mathre DE, Miller JB (1990) Bio-priming seed treatment for biological control of *Pythium ultimum* pre-emergence damping-off in *sh2* sweet corn. Plant Dis 74:368–372

Charpentier CA, Gadille P, Benoit JP (1999) Rhizobacteria microencapsulation: properties of microparticles obtained by spray drying. J Microencapsul 16:215–229

Chet I, Chernin L (2002) Biocontrol, microbial agents in soil. In: Bitton G (ed) Encyclopedia of environmental microbiology. Willey, New York, pp 450–465

Costa E, Teixdo N, Usall J, Atares E, Vinas I (2001) Appl Microbiol Biotechnol 56:367–371

Dandurand LM, Morra MJ, Chaverra MH, Orser CS (1994) Survival of *Pseudomonas* spp in air dried mineral powders. Soil Biol Biochem 26:1423–1430

Dharani-Aiyer PV (2004) Afr J Biotechnol 10:519–522

Glick BR, Patten CL, Holguin G, Penrose DM (1999) Biochemical and genetic mechanisms used by plant growth-promoting bacteria. Imperial College Press, London

Heijnen CE, Burgers SLGE, van Veer JA (1993) Metabolic activity and population dynamics of rhizobia introduced into unamended and bentonite amended loamy sand. Appl Environ Microbiol 59:743–747

Ibrahim L, Butt T, Beckett A, Clark S (1999) The germination of oil-formulated conidia of the insect pathogen, *Metarhizium anisopliae*. Mycol Res 103:901–907

Janisiewicz WJ, Jeffers SN (1997) Efficacy of commercial formulation of two biofungicides for control of blue mold and gray mold of apple in cold storage. Crop Prot 16:629–633

Jeyarajan R, Nakkeeran S (2000) Exploitation of microorganisms and viruses as biocontrol agents for crop disease management. In: Upadhyay et al (eds) Biocontrol potential and their exploitation in sustainable agriculture. Kluwer Academic/Plenum Publishers, USA, pp 95–116

Kevan PG, Al-Mazrawi MS, Sutton JC, Tam L, Boland G, Broadbent B, Thompson SV, Brewer GJ (2003) Using pollinators to deliver biological control agents against crop pests. In: Downer RA, Mueninghoff JC, Volgas GC (eds) Pesticide formulations and delivery systems: meeting the PGPR formulations challenges of the current crop protection. American Society for Testing and Materials International, West Conshohocken

Kings EO, Ward MK, Raney DE (1954) Two simple media for the demonstration of pyocyanin and fluorescein. J Lab Chem Med 44:301–307

Kloepper JW, Schroth MN (1981) Plant growth promoting rhizobacteria and plant growth under gnotobiotic conditions. Phytopathology 71:642–644

Kovach J, Petzoldt R, Harman GE (2000) Use of honey bees and bumble bees to disseminate *Trichoderma harzianum* 1295-22 to strawberries for *Botrytis* control. Biol Control 18:235–242

Krishnamurthy K, Gnanamanickam SS (1998) Biological control of rice blast by *Pseudomonas fluorescens* strain Pf7-14: evaluation of a marker gene and formulations. Biol Control 13:158–165

Larkin RP, Roberts DP, Gracia-Garza JA (1998) Biological control of fungal diseases. In: Hutson D, Miyamoto Y (eds) Fungicidal activity, chemical and biological approaches. Wiley, New York, pp 149–191

Luna CL, Mariano RLR, Souto-Maior AM (2002) Braz J Chem Eng 19:133–140

Mathiyazhagan S, Kavitha K, Nakkeeran S, Chandrasekar G, Manian K, Renukadevi P, Krishnamoorthy AS, Fernando WGD (2004) PGPR mediated management of stem blight of *Phyllanthus amarus* (Schum and Thonn) caused by *Corynespora cassiicola* (Berk and Curt) Wei. Arch Phytopathol Plant Protect 33:183–199

Murphy JF, Zehnder GW, Schuster DJ, Sikora EJ, Polston JE, Kloepper JW (2000) Plant growth promoting rhizobacterial mediated protection in tomato against tomato mottle virus. Plant Dis 84:779–784

Muthukumarasamy R, Revathi G, Lakshminarasimhan C (1999) Diazotrophic associations in sugarcane cultivation in South India. Trop Agric 76:171–178

Nakkeeran S, Kavitha K, Mathiyazhagan S, Fernando WGD, Chandrasekar G, Renukadevi P (2004) Induced systemic resistance and plant growth promotion by *Pseudomonas chlororaphis* strain PA-23 and *Bacillus subtilis* strain CBE4 against rhizome rot of turmeric (*Curcuma longa* L). Can J Plant Pathol 26:417–418

Nakkeeran S, Fernando WGD, Siddiqui ZA (2005) Plant growth promoting rhizobacteria formulations and its scope in commercialization for the management of pests and diseases. In: Siddiqui ZA (ed) PGPR: biocontrol and biofertilization. Springer, Dordrecht, pp 257–296

Nandakumar R, Babu S, Viswanathan R, Sheela J, Raguchander T, Samiyappan R (2001) A new bioformulation containing plant growth promoting rhizobacterial mixture for the management of sheath blight and enhanced grain yield in rice. Biocontrol 46:493–510

Nohata Y, Kurane R (1997) J Ferment Bioeng 83:116–117

Peng G, Sutton JC, Kevan PG (1992) Effectiveness of honey bees for applying the biocontrol agent *Gliocladium roseum* to strawberry flowers to suppress *Botrytis cinerea*. Can J Plant Pathol 14:117–188

Raghuchander T, Shanmugam V, Samiyappan R (2001) Biological control of Panama wilt disease of banana. Madras Agric J 87:320–321

Raupach GS, Kloepper JW (1998) Mixtures of plant growth promoting rhizobacteria enhance biological control of multiple cucumber pathogens. Phytopathology 88:1158–1164

Sabitha Doraisamy, Nakkeeran S, Chandrasekar G (2001) *Trichoderma* – bioarsenal in plant disease management and its scope for commercialization. In: Proceedings of the Indian Phytopathological Society, Southern Zone Meeting, 10–12 Dec 2001, Indian Institute of Spice Research, Calicut, Kerala, pp 43–55

Saleh SA, Mekhemar GAA, Abo El-Soud AA, Ragab AA, Mikhaeel FT (2001) Survival of *Azorhizobium* and *Azospirillum* in different carrier materials: inoculation of wheat and *Sesbania rostrata*. Bull Fac Agric Univ Cairo 52:319–338

Siddiqui ZA, Mahmood I (1996) Biological control of *Heterodera cajani* and *Fusarium udum* on pigeonpea by *Glomus mosseae, Trichoderma harzianum* and *Verticillium chlamydosporium*. Israel J Plant Sci 44:49–56

Taylor AG, Harman GE (1990) Concept and technologies of selected seed treatments. Ann Rev Phytopathol 28:321–339

Vidhyasekaran P, Muthamilan M (1995) Development of formulations of *Pseudomonas fluorescens* for control of chickpea wilt. Plant Dis 79:782–786

Vidhyasekaran P, Sethuraman K, Rajappan K, Vasumathi K (1997) Powder formulation of *Pseudomonas fluorescens* to control pigeonpea wilt. Biol Control 8:166–171

Vivekananthan R, Ravi M, Ramanathan A, Samiyappan R (2004) Lytic enzymes induced by *Pseudomonas fluorescens* and other biocontrol organisms mediate defence against the anthracnose pathogen in mango. World J Microbiol Biotechnol 20:235–244

Wagner B, Katia KC, Antonio DG, Cristiane M, Elida CB (2005) Mass production of *Bacillus subtilis* and *Trichoderma viride* for the control of *Phyllosticta citricarpa* (teleomorph: *Guignardia citricarpa*). Summa Phytopatológica 31:276–278

Weller DM (1983) Colonization of wheat roots by a fluorescent pseudomonad suppressive to take all. Phytopathology 73:1548–1553

Wright B, Rowse HR, Whipps JM (2003) Application of beneficial microorganisms to seeds during drum priming. Biocontrol Sci Tech 13:519–614

Zehnder GW, Murphy JF, Sikora JE, Kloepper JW (2001) Application rhizobacteria for induced resistance. Eur J Plant Pathol 107:39–50

Part II

Pest Management in Fruit Crops

6 Tropical Fruit Crops

India is the second largest producer of fruits next only to China contributing 12.5 % of the total world production. The present area under fruit crops is estimated to be about 6.704 million hectares with an estimated production of 76.424 million MT (Tables 6.1 and 6.2) (Tiwari et al. 2014). Fruit crops and nuts contribute 31.81 % share to the total production of horticultural crops. India is the largest producer of mango, banana, guava and papaya. India produces 45.0 % of the world's mangoes and guavas, 29.0 % of the world's bananas and 38.0 % of the world's papayas. The leading fruit growing states are Maharashtra, Andhra Pradesh, Jammu and Kashmir, Karnataka, Gujarat, Uttar Pradesh, Tamil Nadu, Orissa, Bihar, Kerala, West Bengal, Himachal Pradesh, Uttarakhand, Chhattisgarh, Madhya Pradesh and Assam.

6.1 Banana, *Musa* spp.

6.1.1 Panama Wilt, *Fusarium oxysporum* f. sp. *cubense*

Panama wilt is one of the most devastating diseases of banana in the world. The disease is prevalent in Australia, Coasta Rica, Hawaii, India, Jamaica, Panama, South America, Surinam and West Africa. In India, it became widespread in Tamil Nadu, Kerala, Karnataka, Bihar and Assam. The disease is also prevalent in West Bengal, Maharashtra and Andhra Pradesh, especially where Rasabale, Amritpani, Malbhog and Martban cvs. belonging to Rasthali group are grown.

6.1.1.1 Symptoms

The entry of the fungus is facilitated by root damage caused by the nematodes (*Radopholus similis*, *Meloidogyne incognita*, etc.). The fungus blocks the vascular system and causes wilting (Fig. 6.1). The infected plants show characteristic yellowing of leaf blades developing as a band along the margin and spreading towards midrib. The leaf wilts and the petiole buckles. The leaf hangs between the pseudo stem while the middle of lamina is still green. All leaves eventually collapse, whereas the petioles join the pseudo stem and die. Often the emerging heart leaf gets affected. Four to six weeks after the appearance of the first symptoms, only the pseudo stem with dead leaves hanging round it remains. Young and old plants show dwarfing or stunting. When an affected rhizome is cut transversely, the disease is seen localized in the vascular strands (Fig. 6.1). Individual strand appears yellow. Red or brown dots or streaks are also seen. The cut stem smells like rotten fish. The suckers growing out of diseased corms wilt and eventually the whole mat dies.

6.1.1.2 Biomanagement

Sivamani and Gnanamanickam (1988) reported that application of powder formulation of *Pseudomonas fluorescens* (50 g/pit) before planting and 3, 5 and 7 months after planting was effective for the management of Panama disease.

P.P. Reddy, *Plant Growth Promoting Rhizobacteria for Horticultural Crop Protection*,
DOI 10.1007/978-81-322-1973-6_6, © Springer India 2014

Table 6.1 Major fruit producing states of India (2011–2012)

State	Area ('000 ha)	Production ('000 MT)	Productivity (MT/ha)
Andhra Pradesh	671.69	9841.07	14.65
Assam	142.76	1851.77	12.97
Bihar	299.24	3946.39	13.18
Chhattisgarh	185.19	1569.18	8.47
Gujarat	353.73	7522.43	21.26
Himachal Pradesh	214.57	372.82	1.73
Jammu and Kashmir	455.74	2329.89	5.11
Karnataka	371.80	6428.10	17.28
Kerala	296.14	2429.54	8.20
Madhya Pradesh	159.57	3391.28	21.25
Maharashtra	1560.00	10538.00	6.75
Odisha	328.99	2154.36	6.54
Tamil Nadu	331.97	8535.05	25.71
Uttar Pradesh	337.03	5795.09	17.19
Uttarakhand	200.73	802.12	3.99
West Bengal	216.64	3055.44	14.10
Other states	578.38	5861.68	10.13
Total	*6704.17*	*76424.21*	*11.39*

Table 6.2 Crop-wise area, production and productivity of fruit crops in India (2011–2012)

Fruits	Area ('000 ha)	Production ('000 MT)	Productivity (MT/ha)
Almond	22	4	0.18
Aonla	95	961	10.11
Apple	322	2,203	6.84
Banana	797	28,455	35.70
Ber	34	252	7.41
Citrus	*915*	*7,922*	*8.65*
(i) Lime/lemon	234	2,272	9.70
(ii) Mandarin	329	3,128	9.50
(iii) Sweet orange (mosambi)	162	1,232	7.60
(iv) Others	190	1,290	6.78
Custard apple	19	127	6.68
Grapes	116	2,221	19.14
Guava	220	2,510	11.40
Jackfruit	60	1,042	17.36
Kiwi	3	6	2.00
Litchi	80	538	6.72
Mango	2,378	16,196	6.81
Papaya	117	4,457	38.09
Passion fruit	16	97	6.06
Peach	20	91	4.55
Pear	48	294	6.12
Pineapple	102	1,500	14.70
Plum	26	72	2.76
Pomegranate	112	772	6.89
Sapota	163	1,426	8.74
Walnut	150	284	1.89
Others	889	4,991	5.61
Total fruits	*6,705*	*76,424*	11.39

Fig. 6.1 Panama wilt affected banana plant and vascular browning

Fig. 6.2 Biocontrol of *Fusarium* wilt in banana plantlet treated with *Bacillus pumilus* (*right*) as compared with plantlet not treated (*left*)

Application of the capsule filled with *P. fluorescens* at 60 mg/capsule is recommended at 2, 4 and 6 months after planting. Soil application of banana with *P. fluorescens* strain Pf10 (which produces proteins to detoxify fusaric acid) reduced wilt incidence by 50 % compared with the control (Thangavelu et al. 2001).

Ponnaiah and Subramanian (1994) found that application *P. fluorescens* resulted in 29 % reduction in the incidence of disease. Jahagirdar et al. (2001) reported that seedling dip in *P. fluorescens* (10^8 cells/ml) or *Bacillus subtilis* (10^8 cells/ml) for 2 h was effective for the management of the disease.

The bacterial isolate UPM13B8 (*Pseudomonas* sp.) has been identified as a potential biocontrol agent to manage *Fusarium* wilt disease of banana (*F. oxysporum* f. sp. *cubense* race 4). The application of strain UPM13B8 in formulated forms to the field is highly desirable to ensure its continuous survival in the soil despite the presence of unfavourable environmental conditions. Sodium alginate is an encapsulating material that prolongs shelf life and protects sensitive microorganisms from harsh environments. Addition of UV-absorbing compounds is strongly suggested to develop UV-resistant beads (Adeline and Kuong 2007).

Inoculation of PGPR strain PAB-2 (*Bacillus* sp.) remarkably increased plant growth parameters such as root length (47.6 %) and weight (69.5 %), shoot height (28.9 %), pseudo-stem diameter (18.2 %) and shoot biomass (33.9 %). The inoculated plants showed higher formation of root hair which was visible within 7 days of inoculation; the total biomass (39.8 %) was increased due to PGPR inoculation. A substantial increase in nutrient accumulation (N-51.1 %, P-46.1 %, K-165.2 %, CaO-7.4 % and MgO-32.8 %) was also observed in banana plantlets inoculated with PAB-2. The strain significantly reduced the impact of *Fusarium* wilt on banana, resulting in more than 46.9 % reduction in foliar symptoms (Fig. 6.2) (Table 6.3). The overall

Table 6.3 The disease severity indices of tissue-cultured banana plantlets inoculated with PGPR strains PAB-2 and FOC-4 grown for 56 days

PGPR strains	Disease severity indices				Control effects
	25 days	35 days	45 days	55 days	
FOC-4	36.3	47.9	54.2	66.7	–
FOC-4 + PAB-2	2.6	12.5	29.2	35.4	46.9

growth performance of inoculated seedlings was higher compared to un-inoculated control. This study suggests that PGPR strain PAB-2 (*Bacillus* sp.) could be used as crop enhancer, biofertilizer and biocontrol agent to enhance better plantlet production and control *Fusarium* wilt disease of bananas (Li et al. 2011).

Banana suckers were dipped in talc-based *P. fluorescens* suspension (500 g of the product in 50 l of water) for 10 min after pairing and pralinage. Subsequently it was followed by capsule application (50 mg of *P. fluorescens* per capsule) on the third and fifth month after planting. It resulted in 80.6 % reduction in panama wilt of banana compared to control (Raguchander et al. 1997). Paring and pralinage of banana suckers combined with application of *P. fluorescens* 3 and 5 months after planting recorded low wilt incidence (3.5 %) compared to the control (80.6 %) (Raghuchander et al. 2001).

Antagonistic activity of endophytic *Streptomyces griseorubiginosus* against *F. oxysporum* f. sp. *cubense* has been recorded on banana by Cao et al. (2004).

6.1.1.3 Integrated Management

(a) *Cultural methods and bioagents and botanicals*: Soil application of bioagent *P. fluorescens* along with cassava residue at 10 t/ha and rice bran and fallowing for 21 days before planting was found to check the disease effectively.

6.1.2 Sigatoka Leaf Spot, *Mycosphaerella fijiensis*

The disease mostly attacks leaves and has been reported to cause economic losses to the banana crop. Severe losses have been reported due to Sigatoka leaf spot disease in Java, Guyana, Honduras, Surinam, Tanzania, Fizi, Uganda, Colombia, Costa Rica, Guatemala, Haiti, India, Jamaica, Mexico, Panama and Venezuela. In India, it is distributed in Assam, Andhra Pradesh, Bihar, West Bengal, Kerala, Tamil Nadu, Karnataka, Maharashtra and Gujarat states.

6.1.2.1 Symptoms

The initial symptoms seen in the field are small lesions on leaves which are pale yellow or greenish yellow streaks that appear on both sides of leaves parallel to the leaf veins. The spots on leaves increase in size to form dark brown to black linear oblong areas (Fig. 6.3). The centre of the spots eventually dries out, becoming light grey, but a narrow, dark brown to black border persists giving an eye spot appearance. Rapid drying and defoliation is the characteristic feature of this disease. In severe cases, the petiole collapses and the leaf hangs down from the pseudo stem. The infection may cause complete failure of maturity of the bunch. In some cases, premature ripening of fruits is also reported.

6.1.2.2 Biomanagement

Serratia marcescens isolates showed good antagonism towards black sigatoka pathogen *M. fijiensis* and out yielded the control given by propiconazole in a greenhouse trial. Under field conditions, the effect of these isolates was similar to that of fungicides.

6.1.3 Bacterial Wilt or Moko Disease, *Ralstonia solanacearum*

Bacterial wilt in banana crops is a serious problem in South and Central America, the Caribbean

Fig. 6.3 Sigatoka leaf spot on banana

Fig. 6.4 Moko wilt of banana. *Left* – Wilting of plants. *Right* – Blackening of internal fruit tissues

islands and the Philippines. In India, the disease is reported on banana cvs. Robusta and Poovan from Tamil Nadu causing 75–100 % yield losses. Decline in yield of 74 % in plantain and 64 % in banana were attributed to Moko disease in Guyana from 1963 to 1973. The export of banana in Trinidad was virtually eliminated in the early 1960s due to this disease.

6.1.3.1 Symptoms

The disease is characterized by rapid wilting and collapse of leaves (Fig. 6.4), premature ripening of fruits, discolouration of the vascular strands and wilting and blackening of suckers. In Cavendish cv., the lower leaves develop a yellowish tinge which quickly spread to all the other leaves. Sometimes the second or third leaf or both the leaves collapse near the junction of petiole and lamina. In severe cases, plants collapse before the emergence of bunches. The presence of yellow fingers in otherwise green stem often indicates the occurrence of Moko disease. Internally, a firm dry rot is found in the fruit (Fig. 6.4). Fruit stalks also show discoloured vascular strands.

6.1.3.2 Biomanagement

Anuratha and Gnanamanickam (1990) successfully controlled bacterial wilt of banana by bacterization of suckers with *Pseudomonas fluorescens*.

Fig. 6.5 Symptoms of soft rot of banana

6.1.4 Tip Over/Soft Rot, *Erwinia carotovora*

6.1.4.1 Symptoms

Bananas being propagated from freshly cut rhizome tissue or suckers may show poor sprout emergence, stunting and yellowing (Fig. 6.5). Incidence is highest in wet humid areas where the material has been treated with hot water prior to planting. Affected plants show discolouration and soft rotting of rhizomes and suckers. They have scanty roots with necrotic tips. The plants remain stunted with water-soaked discoloured leaf sheath base, pseudo-stem tip over breaking across the rotted stem.

6.1.4.2 Biomanagement

Bacillus subtilis applied to seedlings as root dip of tissue-cultured banana cv. G-9 followed by three soil drenches at 15 days interval starting from 20 DAP was highly effective and completely controlled the tip over disease in the field. *P. fluorescens* gave 87.5 % protection over control. *B. subtilis* possessed growth-promoting activity as indicated by increased plant height (170.67 cm) followed by *P. fluorescens*-treated plants (157.67 cm), 60 DAP. The highest increase in the yield was obtained in *B. subtilis*-treated plants (85.71 % over control) followed by *P. fluorescens-treated* plants.

6.1.5 Bunchy Top Virus

Bunchy top caused by banana bunchy top virus (BBTV) is one of the most destructive diseases of banana. BBTV infects almost all banana cultivars, retarding the growth of infected plants and causing substantial economic losses. In the hills of the lower Pulney, Western Ghats, Tamil Nadu, India, the prized hill banana 'Virupakshi' (AAB, pome subgroup) is extremely susceptible to BBTV. The area under cultivation of 'Virupakshi' has shrunk from 18,000 to 2,000 ha due to BBTV, despite farmers selecting visually healthy suckers to raise the next crop and spraying plants with systemic pesticides to control *Pentalonia nigronervosa*, the aphid vector of BBTV.

6.1.5.1 Symptoms

The first characteristic symptoms appear as irregular, nodular, green streaks along the secondary veins on the underside of the lower portion of leaf blade, midrib and petiole. The streaks are about 0.75 mm wide and vary in length up to 2.5 cm. The subsequent leaves show the same symptoms and are progressively dwarfed. They also show marginal chlorosis and curling. The leaves are brittle in texture and petioles are incompletely elongated. Pale, white streaks may be seen along the length of lamina. In subsequent leaves, the unfurling process is premature but slow so that several leaves are unrolling at the same time.

Fig. 6.6 Bunchy top symptoms on banana

The leaves become smaller and eventually the crown of the plant becomes composed of stunted leaves, the rosette or bunchy top (Fig. 6.6). The emergence of bunches is choked by the pseudo stem and may split instead of emerging in the normal passion. The bunches are reduced in size and the fruits lose their market value.

6.1.5.2 Biomanagement

The use of microbial inoculants, primarily bacteria, as propagule bio-priming agents in tissue-cultured banana planting material (often called 'bio-hardening') aims at increasing the resistance against bunchy top disease. Tissue-cultured banana plantlets treated with all three bacterial strains [*P. fluorescens* strains Pf1 and CHAO (rhizobacteria) and *B. subtilis* strain EPB22 (endophyte)] in combination both as a 1 % foliar spray and soil drench resulted in a lower incidence of BBTV. Indirect ELISA indicated that bio-hardened plantlets had a reduced viral antigen concentration. Analysis of the plant tissues for oxidative enzymes, pathogenesis-related proteins and phenolic content found that these were higher in plants treated with CHAO, singly or in combination with Pf1 and EPB22, than in an untreated control. Thus, plantlets bio-hardened with beneficial bacterial strains could reduce the damage caused by BBTV through eliciting induced systemic resistance in banana (Kavino et al. 2007b).

Banana plants treated at planting and at the third, fifth and seventh month after planting with two strains of the PGPR *P. fluorescens* (Pf1 and CHAO) formulated with the carrier chitin had significantly reduced bunchy top incidence under field conditions, compared with the control treatment. The reduction in disease incidence was more pronounced with the chitin-amended CHAO strain. The chitin-amended CHAO strain also increased the leaf nutrient status and enhanced growth, bunch yield and the quality of the fruits compared to untreated plants (Kavino et al. 2007a).

Banana plants treated with mixtures of endophytic and rhizobacterial formulations [*P. fluorescens* Pf1, CHAO (rhizobacteria) and *B. subtilis* EPB22 (endophyte)] showed minimum bunchy top infection (20 %) with an 80 % reduction over controls. A low titer of BBTV was recorded in treated plants in ELISA tests. Some treated plants tested negative for BBTV in PCR. The expression of defence-related chemicals and enzymes, such as phenol, PAL, peroxidase and polyphenol oxidase, and pathogenesis-related proteins, such as chitinase and β-1,3-glucanase, were more pronounced in the endophytic bacteria-treated banana plants than in control plants (Harish et al. 2007).

6.1.6 Burrowing Nematode, *Radopholus similis*

The burrowing nematode causes 'rhizome rot' or 'toppling' or 'black head' disease of banana and is becoming a serious problem. *R. similis* is responsible for 30.76–41 % yield loss in banana. Root population of *R. similis* is indirectly correlated with the yield.

6.1.6.1 Symptoms

Wounding of banana roots by the burrowing nematode usually induces reddish-brown cortical lesions which are diagnostic of the disease. These lesions are clearly seen when an affected root is split longitudinally and examined immediately. Root and rhizome necrosis is manifested by varying degrees of retarded growth, leaf yellowing and falling of mature plants (Fig. 6.7).

Fig. 6.7 *Left* – Premature fall of banana plants due to infection of *Radopholus similis*. *Right* – Banana roots infected with *R. similis*

6.1.6.2 Biomanagement

Soil application of *P. fluorescens* at 20 g/plant (2.5×10^8 cfu/g) was effective in increasing the plant growth parameters (pseudo-stem height and girth, number of leaves, root length and weight) and in reducing the nematode population in soil (62.34 %) and roots (60.35 %). Rhizosphere treatment with *P. fluorescens* at 20 g/250 ml water along with soil treatment with 100 g powder/rhizome gave 56.5 % reduction in nematode population. During the second month the nematode reduction was 63.8 % (Nirmal Johnson and Devarajan 2004).

In a glasshouse experiment, maximum increase in plant growth of banana and reduction in nematode population in soil and roots and root lesion index were observed in pots treated with talc-based formulation of *P. fluorescens* (isolate PfB 13) at 10 g/plant which was on par with *P. fluorescens* (isolate Pf1) and carbofuran (Senthilkumar et al. 2008a). Field application of *P. fluorescens* (isolate PfB 13) at 10 g/plant significantly reduced nematode population in soil and roots and root lesion index and increased plant height, pseudo-stem girth, leaf area, number of leaves and fruit yield (Senthilkumar et al. 2008b) (Table 6.4).

Soil application of two isolates of *B. subtilis* (EPB 5 and EPB 31) gave significant increase in plant growth parameters, pseudo-stem girth and number of leaves coupled with reduction in nematode population (*R. similis*) on tissue-cultured banana (Jonathan and Umamaheswari 2006).

Table 6.4 Effect of different isolates of *Pseudomonas fluorescens* on plant growth, yield and management of *Radopholus similis* infecting banana

Treatment	Plant height (cm) (270 DAP)	Pseudo-stem girth (cm) (270 DAP)	Bunch weight (kg)	Root lesion index
P. fluorescens – PfB 13	328.2	57.4	15.6	14
P. fluorescens – PfB 19	287.3	46.4	13.4	32
P. fluorescens – PfB 24	270.3	47.3	11.4	28
P. fluorescens – PfB 32	266.8	41.2	11.2	42
P. fluorescens – PfB 35	287.8	45.2	13.1	46
P. fluorescens – PfB 39	278.6	46.1	10.1	36
P. fluorescens – Pf1	275.4	45.3	14.3	18
Carbofuran	270.9	43.1	13.2	19
Control	210.9	39.7	10.9	54
CD (P = 0.05)	*26.59*	*6.99*	*0.96*	*1.37*

Soil application of biofertilizer *Azospirillum* at 20 g/plant had significant effect in enhancing the plant growth and reduced the population of *R. similis* in soil to an extent of 50.1 % over control (Santhi et al. 2004).

On Cavendish cultivar 'Williams', *Pseudomonas* isolates P-52 and P-58 reduced populations of *R. similis* by 96 % and 94 %, respectively, compared to non-inoculated controls. In the experiments using 'Grande Naine', the results for P-52 and P-58 were 70 % and 83 %, respectively. In the *Bacillus* group, the two most effective bacteria were B-21 and B-50, with 87 % and 84 % control with 'Williams' and 72 % and 68 % control with 'Grand Naine', respectively (Pocasangre et al. 2007).

6.1.6.3 Integrated Management

(a) *Bioagents and botanicals*: Soil application of FYM enriched with *P. fluorescens* (1×10^9 cfu/g) at 2 kg/plant at the time of planting and subsequent application for four times at an interval of 6 months reduced the root population of *R. similis* by 64.5 % and increased the fruit yield by 21 %. Benefit/cost ratio (calculated for marginal cost of biopesticides and returns accrued by application of biopesticides) was 3.6.

6.1.7 Spiral Nematode, *Helicotylenchus multicinctus*

The spiral nematode causes serious decline of banana yield to the tune of 34–56 % with 55.88 % loss in number of fruits per bunch and delayed fruiting by 134 days.

6.1.7.1 Symptoms

The spiral nematode incites discrete, relatively shallow, necrotic lesions on banana roots. *H. multicinctus* causes extensive root necrosis, dieback and dysfunction leading eventually to debility of the entire plant.

6.1.7.2 Biomanagement

Soil application of *P. fluorescens* at 20 g/plant (2.5×10^8 cfu/g) was effective in increasing the plant growth parameters (pseudo-stem height and girth, number of leaves, root length and weight) and in reducing the spiral nematode population in soil (58.87 %) and roots (56.54 %).

Soil application of biofertilizer *Azospirillum* at 20 g/plant had significant effect in enhancing the plant growth and reduced the population of *H. multicinctus* in soil to an extent of 48.9 % over control (Santhi et al. 2004).

6.1.7.3 Integrated Management

(a) *Bioagents and botanicals*: Significant reduction in population of *H. multicinctus* was observed in banana plants treated with *P. fluorescens* at 2 g along with press mud at 3 kg/plant. These treatments also enhanced the plant height, pseudo-stem girth, number of leaves, leaf area and bunch weight (Jonathan and Cannayane 2002).

6.1.8 Lesion Nematode, *Pratylenchus coffeae*

In India, crop loss caused by the root lesion nematode in banana cv. Nendran was reported to be 45 %. The lesion nematode causes most damage to banana in association with *F. oxysporum* f. sp. *cubense*.

6.1.8.1 Symptoms

The symptoms caused by *P. coffeae* are very similar to those caused by *Radopholus similis* (Fig. 6.8). Both *R. similis* and *P. coffeae* are associated with the blackhead toppling disease of bananas.

6.1.8.2 Biomanagement

Soil application of *P. fluorescens* at 20 g/plant (2.5×10^8 cfu/g) was effective in increasing the plant growth parameters (pseudo-stem height and girth, number of leaves, root length and weight) and in reducing the lesion nematode population in soil (67.12 %) and roots (62.00 %).

Soil application of biofertilizer *Azospirillum* at 20 g/plant had significant effect in enhancing the plant growth and reduced the population of *P. coffeae* in soil to an extent of 45.5 % over control (Santhi et al. 2004).

Fig. 6.8 Lesion nematode infection on banana roots

6.1.9 Root-Knot Nematodes, *Meloidogyne* spp.

The root-knot nematodes *M. incognita* and *M. javanica* attack bananas. *Meloidogyne* species have not been reported to cause yield reductions in bananas, but may interact with other nematodes or soil pests. *M. incognita* caused 30.90 % loss in fruit yield of banana.

6.1.9.1 Symptoms

The infected plants were stunted, having small chlorotic leaves and galled roots. Galling of the plant roots is most commonly found in areas previously planted with sugarcane. The galls vary in size and occur at the tips as well as in other areas along the root (Fig. 6.9). Roots with galled tips cease to grow and sometimes develop secondary roots above the gall. Swollen female nematodes were found inside the galled and sometimes non-galled roots.

6.1.9.2 Integrated Management

(a) *Bioagents and botanicals*: Soil application of 2 kg FYM with *P. fluorescens* (with 1×10^9 spores/g) per plant at the time of planting and at an interval of 4 months significantly reduced the root-knot nematode by 76 % compared to control.

6.2 Citrus, *Citrus* spp.

6.2.1 Foot Rot, Root Rot, Crown Rot, Gummosis, Leaf Fall and Fruit Rot, *Phytophthora* spp.

The disease seems to occur especially in the high rainfall areas. Its prevalence has been reported in south India, Maharashtra, Gujarat, Punjab and Assam states. *P. nicotianae* var. *parasitica* is widespread in Assam, while *P. palmivora* is prevalent throughout India.

6.2.1.1 Symptoms

Profuse gumming on the surface of the attacked bark is the main symptom. When gumming occurs on the stem, droplets of gum trickle down the stem. The bark gradually turns brown to dark brown and develops longitudinal cracks. A thin layer of wood tissue is also affected (Fig. 6.10). When gumming starts close to the soil, the disease spreads to the main roots and then around the base of the trunk. As a result of severe gumming, the bark becomes completely rotten and the tree dies owing to girdling effect. The trees usually blossom heavily and die after the fruits mature. In such cases, the disease is called as foot rot or collar rot. The pathogen produces symptoms of decline through rotting of the rootlets, girdling of the trunk and dropping of the blighted leaves. The fruits lying on the ground are liable to invasion by the pathogen and develop brown rot.

6.2.1.2 Biomanagement

Certain strains of *Pseudomonas putida* were found to reduce *Phytophthora* population up to 96 % in greenhouse. This bioagent when tried in field with seedlings reduced *P. citrophthora* population by 41–66 % and controlled 22 % root infection for 3 months (Menge 1993).

6.2.2 Soft Rot, *Aspergillus niger*

The disease has been reported on citrus from the Americas, South Africa, Mediterranean countries and most particularly in India. *Aspergillus* rot is

Fig. 6.9 Root-knot nematode symptoms on banana plants and roots

Fig. 6.10 Gummosis on main stem of Nagpur mandarin tree

an important postharvest disease of citrus in India. The losses due to this fungus were recorded 4–6 % in certain consignments of Nagpur mandarin sent from Nagpur to New Delhi during 1991–1992.

6.2.2.1 Symptoms

A. niger invades through wounds/injuries and causes soft rot. The pathogen infects lemon, lime, sweet orange and mandarin. Preharvest infection may also occur if the fruit is injured by thorns in case of acid lime during windy weather. The infected area of the fruit turns pale yellow with a characteristic halo leaving the shrunk intact cuticle on the macerated peel. In advanced stage, the infected surface is covered with black mass of spores that disseminate to contaminate the whole lot. Nagpur mandarin was found comparatively more susceptible to *Aspergillus* rot in green stage than colour break or orange-coloured fruit.

6.2.2.2 Biomanagement

In lemons treated with *Bacillus subtilis* (AF 1) cells, only 3–4 % fruits showed development of soft rot in comparison to 100 % in control.

Fig. 6.11 *Penicillium* rot of citrus fruits

6.2.3 *Penicillium* Rot, *Penicillium digitatum*, *P. italicum*

Green mould (*P. digitatum*) decay and blue mould (*P. italicum*) decay are important postharvest diseases which occur in all citrus growing areas and often constitute the predominant type of decay. Postharvest losses of citrus fruits caused by these pathogens can account for more than 90 % of all postharvest losses in semi-arid production areas of the world.

6.2.3.1 Symptoms

A soft water-soaked area is developed at the infection site in both diseases. Coloured spore mass is developed at the centre of the lesion surrounded by broad band of white mycelial growth in green mould infection, whereas white mycelial growth around the spore mass of blue mould is usually not more than 2 mm wide (Fig. 6.11). Both pathogens occur frequently, but green mould grows faster at moderate temperature and contaminates the fruit. The spores of green mould are unable to infect healthy uninjured adjacent fruits, while the blue mould develops nesting onto uninjured healthy fruits and may cause serious damage.

6.2.3.2 Biomanagement

Control of green mould was obtained on citrus fruits with several *Pseudomonas* spp. and *B. subtilis*.

Bacillus amyloliquefaciens HF-01 has been shown to be a potential antagonist against *Penicillium digitatum*, isolated from the citrus fruit surface (Hao et al. 2011).

6.2.4 Bacterial Canker, *Xanthomonas axonopodis* pv. *citri*

The disease is known to occur in almost all citrus growing areas. The disease is very severe on acid lime, lemon and grapefruit.

6.2.4.1 Symptoms

The disease affects all the above-ground parts – leaves, twigs, older branches, thorns and fruits (Fig. 6.12). Lesions first appear as small, round, watery, translucent spots on the lower surface of the leaves and then on the upper surface. As the disease progresses, spots become white or greyish and give a rough corky crater appearance. The lesions are surrounded by yellowish halo. Elongated lesions are formed on twigs. On larger branches, the cankers are irregular, rough and more prominent. Cankers on fruits are similar to those on leaves except that the yellow halo is absent and crater-like depressions in the centre are more pronounced. The lesions on fruits remain confined to the rind, but sometimes it causes cracks and fissures on the skin.

Fig. 6.12 Bacterial canker on acid lime leaves, bark and fruit

Fig. 6.13 *Left* – Citrus tree exhibiting symptoms of decline caused by the citrus nematode (note the thinning canopy and bare branches in the upper canopy). *Right* – Citrus root infected with females of *Tylenchulus semipenetrans*

6.2.4.2 Biomanagement

Application of a phylloplane antagonist, i.e. *B. subtilis* isolated from citrus leaves, was effective in reducing the disease incidence (Kalita 1995).

P. fluorescens is reported to be antagonistic to citrus canker pathogen.

6.2.5 Citrus Nematode, *Tylenchulus semipenetrans*

The nematode causes 'slow decline' and is considered to be one of the factors responsible for dieback of citrus trees in India. Yield increases of 40–200 % have been obtained following the control of the nematode over a range of growing conditions. *T. semipenetrans* was responsible for 69 %, 29 % and 19 % loss in fruit yield of sweet orange, lemon and sweet lime, respectively.

6.2.5.1 Symptoms

'Slow decline' of citrus is a disease condition of trees with symptoms similar to those caused by drought and malnutrition. Affected trees exhibit reduced vigour, chlorosis and falling of leaves, twig dieback (Fig. 6.13) and, consequently, reduced fruit production. This decline of the tree is gradual and persists until the crop is so small that tree maintenance may become uneconomical.

Table 6.5 Effect of integrated use of *Pseudomonas fluorescens* and oil cakes on plant growth and population of *Tylenchulus semipenetrans* on acid lime

Treatment	Dose (g)/plant	Plant wt. (g)	CFU/g root	Nematode population	
				Soil (100 ml)	Root (10 g)
P. fluorescens	4×10^9 spores	21.14	2.0×10^7	9,116	2,418
Castor cake	50 g	23.00	–	9,012	2,264
Karanj cake	50 g	22.09	–	9,228	2,426
Neem cake	50 g	23.40	–	8,804	2,310
P. fluorescens + castor cake	½ dose each	25.97	12.4×10^7	6,398	1,218
P. fluorescens + karanj cake	½ dose each	24.42	10.2×10^7	6,548	1,298
P. fluorescens + neem cake	½ dose each	28.97	18.6×10^7	6,034	1,010
Control	–	17.60	–	13,456	6,142
CD (P=0.05)	–	*1.76*	–	*234.64*	*212.34*

Fig. 6.14 Egg and larval instars and adult of lemon butterfly

As the nematode feeds on roots (Fig. 6.13) and reproduces, a large proportion of the feeder roots of citrus trees, particularly in the upper soil layers, is inactivated or destroyed, the uptake of water and minerals from the soil is reduced, and the symptoms appear in the above-ground tree parts. Heavily infested roots are darker in colour with branch rootlets shortened, swollen and irregular in appearance than in normal ones.

Soil particles usually cling tightly, even after washing, to the gelatinous egg masses which cover the protruding part of the nematode body. In heavily infested roots, the cortex separates readily from the vesicular stele.

6.2.5.2 Biomanagement

Application of *P. fluorescens* at 20 g/tree three times in a year at 15 cm depth and 50 cm away from tree trunk helps in reducing the nematode population.

6.2.5.3 Integrated Management

(a) *Bioagents and botanicals*: Integration of neem cake with *P. fluorescens* gave maximum reduction in citrus nematode population both in soil and roots and increased plant growth of acid lime (Parvatha Reddy et al. 2000) (Table 6.5).

6.2.6 Lemon Butterfly, *Papilio demoleus*

An outbreak of the pest during July–August 1996 was reported from the Vidarbha region of Maharashtra State on Nagpur mandarin.

6.2.6.1 Damage

The pest (Fig. 6.14) is active throughout the year except during winter, and the infestation is more pronounced in nurseries and on young trees.

Peak activity coincides with the emergence of new flush. It feeds voraciously on the vegetative growth throughout the year causing severe defoliation of plants. The pest attacks mandarin and acid lime plantations almost throughout the year, but the attack is more severe during July–August.

6.2.6.2 Biomanagement

Serratia marcescens reduce the pest population substantially.

6.3 Papaya, *Carica papaya*

6.3.1 *Fusarium* Wilt, *Fusarium oxysporum*

6.3.1.1 Symptoms

Stem appears water soaked and rots. The pathogen causes wilting and death of papaya seedlings.

6.3.1.2 Integrated Management

(a) *Bioagents and AMF*: Inoculation of PPV3 (*Pseudomonas* sp.) and MTZ01 (consortium of four mycorrhizal fungi *Glomus intraradices*, *G. mosseae*, *G. etunicatum* and *Gigaspora albida*), gave highest disease protection (85 %), and lowest Fusarium wilt in papaya seedlings (10%). Inoculations with MTZ01 or PPV3 showed an efficacy of 54 and 60 %, with a level of disease severity of the 38 and 22 %, respectively. The combination of the AMF complex (MTZ01) with rhizobacterial *Pseudomonas* sp. (PPV3) modified the effects of *F. oxysporum* and provided increased protection for papaya than either acting alone. These results suggest that rhizobacteria and arbuscular mycorrhizal fungi acting together formed a mutualistic relationship that enhances disease control against *F. oxysporum* and stimulates growth in papaya (Hernández-Montiel et al. 2013).

6.3.2 Root-Knot Nematode, *Meloidogyne incognita*, *M. javanica*

Root-knot nematodes are reported to cause 10–20% reduction in papaya fruit yield.

6.3.2.1 Symptoms

General symptoms visible in field include poor growth, yellowing of foliage, dropping of leaves, reduction in leaf production, weak vigour and premature dropping of fruits. Papaya orchards infected with *Meloidogyne* spp. show patches of poor growth with many plants missing in the rows. Roots exhibit typical below-ground symptoms, i.e. galls of varying sizes. The lateral branching of roots is limited. In heavily infected old roots, adjacent galls join together and form large galls. In mild infestation, root tips become swollen and root growth inhibition is distinctly seen (Fig. 6.15).

6.3.2.2 Integrated Management

(a) *Bioagents and AMF*: The combined inoculation of two AMF species (*Glomus mosseae* or *G. manihotis*) and a *Bacillus* consortium based on three strains previously described as PGPR in other crops were able to reduce nematode infection and damage on papaya.

(b) *Two bioagents*: Seed treatment with *P. fluorescens* (10^8 spores/g) combined with soil application of *T. harzianum* (10^6 spores/g) and *P. fluorescens* (10^8 spores/g) each at 5 g/kg soil gave significant reduction in *M. incognita* population both in soil and roots, number of eggs per egg mass and hatching of eggs and increased root colonization by both the bioagents in papaya under nursery conditions (Rao 2007) (Table 6.6).

(c) *Bioagents and botanicals*: Application of 2 kg FYM enriched with *P. fluorescens* (10^9 spores/g) per plant at the time of planting and at an interval of 6 months significantly reduced reniform and root-knot nematodes on roots by 66 and 70 %, respectively. Significant increase in fruit yield (28 %) was also observed. Benefit/cost ratio (calculated for marginal cost

Fig. 6.15 Papaya plant infected with *Meloidogyne incognita*

Table 6.6 Effect of integration of bioagents for the management of *Meloidogyne incognita* infecting papaya

Treatment[a]	No. of nemas/ 10 g roots	No. of J_2/100 ml soil	No. of eggs/ egg mass	% egg hatch suppression	Root colonization (cfu/g)	
					T. harzianum	*P. fluorescens*
T1	47	78	389	32	–	8,769
T2	42	70	364	45	–	18,457
T3	50	65	375	43	22,649	–
T4	35	52	358	65	21,895	18,249
T5	38	56	326	55	–	24,531
T6	45	63	347	54	22,412	8,566
T7	31	45	310	67	20,874	24,124
T8	68	126	412	–	–	–
CD at 5 %	*6.6*	*8.5*	*33.2*	*7.4*	*2365.7*	*2487.2*

[a]T1 – seed treatment with *P. fluorescens*, T2 – nursery soil mixed with *P. fluorescens* (5 g/kg), T3 – nursery soil mixed with *T. harzianum* (5 g/kg), T4 – nursery soil mixed with *T. harzianum* (5 g/kg) and *P. fluorescens* (5 g/kg), T5 – seed treatment with *P. fluorescens* + nursery soil mixed with *P. fluorescens* (5 g/kg), T6 – seed treatment with *P. fluorescens* + nursery soil mixed with *T. harzianum* (5 g/kg), T7 – seed treatment with *P. fluorescens* + nursery soil mixed with *T. harzianum* (5 g/kg) and *P. fluorescens* (5 g/kg), T8 – control

of biopesticides and returns accrued by application of biopesticides) was 3.2 (Anon 2012).

6.4 Mulberry, *Morus* spp.

6.4.1 Leaf Spot, *Cercospora moricola*

6.4.1.1 Symptoms

The disease appears as brownish necrotic spots (Fig. 6.16) causing severe loss to mulberry yield and quality.

6.4.1.2 Biomanagement

Pseudomonas maltophilia reduced the incidence of brown leaf spot disease in mulberry (Sukumar and Ramalingam 1986).

6.4.2 Leaf Rust, *Aecidium mori*, *Cerotelium fici*

6.4.2.1 Symptoms

The disease is characterized by small pinhead-sized spots giving rusty appearance on both surfaces of leaves.

Fig. 6.16 Mulberry *Cercospora* leaf spot

Fig. 6.17 Root-knot galls on mulberry roots

6.4.2.2 Biomanagement

PGPR isolates, *Azotobacter chroococcum*, *B. subtilis* and *P. fluorescens*, have shown their potential in inducing systemic resistance in mulberry against leaf rust disease in addition to promoting the plant growth (Gupta et al. 2003).

6.4.3 Root-Knot Nematodes, *Meloidogyne incognita*

M. incognita is responsible for 11.8 % loss in herbage yield of mulberry.

6.4.3.1 Symptoms

Above-ground symptoms include stunted growth, poor and delayed sprouting, reduced leaf size and yield, chlorosis and marginal necrosis of leaves, yellowing and wilting of leaves in spite of adequate soil moisture availability and death of plants in severe cases. These symptoms first appear in isolated patches, slowly spreading over the entire garden.

Below-ground symptoms include formation of galls or knots on roots (Fig. 6.17), reduced and stubby root system and necrotic lesions on the root surfaces and death of roots.

6.4.3.2 Integrated Management

(a) *Bioagents and AMF*: Combined inoculation of AMF and *P. fluorescens* had positive effect on root-knot nematode control on mulberry (Kumutha 2001).
(b) *Two bioagents*: Combined soil application of *P. fluorescens* and *Trichoderma viride* each at 10 g/plant was effective to check the nematode population and improve growth of mulberry with increased leaf yield (Muthulakshmi et al. 2010).

6.5 Avocado, *Persea americana*

6.5.1 *Cercospora* Leaf Spot and Anthracnose

6.5.1.1 Symptoms

Individual spots on leaves are very small, less than 2.5 mm in diameter, and brown to purple in colour. Angular appearance of leaf spots is highly diagnostic. Many of these leaf spots are surrounded by yellow halos. Individual leaf spots may coalesce to form irregular areas of brown tissue. On fruit, damage begins as small, irregular, brown spots that enlarge and coalesce. Fissures often appear in these spots (Fig. 6.18).

6.5.1.2 Biomanagement

Avogreen is a commercial product of *B. subtilis* isolated from avocado phylloplane to control diseases caused by *Cercospora* leaf spot and anthracnose of avocado (Janisiewiez and Kortsen 2002).

Fig. 6.18 *Cercospora* spots on leaf and fruit of avocado

References

Adeline TSY, Kuong CS (2007) Viability of alginate-encapsulated isolate UPM13B8, a potential biological agent for the control of Fusarium wilt of banana, after exposure to "sun rays". In: ISHS/ProMusa symposium, White River, South Africa, p 66

Anon (2012) Annual report 2011–2012. Indian Institute of Horticulture Research, Bangalore, pp 23–24.

Anuratha CS, Gnanamanickam SS (1990) Biological control of bacterial wilt caused by *Pseudomonas solanacearum* in India with antagonistic bacteria. Plant Soil 124:109–116

Cao L, Qiu Z, Dai X, Tan H, Lin Y, Zhou S (2004) Isolation of endophytic actinomycetes from roots and leaves of banana (*Musa acuminata*) plants and their activities against *Fusarium oxysporum* f. sp. *cubense*. World J Microbiol Biotechnol 20:501–504

Gupta VP, Sharma DD, Sarkar A, Rajan MV (2003) Plant growth-promoting rhizobacteria as inducers of systemic resistance in mulberry against leaf spot and leaf rust diseases (Abstr. MPP/P-23). In: National conference on tropical sericulture for global competitiveness, Mysore, India,p 51

Hao W, Li H, Hu M, Yang L, Rizwan-ul-Haq M (2011) Integrated control of citrus green and blue mold and sour rot by *Bacillus amyloliquefaciens* in combination with tea saponin. Postharvest Biol Technol 59:316–323

Harish S, Kavino M, Kumar N, Samiyappan R (2007) Biopriming banana with plant-growth promoting endophytic bacteria induces systemic resistance against Banana bunchy top virus. In: ISHS/ProMusa symposium, White River, South Africa, p 40

Hernández-Montiel LG, Rueda-Puente EO, Cordoba-Matson MV, Holguín-Peña JR, Zulueta-Rodríguez R (2013) Mutualistic interaction of rhizobacteria with arbuscular mycorrhizal fungi and its antagonistic effect on *Fusarium oxysporum* in *Carica papaya* seedlings. Crop Prot 47:61–66

Jahagirdar S, Siddaramaiah AL, Ramaswamy GR (2001) Influence of biocontrol agents and MPG-3 on *Fusarium oxysporum* f. sp. *cubense*, incitant of Panama disease of banana. Plant Dis Res 16:68–72

Janisiewicz WJ, Korsten L (2002) Control of postharvest diseases of vegetables by application of microbes. In: Bredt J, Bartz J (eds) Postharvest pathology of vegetables. Marcel Dekker, New York, pp 543–562

Jonathan EI, Cannayane I (2002) Field application of biocontrol agents for management of spiral nematode, *Helicotylenchus multicinctus* in banana (Abstr.). In: Global conference on banana and plantain, Bangalore, p 193

Jonathan EI, Umamaheswari R (2006) Biomanagement of nematodes infesting banana by bacterial endophytes. Indian J Nematol 36:234–240

Kalita PK (1995) Epidemiology and management of citrus canker caused by Xanthomonas campestris *pv.* citri. M. Sc. thesis, Assam Agricultural University, Jorhat

Kavino M, Kumar N, Harish S, Samiyappan R (2007a) Rhizobacteria-mediated growth promotion of banana leads to protection against Banana bunchy top virus under field conditions. In: ISHS/Promusa symposium on recent advances in banana crop protection for sustainable production and improved livelihoods, White River, South Africa, p 14

Kavino M, Kumar N, Harish S, Samiyappan R (2007b) Bio-hardening of tissue-cultured plantlets of 'Robusta' (AAA, Cavendish subgroup) with rhizosphere and endophytic bacteria to resist infection by banana bunchy top virus. In: ISHS/Promusa symposium on recent advances in banana crop protection for sustainable production and improved livelihoods, White River, South Africa, pp 60–61

Kumutha K (2001) Symbiotic influence of AM fungi and rhizobacteria on biochemical and nutritional changes in Mulberry (Morus alba). Ph. D. thesis, Tamil Nadu Agricultural University, Coimbatore

Li Wen-ying, Peng Zhi-ping, Yang Shao-hai, Yu Jun-hong, Huang Ji-chuan (2011). Effects of PGPR strain PAB-2 on growth promotion and control of Fusarium-wilt of banana plantlets under greenhouse conditions. In: ISHS/ProMusa symposium, White River, South Africa, pp 479–484

Menge JA (1993) Improved root rot control with mulches. Calif Citrog 79:16–17

Muthulakshmi M, Devarajan K, Jonathan EI (2010) Biocontrol of root-knot nematode, *Meloidogyne incognita* (Kofoid and White) Chitwood in mulberry (*Morus alba* L.). J Pestic 3:479–482

Nirmal Johnson SB, Devarajan K (2004) Effect of *Pseudomonas fluorescens* on banana nematodes. In: National symposium on paradigms in nematological research for biodynamic farming,. University of Agricultural Science, Bangalore, pp 81–82

Parvatha Reddy P, Nagesh M, Rao MS, Rama N (2000) Management of *Tylenchulus semipenetrans* by integration of *Pseudomonas fluorescens* with oil cakes. In: Proceedings of international symposium on citriculture, Nagpur, pp 830–833

Pocasangre L.E, Nuñez C, Quesada C, Sikora R.A (2007) The use of banana rhizosphere endophytic bacteria to control the burrowing nematode *Radopholus similis*. In: ISHS/ProMusa symposium, White River, South Africa, p 20

Ponnaiah S, Subramanian KS (1994) Biocontrol of banana wilt. Kisan World

Raghuchander T, Jayashree K, Samiyappan R (1997) Management of Fusarium wilt of banana using antagonistic microorganisms. J Biol Control 11:101–105

Raghuchander T, Shanmugam V, Samiyappan R (2001) Biological control of Panama wilt disease of banana. Madras Agric J 87:320–321

Rao MS (2007) Papaya seedlings colonized by the bio-agents *Trichoderma harzianum* and *Pseudomonas fluorescens* to control root-knot nematodes. Nematol Mediterr 35:199–203

Santhi A, Rajendran G, Sivakumar M (2004) Influence of biofertilizers on lesion nematodes in banana. In: National symposium on paradigms in nematological research for biodynamic farming, University of Agricultural Science, Bangalore, p 97

Senthilkumar M, Jonathan EI, Samiyappan R (2008a) Bioefficacy of *Pseudomonas fluorescens* on burrowing nematode. *Radopholus similis* in banana. Indian J Nematol 38:46–52

Senthilkumar M, Jonathan EI, Samiyappan R (2008b) Field application of biocontrol agent, *Pseudomonas fluorescens* for the management of burrowing nematode. *Radopholus similis* in banana. Indian J Nematol 38:57–61

Sivamani E, Gnanamanickam SS (1988) Biological control of *Fusarium oxysporum* f. sp. *cubense* in banana by inoculation of *Pseudomonas fluorescens*. Plant Soil 107:3–9

Sukumar J, Ramalingam A (1986) Antagonistic effects of phylloplane microorganisms against *Cercospora moricola* Cooke. Curr Sci 55:1208–1209

Thangavelu R, Palaniswami A, Ramakrishnan G, Doraiswami S, Muthukrishnan S, Velazhahan R (2001) Involvement of fusaric acid detoxification by *Pseudomonas fluorescens* strain Pf 10 in the biological control of Fusarium wilt of banana caused by *Fusarium oxysporum* f. sp. *cubense*. Zeitschrift fur Pflanzenkrankheiten und Pflanzenchutz 108:433–445

Tiwari RK, Mistry NC, Singh B, Gandhi GP (2014) Indian horticulture database-2012. National Horticulture Board, Gurgaon

7 Subtropical and Temperate Fruit Crops

7.1 Subtropical Fruit Crops

7.1.1 Mango, *Mangifera indica*

7.1.1.1 Anthracnose, *Colletotrichum gloeosporioides*

Anthracnose, also known as blossom blight, leaf spot and fruit rot, is a widespread and destructive disease of mango both in field and storage. Young plantations of Bombay Green cv. were completely wiped out from the Terai region of Uttar Pradesh as a result of severe wither tip.

i) Symptoms

The pathogen infects leaves, twigs, inflorescence and fruits (Fig. 7.1). On leaves, symptoms appear as brownish black sunken spots which later increase in size and coalesce, giving blighted appearance leading to shedding of leaves. On the twigs, brownish black necrotic lesions appear which later coalesce, showing twig blight and dieback symptoms. The fruit infection at early stage appears as black necrotic spots. These spots later coalesce resulting in reduced fruit growth and size. On mature fruits, the spots become more conspicuous resulting in reduced market value of the produce.

ii) Biomanagement

Bacillus subtilis was found relatively effective by controlling 50 % of the disease on fruits at postharvest. Anthracnose disease of mango is biologically controlled by *B. subtilis* LB5 through inhibiting conidial germination (Ruangwong et al. 2012). *B. subtilis* isolated from the avocado phylloplane has been successfully exploited for control of preharvest (Korsten et al. 1992) and postharvest diseases of mango (De Villiers and Korsten 1994).

Koomen and Jeffries (1993) found that among 121 isolates antagonistic to the anthracnose pathogen, only *P. fluorescens* gave significant reduction of mango anthracnose. Pre-harvest foliar application of talc-based fluorescent pseudomonad strain FP7 supplemented with chitin at fortnightly intervals (5 g/l; spray volume 20 l/tree) on mango trees from pre-flowering to fruit maturity stage induced flowering to the maximum extent, reduced the latent infection by *C. gloeosporioides* besides increasing the fruit yield and quality (Vivekananthan et al. 2004).

7.1.1.2 Canker, *Xanthomonas campestris* pv. *mangiferaeindicae*

Bacterial canker is known to be present in Australia, Brazil, Japan, Kenya, Malaysia, Pakistan, South Africa, Sudan, Venezuela, Taiwan, Reunion Island, Somali Land and many other countries. In India, it is widely prevalent throughout the country, especially severe in North India. The loss ranges from 10 % to 80 % depending upon the severity and stage of infection. Its

P.P. Reddy, *Plant Growth Promoting Rhizobacteria for Horticultural Crop Protection*,
DOI 10.1007/978-81-322-1973-6_7, © Springer India 2014

Fig. 7.1 Anthracnose on mango leaves and fruit

incidence and intensity ranged from 0.52 % to 60.0 % and 15.0 % to 90.0 %, respectively.

i) Symptoms

The disease attacks leaves, petioles, twigs, stems and fruits. Small water-soaked lesions appear in groups on any part of the leaf blade. They increase in size and turn brown to black surrounded by a yellow halo (Fig. 7.2). Several lesions may coalesce to form large necrotic patches which are often rough and raised. As the disease further advances, the infected leaves drop off. On young fruits, water-soaked lesions develop which also turn dark brown to black. Infected fruits may develop cracks in the skin, and those badly affected drop prematurely. The canker stage also invites the secondary infection from saprophytic fungi.

ii) Biomanagement

Bacillus coagulans is a potential bioagent which reduced the disease incidence to the extent of 75 % (Kishun 2003). Antagonistic bacteria like *B. coagulans*, *B. subtilis* and *B. amyloliquefaciens* (Pruvost and Luisetti 1991) and fluorescent pseudomonads (Tzeng et al. 1994) have been found to be effective biocontrol agents against the disease.

7.1.2 Grapevine, *Vitis vinifera*

7.1.2.1 Crown Gall, *Agrobacterium tumefaciens*

Crown gall is an important disease of wine grapes when they are grown in cold climates. Incidence may range from a few vines in a vineyard to 100 % of the vines.

i) Symptoms

Gall formation on the aerial parts of the vine is the most common symptom associated with this disease. The bacterium which causes crown gall may be present in plants that do not show any symptoms. Galls are usually noticed as swellings near the base of the vine and up the trunk (Fig. 7.3). Galls on roots of grape are not typical; however, the bacteria can induce a localized necrosis of roots. Young galls are soft, creamy to greenish in colour, with no bark or covering. As they age, the tissue darkens to brown. The surface becomes open and the texture becomes

Fig. 7.2 Bacterial black spot on mango leaf and fruit

Fig. 7.3 Aerial crown gall on grapevine

moderately hard and very rough. The surface tissue of the galls turns black as it dies, but the bacterium remains alive in the vine.

ii) Biomanagement

Control of *A. tumefaciens* by its relative *Agrobacterium radiobacter* strain K84 (Moore and Warren 1979) is achieved by immersing the roots of cuttings into a bacterial suspension (*A. radiobacter*) before transplanting them in the fields. *A. radiobacter* actively colonizes the roots of treated cuttings, preventing infection by *A. tumefaciens*. *A. radiobacter* produces a bacteriocin called Agrocin 84, which is thought to be responsible for the antagonistic effect (Slota and Farrand 1982). Commercial formulations of *A. radiobacter* include Galltrol-A, Nogall, Diegall and Norbac 84C.

7.1.2.2 Grey Mould, *Botrytis cinerea*

It is one of the principal causes of postharvest spoilage in storage.

i) Symptoms

In early stage, tissue just beneath the surface of fruit is infected loosening the skin from the flesh. The affected area turns light brown. The fungal infection advances into the inner flesh resulting in a soft watery mass of decayed tissue. Under

Fig. 7.4 Grey mould symptoms on grape berries

moist atmosphere, the fungus sporulates on the surface of the fruit and the typical powdery grey mould stage becomes evident. Infected fruits shrivel and turn dark brown (Fig. 7.4). The disease starts in midseason and continues to develop until harvest in the absence of rain. The fungus infects stigma and style and becomes latent in the necrotic stigma and style tissues at the stylar end of the berry. In a compact infected bunch, fruits inside may split during growth. Infection of such fruits results in bunch rot. Even a single field infected berry may cause 'nest rot' in transit and storage.

ii) Biomanagement

An endophytic plant growth-promoting rhizobacterium, *Burkholderia phytofirmans* strain *PsJN* colonizes the roots of grapevine and protects it from the grey mould disease. *B. phytofirmans* is the host bacterium of grapevine that protects the host plant by tightly attaching to the plant and extracellular alkalization. However, a non-host bacterium *Pseudomonas syringae* pv. *pisi* attached tightly primes plant with extracellular alkalization and by a two-phase oxidative burst, with a Hypersensitive Reaction (HR)-like response (Bordiec et al. 2011).

7.1.2.3 Root-Knot Nematodes, *Meloidogyne incognita*, *M. javanica*

In India, *M. javanica* is most prevalent in the northern part of the country and *M. incognita* in the southern part. *M. incognita* is also reported in some parts of Haryana. *M. incognita* was responsible for 55 % loss in fruit yield of grapes, while *M. javanica* caused 53 % loss in yield.

i) Symptoms

Patches of poorly branched vines with scant foliage, pale and small leaves and poor bearing are the indications of root-knot nematode damage. In young plants, premature decline and weak vegetative growth are commonly associated with nematode attack. The root system shows typical localized swellings particularly on feeder roots and young secondary roots, and females may be found on internodal trunk just below the ground level. *M. incognita* has been reported to stimulate the production of many new fine rootlets above the site of nematode infection resulting in 'hairy root' condition. Depending upon the variety of grape, *M. javanica* forms galls of varying size and shape and distorts the normal appearance of roots (Fig. 7.5).

Fig. 7.5 Root-knot nematode on grape roots

ii) Biomanagement

Sundarababu et al. (1999) observed that the commercial formulation of *P. fluorescens* reduced root-knot nematode population effectively and enhanced yield in grapevine.

iii) Integrated Management

(a) *Cultural methods and bioagents*: Pruning (during July) and soil application of 4 g talc formulation of *P. fluorescens* (containing 15 × 10^8 cfu/g)/vine around root-knot-infested grapevine roots at 15 cm depth in the basin significantly reduced root galling due to *M. incognita* (39 %), number of egg masses (250 %) and increased fruit yield (166 %) (Santhi et al. 1998) (Table 7.1).

7.1.2.4 Root-Knot, *Meloidogyne incognita*, and Wilt, *Fusarium moniliforme*, Disease Complex

i) Symptoms

Simultaneous inoculation of nematode and fungus resulted in the greatest reduction of shoot length (40.6 %) and shoot weight (63.9 %). Wilt disease was observed when both *M. incognita* and *F. moniliforme* were inoculated or when the fungus was inoculated alone. The incidence of wilt disease (35 %) was higher in vines inoculated with nematodes a week prior to fungal inoculation.

ii) Integrated Management

(a) *Bioagents and botanicals*: Soil application of *P. fluorescens* at 100 g and FYM at 20 kg/vine gave effective management of disease complex and improved the plant stand by reducing the final soil nematode population (56.9 %), root gall index (1.8) and per cent disease incidence (15.67 %). This treatment also increased the number and weight of fruit bunches (17.83 and 155.40 %, respectively) and fruit quality (more TSS 13.53 Brix, TSS/acid ratio 14.87, lower acidity 0.91 %). The bunch weight of grapevine increased by 155.4 % compared to untreated control (Senthil Kumar and Rajendran 2004).

7.1.3 Guava, *Psidium guajava*

7.1.3.1 Anthracnose, *Gloeosporium psidii*

The disease is a serious problem in Uttar Pradesh, Punjab and Karnataka. This is a serious postharvest disease of guava.

Table 7.1 Field evaluation of biocontrol potential of *Pseudomonas fluorescens* against *Meloidogyne incognita* infecting grapevine

Treatment (dose/vine)	No. of galls/5 g roots	No. of egg masses/5 g roots	Root colonization (cfu/g + 10^8)	Yield (MT/ha)
P. fluorescens – 1 g	428c	94c	24	12.07b
P. fluorescens – 2 g	390b	85b	36	15.41c
P. fluorescens – 4 g	326a	72a	58	22.07d
Carbofuran 3G – 60 g	300a	67a	–	31.65e
Control (untreated)	535d	180d	–	8.33a

Figures with different letters are significantly different from each other at 5 % level by analysis of variance test

i) Symptoms

The plant shows dieback symptoms. Pinhead spots are first seen on the unripe fruits which gradually enlarge measuring 5–6 mm in diameter (Fig. 7.6). They are dark brown to black in colour, sunken and circular and have minute, black stromata in the centre of the lesions which produce creamy spore masses in moist weather. Several spots coalesce to form bigger lesions. The infected area in unripe fruits becomes corky and hard and often develops cracks in case of severe infection. On ripe fruit, the infection causes softening of tissue, and lesions attain a diameter of 10–20 mm. Unopened buds and flowers are also attacked and cause their shedding. Spread of infection is very rapid on fully mature green fruits, while young fruits do not normally take infection, perhaps due to the differences in the concentration of K ions in the tissues.

ii) Biomanagement

P. fluorescens was found effective in checking the spread of pathogens on fruits compared with the pathogen-inoculated control.

B. subtilis was found relatively effective by controlling 50 % of the anthracnose disease on fruits at postharvest stage.

Actinomycetes of cow dung have been identified as having potential against anthracnose of guava (Garg et al. 2003).

Fig. 7.6 Anthracnose on guava fruits

7.1.4 Pomegranate, *Punica granatum*

7.1.4.1 Root-Knot Nematode, *Meloidogyne incognita*

The root-knot nematodes were responsible for 24.64–27.45 % loss in fruit yield of pomegranate.

i) Symptoms

Heavy root galling and visible damage to pomegranate trees in young orchards under irrigation is frequently encountered (Fig. 7.7).

ii) Biomanagement

P. fluorescens at 20 kg/ha was found to be most effective in reducing the root-knot nematode population (*M. incognita* race 2) (31.28 %) and number of root galls/5 g roots (29.28 %) and increasing the fruit yield (18.99 %) with benefit/cost ratio of 2.37 (Table 7.2) (Pawar et al. 2013).

7.2 Temperate Fruit Crops

7.2.1 Apple, *Malus pumila*

7.2.1.1 Scab, *Venturia inaequalis*

Scab is the most destructive disease of apples and is present in all the countries of the world where apples are grown. The disease is particularly severe in high rainfall and humid areas.

Fig. 7.7 Pomegranate roots infected with *Meloidogyne incognita*

Table 7.2 Effect of bioagents and chemicals for the management of root-knot nematodes infecting pomegranate

Treatment	% decline in nema popn	% decline in root galling	Yield (MT/ha)	Benefit/cost ratio
Pseudomonas fluorescens at 20 g/m^2	31.2	29.2	18.4	2.37
Trichoderma viride at 20 g/m^2	28.3	23.9	18.0	2.33
Cartap hydrochloride at 0.3 g a.i./m^2	27.2	20.4	17.6	2.27
Carbofuran at 0.3 g a.i./m^2	29.8	25.1	17.8	2.27
Untreated control	–	–	15.4	–
CD (P=0.05)	2.99	2.30	1.17	–

Fig. 7.8 Apple leaf and fruits infected with scab

In India, it is present in all the apple growing areas of Jammu and Kashmir, Himachal Pradesh, Uttar Pradesh, Sikkim, Arunachal Pradesh and Nilgiris. The first epidemic of scab in Kashmir valley in 1973 completely ruined the apple crop worth US$ 540,000. During 1983, it rendered 10 % apple crop (30,000 MT) unfit for consumption in Himachal Pradesh which has to be destroyed, and as a result the state suffered a loss of Rs. 15 million.

i) Symptoms

Scab may appear on leaves, fruits, petioles and green twigs. The most striking symptoms of scab are commonly observed on leaves and fruits. Fruits may show small, rough, black, circular lesions on their skin (Fig. 7.8) while on the tree or after keeping in cold storage. Fruits picked from infected trees appear apparently healthy but are too sticky to keep even in cold storage. Such fruits soon develop scab symptoms even at low temperature and may not last long in storage. The affected fruits rot due to secondary infection of the lesions. Secondary infections on leaves are so numerous that the entire leaf surface appears covered with scab, commonly referred as sheet scab. Lesions on young fruits resemble those on leaves but turn dark brown to black and become corky or scab-like with time. Infections are often limited to one or two spots per fruit. Secondary infections are clumped together.

ii) Biomanagement

Flavobacterium sp., *Cryptococcus* sp. and two unidentified actinomycetes were found to be most antagonistic against scab (Heye and Andrews 1983).

Fig. 7.9 Apple root showing signs of the white root rot fungus

iii) Integrated Management

(a) *Bioagents and chemicals*: Integrated control of apple scab by modifying the nutrient status of overwintering leaves is well established. Treatment of senescent apple leaves on trees shortly before leaf fall or of fallen leaves on the orchard floor with a solution of 2 % urea greatly reduced ascospore production in the spring. Urea treatment of fallen apple leaves greatly increased the antagonistic bacterial populations found on decaying leaves. Fluorescent pseudomonads increased greatly in number on urea treated leaves. Pseudothecial development may be reduced directly by antibiotics produced by fluorescent pseudomonads on decaying leaves or indirectly by degeneration of leaf structure due to enhanced degradation of leaf material in the presence of higher bacterial populations induced by treatment with a good nitrogen source (Burchill and Cook 1971).

7.2.1.2 White Root Rot, *Dematophora necatrix*

i) Symptoms

This is a soilborne disease which is most important as the fungus can cause death of plants (Fig. 7.9). The disease appears on underground plant parts and causes complete rotting of the roots. The fine roots are attacked first which are completely devoured and infection spreads to the main root through the secondary root system. Lateral roots turn dark brown and become infected with white flocculent fungus during monsoon months. Bronzing of leaves and stunted growth and size are important above-ground symptoms. Root rot affected trees are usually associated with a heavy blossom and fruiting next year. However, in succeeding years, few leaves emerge and much of the immature fruits fail to reach maturity. Infected trees often persist for 2–3 years depending upon infestation of fungus.

ii) Biomanagement

Charcoal-based seed treatment (200 g of charcoal-based *Bacillus licheniformis* CKA1 culture per kg of seed) followed by seedling dip (at 1 l/500 seedlings) gave 20–25 % increase in apple seedling biomass and significant improvement in plant root and shoot parameters and controlled white root rot and gave 20–35 % increase in apple fruit yields.

Using antagonists like *Enterobacter aerogenes* and *B. subtilis* is effective for the management of white root rot.

iii) Integrated Management

(a) *Physical methods and bioagents*: Use of antagonists like *Enterobacter aerogenes* and *B. subtilis* along with soil solarization has been found to protect the plants from root rot infection.

(b) *Bioagents and chemicals*: Application of *E. aerogenes* + carbendazim (0.1 or 0.05 %) showed more than 92 % disease control when applied as pre-inoculation to the pathogen. In simultaneous inoculation, 0.1 % carbendazim in combination with *E. aerogenes*, completely prevented the appearance of the disease; however, 0.05 % carbendazim in combination with *E. aerogenes* gave 90.4 % disease control (Gupta and Sharma 2004).

(c) *Bioagents/botanicals and physical methods*: Soil solarization in combination with organic amendments and biocontrol agents in general gave good control of the disease in nursery. However, cent per cent control of the disease was achieved in plots where *Bacillus* sp. was incorporated in combination with soil solarization (Table 7.3).

7.2.1.3 Collar Rot, *Phytophthora cactorum*

The disease is also known as crown rot and trunk canker is universally present in all the apple growing regions of the world including the USA, the UK, Canada, New Zealand, the Netherlands, Germany, Australia and India. On susceptible varieties, it causes extensive losses even resulting in death of apple trees within a few years.

i) Symptoms

The above-ground symptoms are often confused with white root rot. The infection starts from the collar region and spreads to the underground parts. Bark at the soil level becomes slimy and rots resulting in cankered areas (Fig. 7.10). The attacked trees are recognized by chlorotic foliage with red colouration of veins and margins.

Table 7.3 Effect of soil solarization, organic amendments and biocontrol agents against white root rot disease in apple nursery

Treatment	White root rot (%)
Soil solarization	1.15
Soil solarization + neem cake	0.00
Soil solarization + deodar needles	3.75
Soil solarization + *Bacillus* sp.	0.00
Unsterilized soil + neem cake	4.50
Unsterilized soil + deodar needles	7.75
Unsterilized soil + *Bacillus* sp.	2.01
Unsterilized soil	22.82

Fig. 7.10 Crown rot on apple

ii) Biomanagement

Biological control of collar rot on apple with *B. subtilis* and *Enterobacter aerogenes* is feasible and requires commercial exploitation (Sharma et al. 1999).

iii) Integrated Management

(a) *Physical methods and bioagents*: Use of rhizobacteria like *E. aerogenes* and *B. subtilis* along with soil solarization has been found to protect the plants from collar rot infection.

(b) *Physical methods and botanicals/bioagents*: Cent per cent control of the disease was achieved in plots where *Bacillus* sp. was incorporated in combination with soil solarization. Collar rot pathogen was highly sensitive to the biocontrol agents and organic amendments both in solarized and unsolarized soil (Table 7.4).

7.2.1.4 Fruit Rot, Blue Mould, *Penicillium expansum*, and Grey Mould, *Botrytis cinerea*

i) Symptoms

In early stages, the rotted area is light in colour and soft and watery in texture. The rotted area does not become sunken although it is very soft in texture. The rot advances rapidly. In humid

Table 7.4 Effect of soil solarization, organic amendments and biocontrol agents against collar rot disease in apple nursery

Treatment	Collar rot (%)
Soil solarization	0.00
Soil solarization + neem cake	0.00
Soil solarization + deodar needles	0.00
Soil solarization + *Bacillus* sp.	0.00
Unsterilized soil + neem cake	1.00
Unsterilized soil + deodar needles	0.75
Unsterilized soil + *Bacillus* sp.	0.00
Unsterilized soil	2.05

atmosphere, grey blue cushions of fruiting bodies of the pathogen appear on the diseased skin (Fig. 7.11).

ii) Biomanagement

A saprophytic strain of *Pseudomonas syringe* was shown to provide biological control of all pathogens responsible for fruit rot on wounded Red and Golden Delicious apples in co-inoculations with individual pathogens (Jeffers and Wright 1994). The strain (*P. syringe* ESC-11) is currently registered for postharvest application to apples and is marketed as the product Bio-save 110.

Treatment of wounded fruits of apple (inoculated with *Penicillium expansum*) with most effective antagonist, *Pseudomonas syringae* pv. *lachrymans* isolate L-22-64 (increased in population at the wound site during 30 days of storage at 2° or 24 °C), controlled *P. expansum* (Fig. 7.12) (Janisiewicz 1988).

P. syringae (10 % wettable powder) in the modified packing line was sprayed at the rate of 10 g/l over apple fruit to control blue and grey moulds of apple. The population of antagonist increased in the wounds more than tenfold during 3 months in storage (Janisiewicz and Jeffers 1997).

B. subtilis applied to wounded apples reduced fruit rot (Leibinger et al. 1997).

iii) Integrated Management

(a) *Two bioagents*: Combined application of *Pseudomonas* sp. and *Acromonium breve* gave complete control of *P. expansum* and *B. cinerea* on apple (Janisiewicz 1988). A co-application involving the bacterial antagonist *P. syringe* and the yeast *S. roseus* applied in equal biomass provided control of blue mould that was superior to that obtained by treatment with the individual agents applied separately (Janisiewicz and Bors 1995).

(b) *Bioagents and chemicals*: Combination of *Pseudomonas syringae* MA-4 at $1–3 \times 10^7$ cfu/ml with cyprodinil at 5–10 μg/ml controlled both blue and grey mould by more than 90 % on apple, demonstrating that the integration could not only improve disease control efficacy but also extended the degree of control to more than one important disease (Zhou et al. 2002).

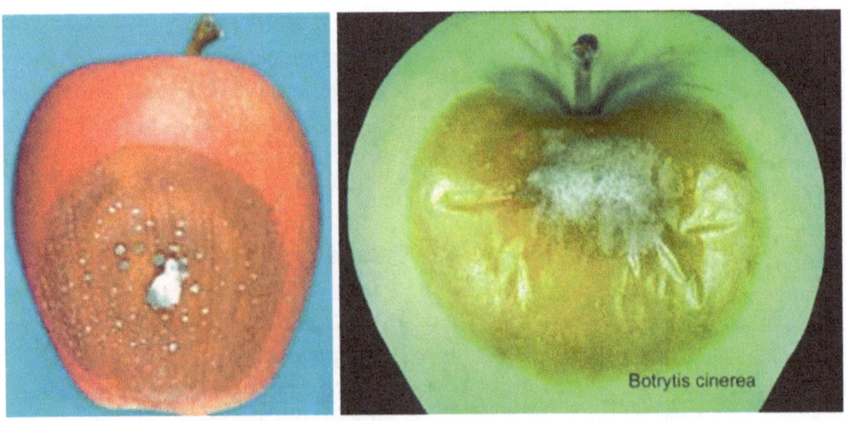

Fig. 7.11 Blue and grey mould on apple fruits

Fig. 7.12 Management of blue and grey mould on apples by *Pseudomonas syringe. Left* – control, *right* – treated

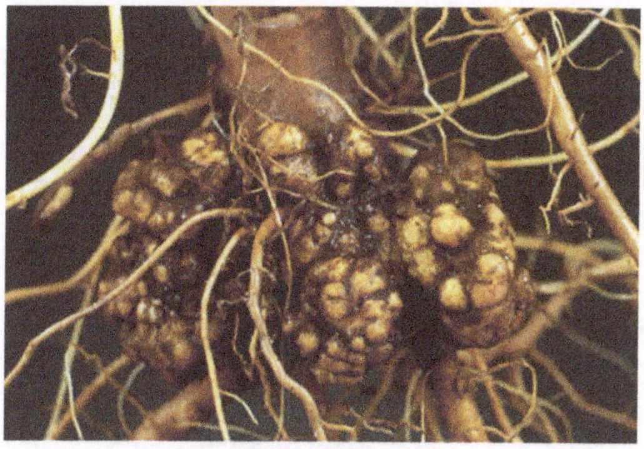

Fig. 7.13 Hairy root symptoms on apple

Table 7.5 Effect of soil solarization, organic amendments/biocontrol agents against hairy root disease in apple nursery

Treatment	Hairy root (%)
Soil solarization	2.25
Soil solarization + neem cake	4.15
Soil solarization + deodar needles	2.00
Soil solarization + *Bacillus* sp.	0.00
Unsterilized soil + neem cake	7.00
Unsterilized soil + deodar needles	9.45
Unsterilized soil + *Bacillus* sp.	3.15
Unsterilized soil	16.00

7.2.1.5 Hairy Root, *Agrobacterium rhizogenes*

i) Symptoms

At the union between scion and root piece, an enlargement somewhat resembling a newly formed crown gall appears. From this arise numerous roots, fleshy or fibrous in texture, with many containing numerous branches (Fig. 7.13). The surface of these enlargements bears numerous convolutions with fissures extending deep into the interior of the enlargements.

ii) Integrated Management

(a) *Physical methods and botanicals/bioagents*: Soil solarization in combination with organic amendments and biocontrol agents in general gave good control of hairy root disease in nursery. However, cent per cent control of the disease was achieved in plots where *Bacillus* sp. was incorporated in combination with soil solarization (Table 7.5).

7.2.1.6 Crown Gall, *Agrobacterium tumefaciens*

i) Symptoms

Crown gall bacterium enters the plant through wounds in roots or stems and stimulates the plant tissues to grow in a disorganized way, producing swollen galls. Galls are present all year round. Crown gall is identified by overgrowths appearing as galls on roots and at the base or 'crown' of apple (Fig. 7.14).

Fig. 7.14 *Agrobacterium tumefaciens* induced galls on apple roots (crown gall)

ii) Biomanagement

Satisfactory and nearly cent per cent control of crown gall is achieved by dipping seeds or root system of nursery plants or the rootstocks up to the crown region in *Agrobacterium radiobacter* strain 84 suspension (Kerr 1980). This biocontrol agent occupies the infection sites and produces a bacteriocin Agrocin 84 which is inhibitory to the pathogen. A derivative strain of K 84, designated as K 1026, is superior to its original K 84 strain (Jones and Kerr 1989).

iii) Integrated Management

(a) *Physical methods and botanicals/bioagents*: Cent per cent control of the disease was achieved in plots where *Bacillus* sp. was incorporated in combination with soil solarization (Table 7.6).

Table 7.6 Effect of soil solarization, organic amendments and biocontrol agents against soilborne diseases in apple nursery

Treatment	Crown gall (%)
Soil solarization	4.00
Soil solarization + neem cake	1.50
Soil solarization + deodar needles	0.50
Soil solarization + *Bacillus* sp.	0.00
Unsterilized soil + neem cake	3.00
Unsterilized soil + deodar needles	6.55
Unsterilized soil + *Bacillus* sp.	5.25
Unsterilized soil	8.00

7.2.1.7 Replant Problem

The living pathogens may include fungi, nematodes and bacteria; the nonliving factors can range from nutrient imbalances to herbicide residues to compaction. Synergistic interactions of fungi, bacteria and nematodes were found responsible for causing apple replant problem.

i) Symptoms

The symptoms commonly seen are decreased growth both above and below ground, delayed productivity and even tree death.

ii) Biomanagement

Application of Bact-I, EBW4, *B. subtilis* and B8 soil drenches increased tree growth in apple replant disease sick soils.

7.2.1.8 Fire Blight, *Erwinia amylovora*

i) Symptoms

The intensity of blossom infection with consequent loss of crop and the possible loss of some branches as a result of cankering would be very serious (Fig. 7.15). Primary infection frequently occurs as a result of transfer of the bacteria by pollinating insects to open blossom. At that time (April) in North India, temperature above 24 °C and plentiful rains occur which favour infection and rapid spread of the disease. The presence of naturally occurring hosts (pear) in the vicinity of the apple orchard constitutes a permanent reservoir of infection.

ii) Biomanagement

Biological control of the blossom blight phase of fire blight has been obtained in the field using *P. fluorescens*, Pf-A506 (Blight Ban) (Vanneste and Yu 1996). This agent multiplies rapidly and colonizes open flowers to the extent that it excludes any significant subsequent colonization by the fire blight organism. If this antagonist is applied after *E. amylovora* is already

Fig. 7.15 Fire blight on apple

present or even as a mixture with the pathogen, it is not effective.

The second promising bioantagonist is another bacterium, *Erwinia herbicola*, strain C9-1, which is a common epiphyte on apples. In addition to the competition for space that occurs with Pf-A506, *E. herbicola* C9-1 also produces an antibiotic of its own that inhibits the multiplication of the pathogen. Like its A506 counterpart, this second bioantagonist must also be present in the flower before the arrival of the pathogen for it to be effective.

E. amylovora causing fire blight of apple infects through flower and develops extensively on stigma. Colonization by antagonist at the critical juncture is necessary to prevent flower infection. Since flowers do not open simultaneously, the biocontrol agent *P. fluorescens* has to be applied to flowers repeatedly to protect the stigma. Nectar seeking insects like *Aphis mellifera* can be used to deliver *P. fluorescens* to stigma. Bees deposit the bacteria on the flowers soon after opening due to their foraging habits.

iii) Integrated Management

(a) *Bioagents and chemicals*: Studies conducted with *P. fluorescens* A506 in combination with antibiotic (streptomycin/oxytetracycline) applications (7 days after application of the antagonist) suggest that the control achieved is likely to be additive in nature (Lindow et al. 1996).

7.2.2 Peach, *Prunus persica*, and Plum, *Prunus salicina*

7.2.2.1 Brown Rot of Fruits, *Monilinia fructicola*, *M. fructigena*, *M. laxa*, *M. laxa* f. sp. *mali*

i) Symptoms

Symptoms of the disease are blighting of blossom and leaves, canker production on woody tissues and rotting of fruits. Blossom blight is the first symptom during spring, and attacked parts turn grey to dark brown. The fungus spreads through the peduncle and reach branches causing twig blight. Stem cankers usually develop from blighted twigs or fruit spurs. Fruit rot is the most destructive phase of the disease. Rotted fruits either fall down or hang as firm mummies. Conidia and ascospores produced on the mummified fruits and canker spots serve as a source of primary infection.

ii) Biomanagement

Brown rot of peaches in storage was controlled under simulated commercial conditions by incorporating the antagonist *B. subtilis* into wax used in the packing process (Wilson and Pusey 1985).

B. subtilis gave excellent control of brown rot caused by *Monilinia fructicola* in storage in peach and plum. No fruit treated with 10 cfu/ml of suspension developed brown rot (Fig. 7.16) (Wilson and Pusey 1985).

iii) Integrated Management

(a) *Bioagents and chemicals*: Biocontrol agents can be used in combination with fungicides in controlling rots on peach, in cases where *B. subtilis* (B3), effective against brown rot (incited by *M. fructicola*), was combined with dicloran used for the control of *Rhizopus* rot (Pusey et al. 1988).

Zhou et al. (1999) reported that the addition of 0.5 % calcium to cell suspension of *Pseudomonas*

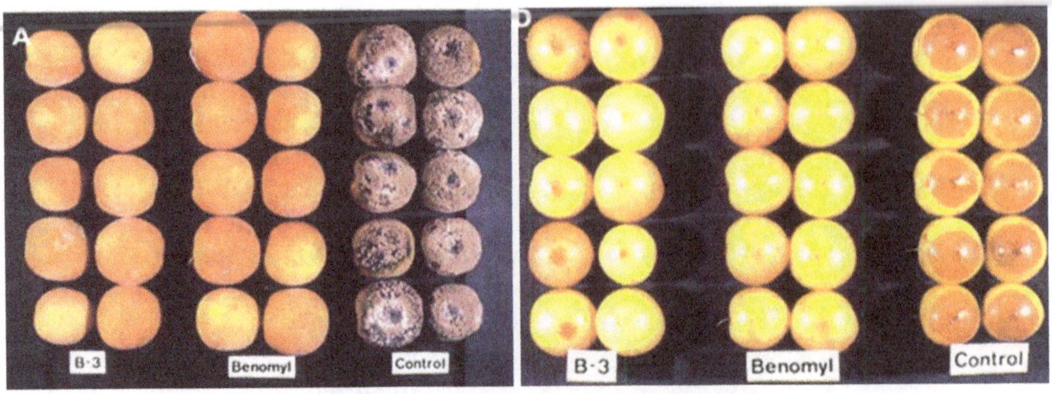

Fig. 7.16 Wounded peaches and plums after treatment with *B. subtilis* (B-3), benomyl and water (control)

syringae MA-4 resulted in a greater reduction of peach brown rot incidence when sprayed on peaches naturally infected with *M. fructicola*. Preharvest application of *P. syringae* MA-4 with a foliar calcium fertilizer also significantly increased biocontrol efficacy against peach brown rot (Zhou and Schneider 1998).

7.2.2.2 Crown Gall, *Agrobacterium radiobacter*

i) Symptoms

Small outgrowth appears on the stem and roots. At the young stage, the galls are soft, spherical, white or flesh coloured. The galls vary in size from 7 to 10 mm in diameter. On woody stems, the galls are hard and corky (Fig. 7.17). They are generally knobby and knotty and become cleftier as they grow older. The affected plant may become stunted with chlorotic leaves.

ii) Biomanagement

A nonpathogenic strain of *A. radiobacter* (strain 84) is employed to control crown gall, and 99 % success has been achieved in case of peach crown gall.

7.2.2.3 Ring Nematode, *Criconemella xenoplax*

It is ectoparasitic and is an economically important nematode which is one of the factors in bacterial canker and peach tree short life (PTSL) disease. The estimated losses that occurred in peach orchards suffering from PTSL in Georgia were 30–70 % after 5 years. *C. xenoplax* was responsible for 33 % loss in fruit yield of peach. It is distributed in Himachal Pradesh and Haryana.

Fig. 7.17 Crown gall on peach seedlings

i) Symptoms

Parasitism of *C. xenoplax* causes chlorosis, leaf drop, stunting and loss of vigour on top of peach trees. Peach roots showed lesions, pits and fewer feeder roots (Fig. 7.18). The roots are predisposed to bacterial canker caused by secondary infection of *Pseudomonas syringae*.

ii) Biomanagement

Treatment of peach seedlings with a strain of *Pseudomonas aureofaciens* isolated from suppressive soil inhibited infection by the ring nematode in soil (McInnis et al. 1990).

Fig. 7.18 Peach roots infected with *Criconemella xenoplax*

7.2.3 Pear, *Pyrus communis*

7.2.3.1 Grey Mould, *Botrytis cinerea*

i) Symptoms

Grey mould and its many strains cause death of flower parts, leaves, buds, shoots, seedlings and fruits (Fig. 7.19). The disease needs moisture as one of its criteria for infection. The wetter the plant is, the more likely the grey mould will show up on plants. Not only are the numbers of infected areas increased but also are the numbers of plants attacked as well as the severity of the infections (quicker growth of the disease and death of tissue).

ii) Integrated Management

(a) *Bioagents and chemicals*: In a packing house trial, combination of Bio-Save 110 (*Pseudomonas syringae*) or Aspire (*Candida oleophila* strain 1-18) with TBZ at 100 μg/ml (about 17.6 % of the label rate) provided control of blue mould and grey mould of pears, similar to that of TBZ alone used at the label rate (569 μg/ml) (Sugar and Spotts 1999).

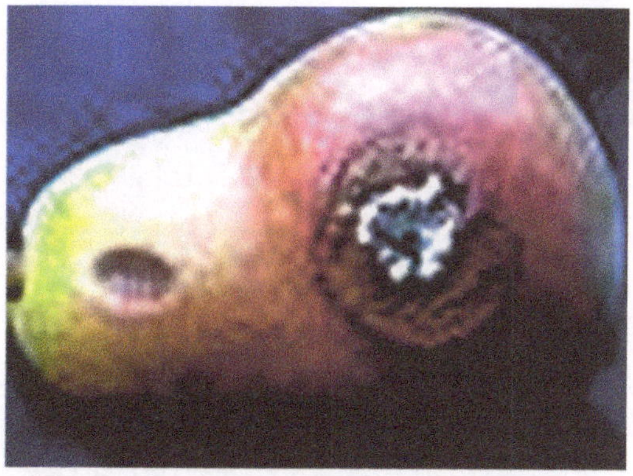

Fig. 7.19 Grey mould on pear fruit

7.2.3.2 Blue Mould, *Penicillium expansum*

i) Symptoms

The rotted areas are soft, watery and light brown in colour. The surface of older lesions may be covered by bluish-green spores that initially are nearly snow white in colour. The lesions are of varying shades of brown, being lighter on the yellow or green varieties. Two characteristics are of importance in the recognition of *P. expansum*, the most common species, namely, the musty odour and the formation of conidial tufts or coremia on the surface of well-developed lesions.

ii) Biomanagement

Application of *P. syringe* at 2.7×10^6 cfu/ml gave effective control of blue mould on pear (Fig. 7.20).

iii) Integrated Management

(a) *Bioagents and chemicals*: In a packing house trial, combination of Bio-Save 110 (*Pseudomonas syringae*) or Aspire (*Candida oleophila* strain 1-18) with TBZ at 100 μg/ml (about 17.6 % of the label rate) provided control of blue mould and grey mould of pears,

Fig. 7.20 Management of blue mould rot on pear by *Pseudomonas syringe*. *Left* – control, *right* – treated

Fig. 7.21 Brown spot on pear leaf

Fig. 7.22 Fire blight infection on pear tree and fruit

similar to that of TBZ alone used at the label rate (569 μg/ml) (Sugar and Spotts 1999).

7.2.3.3 Brown Spot, *Stemphylium vesicarium*

i) Symptoms

The pathogen attack leaves (Fig. 7.21), fruits and occasionally twigs of pear.

ii) Biomanagement

Montesinos et al. (1996) showed the antagonism of *Erwinia herbicola* and *P. fluorescens* to brown spot of pear. They found that *P. fluorescens* strain EPS 288 was the most effective and could reduce disease incidence by 57 %.

7.2.3.4 Fire Blight, *Erwinia amylovora*

i) Symptoms

The term 'fire blight' describes the appearance of the disease, which can make affected areas appear blackened, shrunken and cracked, as though scorched by fire (Fig. 7.22).

ii) Biomanagement

P. fluorescens strain A506 effectively controlled fire blight of pear (Wilson and Lindow 1993).

iii) Integrated Management

(a) *Bioagents and chemicals*: Lindow et al. (1996) reported that use of *P. fluorescens* strain A506 in combination with streptomycin and oxytetracycline could reduce pear fire blight by 40–50 %.

7.2.4 Almond, *Amygdalus communis*

7.2.4.1 Crown Gall, *Agrobacterium tumefaciens*

i) Symptoms

Rough, abnormal galls appear on roots or trunk. Galls are soft and spongy (Fig. 7.23). The centres of older galls decay. Young trees become stunted; older trees often develop secondary wood rots.

ii) Biomanagement

Dipping of younger (2 months old) and older (10 months old) almond seedlings in a suspension of either *Agrobacterium radiobacter* strain K 1026 or strain K 84 and planted in soil infested with a pathogenic *A. tumefaciens* biovar 2 gave effective control of crown gall disease (Jones and Kerr 1989) (Table 7.7).

7.2.5 Apricot, *Prunus armeniaca*

7.2.5.1 Brown Rot, *Monilinia fructicola*

i) Symptoms

Fruit rot first appears as small, circular brown spots that increase rapidly in size causing the entire fruit to rot. Greyish spores appear in tufts on rotted areas. Infected fruit eventually turn into shrivelled, black mummies that may drop or remain attached to the tree through the winter.

ii) Biomanagement

B. subtilis gave excellent control of apricot brown rot in storage. No fruit treated with 10 cfu/ml of suspension developed brown rot (Fig. 7.24) (Wilson and Pusey 1985).

Fig. 7.23 Crown gall on the crown and roots of a young almond tree

Table 7.7 Effect of *Agrobacterium radiobacter* strains on crown gall of almond

Plant age (months)	Treatment	No. of plants surviving	% plants galled	No. of galls/plant
2	Water	12	100	9.33
	K 84	14	14	0.21
	K 1026	12	25	0.33
10	Water	15	100	46.33
	K 84	15	20	0.20
	K 1026	15	27	0.67

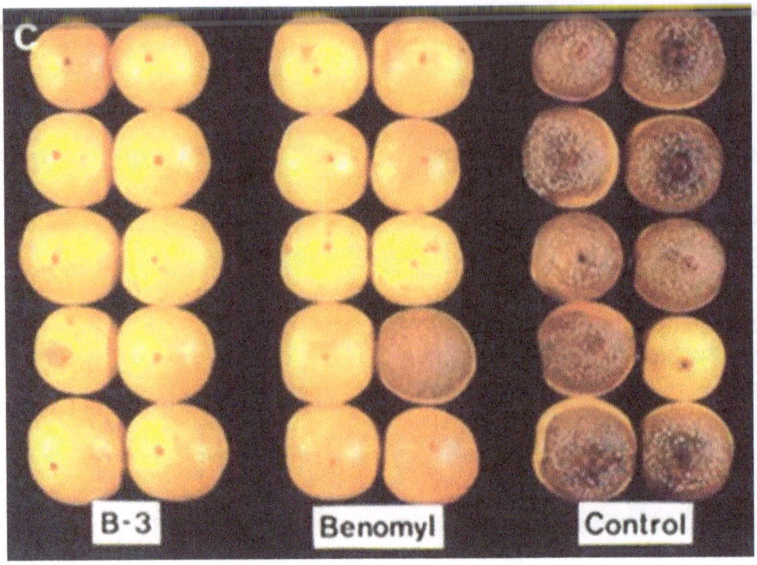

Fig. 7.24 Wounded apricots after treatment with *B. subtilis* (B-3), benomyl and water (control)

Fig. 7.25 Symptoms of strawberry *Verticillium* wilt

7.2.6 Strawberry, *Fragaria vesca*

7.2.6.1 Wilt, *Verticillium dahliae*

i) Symptoms

Symptoms resemble drought stress. Infection and severe disease occurs mostly in first year of growth. Outer leaves develop brown necrotic areas between veins and on the edges during early spring. Interior leaves of diseased plants usually remain alive and blue green in colour. Infected plants wilt rapidly under stress (Fig. 7.25). Diseased plants may occur singly or in patches.

ii) Biomanagement

Dipping of strawberry roots for 15 min in bacterial suspension of *P. putida* (2×10^9 cfu/ml) isolated from strawberry rhizosphere reduced *Verticillium* wilt of strawberry by 11 % compared to untreated control (Berg et al. 2001).

Significant reduction of *Verticillium* wilt with root dip treatment in *Streptomyces albidoflavus* or *P. fluorescens* was recorded in the greenhouse and fields naturally infected by *V. dahliae*. The relative increase in yield ranged from 113 % by *S. albidoflavus* S1 to 247 % by *P. fluorescens* P6 and correlated with a suppression of pathogen in

Fig. 7.26 *Fusarium* wilt affected crown with brown discolouration of water conducting tissues

field trials integrated in commercial strawberry production (Berg et al. 2000).

Dipping plants in a suspension of *Serratia plymuthica* prior to planting reduced *Verticillium* wilt by 24.2 % and increased the yield by 296 % (Berg et al. 1999).

The *B. subtilis* strain TS06 has a broad antifungal application to severe plant disease of strawberry (*Fragaria ananassa*), which is caused by *Verticillium* wilt (Zhang et al. 2012).

7.2.6.2 *Fusarium* Wilt

i) Symptoms

Symptoms consisted of wilting of foliage, drying and withering of older leaves, stunting of plants and reduced fruit production. Plants eventually collapsed and died. Internal vascular and cortical tissues of plant crowns showed a brown to orange-brown discolouration (Fig. 7.26).

ii) Biomanagement

The *B. subtilis* strain TS06 has a broad antifungal application to severe plant disease of strawberry (*Fragaria ananassa*), which is caused by *Fusarium* wilt (Zhang et al. 2012).

iii) Integrated Management

(a) *Bioagents, botanicals and AMF*: Combined use of AMF *Gigaspora margarita* and 3–15 % charcoal compost (which contained antagonistic microorganisms such as *B. subtilis*, *Thermomonospora* sp. and *Thermoactinomyces* sp.) drastically reduced *Fusarium* wilt in strawberry. Moreover, AMF and charcoal compost stimulated rooting and increased the root volume and hence plant growth (Kabayashi 1989).

Fig. 7.27 Grey mould on strawberry fruits

7.2.6.3 Grey Mould, *Botrytis cinerea*

i) Symptoms

Infection usually begins on berries touching the soil. However, infection may start in that part of a berry that touches another decayed berry or dead leaf (Fig. 7.27). Grey mould often starts on blossoms and green fruit injured by frost. Sometimes disease affects flower stalks enough to prevent the development of fruit.

ii) Integrated Management

Combined application of *Pichia guilliermondii* (yeast) and *Bacillus mycoides* (B16) reduced the infection of *B. cinerea* by 75 % on fruits in strawberry plants grown commercially under greenhouse conditions. But the individual application of either antagonist resulted in 50 % reduction of strawberry fruit infection. Population of yeast increased when applied as mixture rather than single application (Guetsky et al. 2002).

References

Berg G, Kunze S, Dahl R (1999) Rhizobacterial isolates for use against phytopathogenic soil fungi and process for applying these rhizobacterial isolates. European Patent no. 98124694

Berg G, Kunze S, Buchner A, Wellington EM, Smalla K (2000) Successful strategy for the selection of new strawberry associated rhizobacteria antagonistic to Verticillium wilt. Can J Microbiol 46:1128–1137

Berg G, Fritze A, Roskot N, Smalla K (2001) Evaluation of potential biocontrol rhizobacteria from different host plants of *Verticillium dahliae*. Kleb J Appl Microbiol 91:963–971

Bordiec S, Paquis S, Lacroix H, Dhondt S, Barka EA, Kauffmann S, Jeandet P, Mazeyrat-Gourbeyre F, Cle'ment C, Baillieul F, Dorey S (2011) Comparative analysis of defence responses induced by the endophytic plant growth-promoting rhizobacterium *Burkholderia phytofirmans* strain PsJN and the non-host bacterium *Pseudomonas syringae* pv. *pisi* in grapevine cell suspensions. J Exp Bot 62:595–603

Burchill RT, Cook RTA (1971) The interaction of urea and microorganisms in suppressing development of perithecia of *Venturia inequalis* (Cke.) Wint. In: Preace TF, Dickensen CH (eds) Ecology of leaf surface microorganisms. Academic, New York, pp 471–483

De Villiers EE, Korsten L (1994) Biological treatments for the control of mango post-harvest diseases. S Afr Mango Grow Rsquo Assoc Yearb 14:48–51

Guetsky R, Elad Y, Shtienberg D, Dinoor A (2002) Establishment, survival and activity of the biocontrol agents *Pichia guilliermondii* and *Bacillus mycoides* applied as a mixture on strawberry plants. Biocont Sci Tech 12:705–714

Gupta VK, Sharma K (2004) Integration of chemicals and biocontrol agents for managing white root rot of apple. Acta Hort 635:141–149

Heye CC, Andrews JH (1983) Antagonism of *Athelia bombacina* and *Chaetomium globosum* to the apple scab pathogen, *Venturia inaequalis*. Phytopathology 73:650–654

Janisiewicz WJ (1988) Biocontrol of postharvest diseases of apples with antagonist mixtures. Phytopathology 78:194–198

Janisiewicz WJ, Bors B (1995) Development of a microbial community of bacterial and yeast antagonists to control wound-invading postharvest pathogens of fruits. Appl Environ Microbiol 61:3261–3267

Janisiewicz WJ, Jeffers SN (1997) Efficacy of commercial formulation of two biofungicides for control of blue mold and gray mold of apple in cold storage. Crop Prot 16:629–633

Jeffers SN, Wright TS (1994) Comparison of four promising biological control agents for managing postharvest diseases of apples and pears. Phytopathology 84:1082

Jones DA, Kerr A (1989) *Agrobacterium radiobacter* K1026, a genetically engineered derivative of strain K84, for biological control of crown gall. Plant Dis 73:15–18

Kabayashi N (1989) Charcoal compost fixed antagonistic microorganisms and its effect to soil-borne diseases (Abstr.). Ann Phytopathol Soc Jpn 55

Kerr A (1980) Biological control of crown gall through production of agrocin 84. Plant Dis 64:25–30

Kishun R (2003) Mango bacterial canker disease and its management strategies. In: Prasad D, Puri SN (eds) Crop pest and disease management: challenges for the millennium. Jyothi Publications, New Delhi, pp 253–259

Koomen I, Jeffries P (1993) Effect of antagonistic microorganisms on the post harvest development of *Colletotrichum gloeosporoides* on mango. Plant Pathol 42:230–237

Korsten L, Lonsdale JH, De Villiers EE, De Jager ES (1992) Pre-harvest control of mango diseases. S Afr Mango Grow Rsquo Assoc Yearb 12:72–78

Leibinger W, Breuker B, Hahn M, Mendgen K (1997) Control of post-harvest pathogens and colonization of the apple surface by antagonistic microorganisms in the field. Phytopathology 87:1103–1110

Lindow SE, McGourty G, Elkins R (1996) Interactions of antibiotics with *Pseudomonas fluorescens* strain A506 in the control of fire blight and frost injury to pear. Phytopathology 86:841–848

McInnis T, Klupfel D, Zehr E (1990) Suppression of *Criconemella xenoplax* on peach by rhizosphere bacteria. In: Proceedings of the 2nd international nematology congress. Veldhoven

Montesinos E, Bonaterra A, Ophir A, Beer SV (1996) Antagonism of selected bacterial strains to *Stemphylium vesicarium* and biological control of brown spot of pear under controlled environmental conditions. Phytopathology 86:856–863

Moore LW, Warren G (1979) *Agrobacterium radiobacter* strain 84 and biological control of crown gall caused by *Agrobacterium tumefaciens*. Ann Rev Phytopathol 17:163–179

Garg N, Prakash O, Pathak RK (2003) Biocontrol of anthracnose disease of guava using actinomycete from cow dung. In: Seminar on plant diseases of national importance with special reference to guava wilt and mango malformation. Central Institute of Sub-tropical Horticulture, Lucknow, Uttar Pradesh, India, p 55

Pawar SA, Mhase NL, Kadam DB, Chandlee AG (2013) Management of root-knot nematode, *Meloidogyne incognita,* Race-II infesting pomegranate by using bioinoculants. Indian J Nematol 43:92–94

Pruvost O, Luisetti J (1991) Acta Hort 291:314–327

Pusey PL, Hotchkiss MW, Dulmage HT, Baumgardner RA, Zehr EI, Reilly CC, Wilson CL (1988) Pilot tests for commercial production and application of *Bacillus subtilis* (B-3) for postharvest control of peach brown rot. Plant Dis 72:622–626

Ruangwong OU, Chang CI, Lamine SA, Liang WJ (2012) Identification of antifungal compound produced by *Bacillus subtilis* LB5 with ability to control anthracnose disease caused by *Colletotrichum gloeosporioides*. Afr J Microbiol Res 6:3732–3738

Santhi A, Rajeswari S, Sivakumar CV (1998) Soil application of *Pseudomonas fluorescens* (Migula) for the control of root-knot nematode (*Meloidogyne incognita*) on grapevine (*Vitis vinifera* Linn.). In: Mehta UK (ed) Nematology – challenges and opportunities in 21st century. Sugarcane Breeding Institute, Coimbatore, pp 203–206

Senthil Kumar T, Rajendran G (2004) Biocontrol agents for the management of disease complex involving root-knot nematode, *Meloidogyne incognita* and *Fusarium moniliforme* on grapevine (*Vitis vinifera*). Indian J Nematol 34:49–51

Sharma RC, Sharma S, Sharma JN (1999) Present concepts and strategies of disease management in pome and stone fruits. Int J Trop Plant Dis 17:69–80

Slota JE, Farrand SK (1982) Genetic isolation and physical characterization of pAgK84, the plasmid responsible for agrocin 84 production. Plasmid 8:175–186

Sugar D, Spotts RA (1999) Control of postharvest decay in pear by four laboratory- grown yeasts and two registered biocontrol products. Plant Dis 83:155–158

Sundarababu R, Mani MP, Sivakumar CV (1999) A new biocontrol agent for plant parasitic nematodes. Indian Hort 12:25

Tzeng KC, Lin YC, Hsu ST (1994) Place Path Bull 3:24–33

Vanneste JL, Yu J (1996) Biological control of fire blight using *Erwinia herbicola* Eh252 and *Pseudomonas fluorescens* A506 separately or in combination. Acta Hort 411:351–353

Vivekananthan R, Ravi M, Ramanathan A, Samiyappan R (2004) Lytic enzymes induced by *Pseudomonas fluorescens* and other biocontrol organisms mediate defence against the anthracnose pathogen in mango. World J Microbiol Biotechnol 20:235–244

Wilson M, Lindow SE (1993) Interactions between the biological control agent *Pseudomonas fluorescens* strain A 506 *Erwinia amylovora* in pear blossoms. Phytopathology 83:117–123

Wilson CL, Pusey PL (1985) Potential for biological control of postharvest plant diseases. Plant Dis 69:375–378

Zhang Y, Fan T, Jia W, Zhang W, Liu Q, Li B, Zhang L (2012) Identification and characterization of a *Bacillus subtilis* strain TS06 as bio-control agent of strawberry replant disease (*Fusarium* and *Verticillium* wilts). Afr J Biotechnol 11(3):570–580

Zhou T, Schneider K (1998) Control of peach brown rot by preharvest applications of an isolate of *Pseudomonas syringae.* In: Abstracts of 7th international congress of plant pathology 3:20

Zhou T, Northover J, Schneider K (1999) Biological control of postharvest diseases of peach with phyllosphere isolates of *Pseudomonas syringae*. Can J Plant Pathol 21:375–381

Zhou T, Northover J, Schneider K, Lu X (2002) Interaction between *Pseudomonas syringae* MA-4 and cyprodinil in the control of blue mold and gray mold of apple. Can J Plant Pathol 24:154–161

Part III

Pest Management in Vegetable Crops

8 Solanaceous Vegetable Crops

India is the second largest producer of vegetables next only to China contributing 14.0 % of the total world production. Vegetables occupy an area of 8.989 million ha with a production of 156.325 million MT (Tables 8.1 and 8.2) (Tiwari et al. 2014). The major vegetable-producing states are West Bengal, Bihar, Uttar Pradesh, Orissa, Andhra Pradesh, Maharashtra, Gujarat, Madhya Pradesh and Karnataka. India produces 18.0 % of the world onions. In vegetables, India occupies the first position in the production of okra (73 %) and second in onion (19.9 %), brinjal (27.55 %), cabbage (13 %), cauliflower (36 %), potato (13 %) and tomato (11 %). The per capita consumption of vegetables has increased from 96 to 175 g/day within the last one decade.

8.1 Potato, *Solanum tuberosum*

8.1.1 Black Scurf, *Rhizoctonia solani*

Black scurf, also called as *Rhizoctonia* stem and stolen canker, is an occasionally serious disease of potato found in all production areas of the world.

8.1.1.1 Symptoms

Lesions characteristic of *Rhizoctonia* on stems and stolons are brown to black and sunken. These cankers can continue to expand and are capable of girdling stems and stolons of young developing plants (Fig. 8.1). *Rhizoctonia* infection of older plants very seldom leads to girdled stems that die. However, the health of these plants can be severely compromised and they can frequently become more susceptible to other diseases, particularly early blight.

Perhaps the most readily apparent phase of *Rhizoctonia* disease is the black scurf or sclerotia present on tuber surfaces (Fig. 8.1). These sclerotia can vary in size from very small, flat, superficial black specks to large, raised, irregularly shaped masses that can cover a major portion of the tuber. *Rhizoctonia* can seriously limit marketable yield, however, by causing infected plants to produce an abnormally high percentage of tubers that are misshapen, knobby and cracked while also adversely affecting size distribution.

8.1.1.2 Biomanagement

The strain of *Enterobacter* significantly reduced the infection of young stems and stolons by *R. solani.*

Kodiak (*Bacillus subtilis*) is a seed or soil treatment that significantly reduced black scurf and stem canker on potato. *B. subtilis* GBO3 was most effective in reducing stem canker severity (40–49 % reduction) relative to the infested controls over all trials.

Seed treatment of potatoes with *B. subtilis* successfully reduced the incidence of *R. solani* in greenhouse and field trials. Reduction of the disease incidence in field trials reached to a level of up to 50 % (Schmiedeknecht et al. 1998).

P.P. Reddy, *Plant Growth Promoting Rhizobacteria for Horticultural Crop Protection*,
DOI 10.1007/978-81-322-1973-6_8, © Springer India 2014

Table 8.1 Major vegetable-producing states of India (2011–2012)

State	Area ('000ha)	Production ('000MT)	Productivity (MT/ha)
Andhra Pradesh	660.97	12025.28	18.19
Assam	266.00	3045.56	11.44
Bihar	857.01	15552.38	18.14
Chhattisgarh	351.55	4582.63	19.03
Gujarat	517.63	10049.81	19.41
Haryana	356.77	5068.42	14.20
Jharkhand	261.24	3902.63	26.18
Karnataka	454.70	7662.50	16.85
Kerala	149.05	3626.00	24.32
Madhya Pradesh	506.99	10084.01	19.88
Maharashtra	591.00	8778.00	14.85
Odisha	690.07	9520.56	13.40
Punjab	178.22	3674.53	20.61
Rajasthan	181.71	1287.41	7.08
Tamil Nadu	306.66	9068.49	29.57
Uttar Pradesh	852.09	18563.75	21.78
West Bengal	1330.94	23415.69	17.59
Other states	476.94	6417.83	13.45
Total	**8989.54**	**156325.48**	**17.38**

Table 8.2 Crop-wise area, production and productivity of vegetable crops in India (2011–2012)

Vegetables	Area ('000ha)	Production ('000MT)	Productivity (MT/ha)
Beans	118	1,151	9.75
Bitter gourd	77	866	11.24
Bottle gourd	105	1,984	18.89
Brinjal	692	12,634	18.25
Cabbage	390	8,412	21.56
Capsicum	10	127	12.70
Carrot	62	1,153	18.59
Cauliflower	391	7,349	18.79
Cucumber	40	607	15.17
Muskmelon	38	791	20.81
Okra	518	6,259	12.08
Onion	1,087	17,511	16.10
Peas	408	3,745	9.17
Potato	1,907	41,483	21.75
Radish	160	2,286	14.28
Pumpkin	11	278	25.27
Sweet potato	110	1,073	9.75
Tapioca	227	8,747	38.53
Tomato	907	18,653	20.56
Watermelon	71	1,727	24.32
Others	1,661	19,487	11.73
Total	**8,989**	**156,325**	**17.39**

Fig. 8.1 Black scurf on collar region and tubers of potato

8.1.1.3 Integrated Management

(a) *Bioagents and botanicals:* Compete Plus (six species of *Bacillus*, *Streptomyces griseoviridis* and *Trichoderma harzianum* plus organic nutrients) significantly reduced incidence of potato tubers with both black scurf (*R. solani*) and common scab (*Streptomyces scabies*).

(b) *Two bioagents*: One combination of biocontrol organisms, *B. subtilis* and *Trichoderma virens*, demonstrated somewhat better control of stem canker than each organism alone, suggesting that this approach may provide improved biocontrol efficacy (Brewer and Larkin 2005).

8.1.2 Dry Rot, *Fusarium roseum* var. *sambucinum*

8.1.2.1 Symptoms

Tubers showing typical symptoms of dry rot of potato (Fig. 8.2) plants include partial or total wilting.

8.1.2.2 Biomanagement

The best isolates to inhibit *F. roseum* var. *sambucinum* belonged to the species *B. cereus*, *B. lentimorbus* and *B. licheniformis* (Sadfi et al. 2001). The antifungal activity of the selected isolates

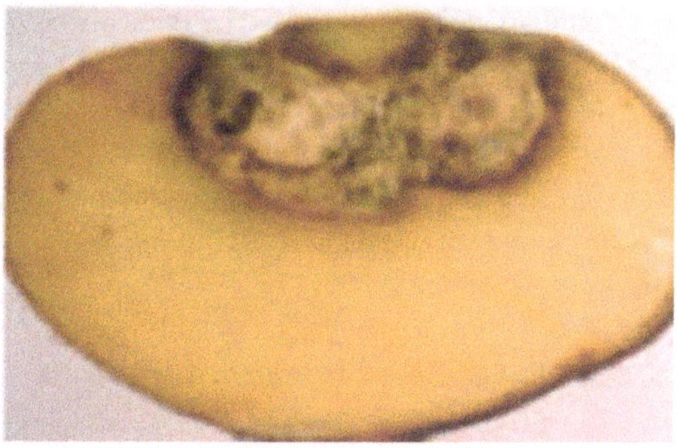

Fig. 8.2 Symptoms of potato dry rot

was associated with their ability to produce inhibitory volatile substances and diverse and complex lytic chitinases.

Greenhouse trials undertaken in pots demonstrated that *Bacillus* spp. stimulated the emergence of potato tubers. Evaluation of yield parameters showed the efficiency of isolates I32 of *B. licheniformis* and X16 of *B. cereus* to reduce disease severity. Field experiments showed that the isolate X16 of *B. cereus* seems to be the most effective to control *Fusarium* rot on seed tubers and increase yield parameters (Sadfi et al. 2002).

Burkholderia cepacia OSU-7 has great potential to be used as biocontrol agent for the management of *Fusarium* dry rot of potato. *B. cepacia* strain OSU-7 had significant effects on

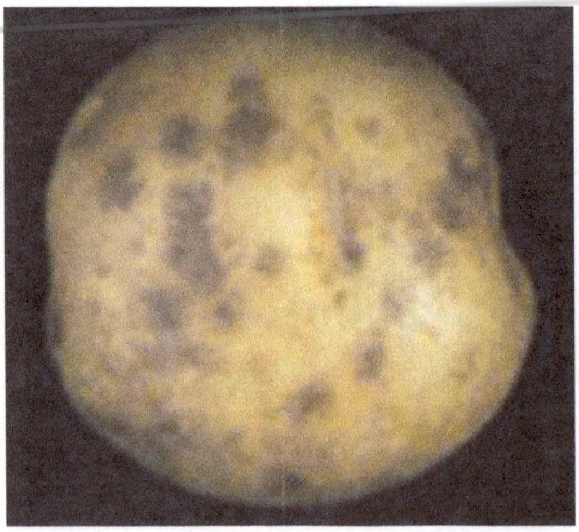

Fig. 8.3 Symptoms of silver scurf on potato tuber

controlling potato dry rot caused by three different fungi (*Fusarium sambucinum*, *F. oxysporum* and *F. culmorum*) (Al-Mughrabi 2010).

The strains of *Pseudomonas putida* and *Serratia grimesii* significantly decreased *Fusarium* dry rot of potato at 10 cfu/ml.

8.1.3 Silver Scurf, *Helminthosporium solani*

8.1.3.1 Symptoms

The pathogen infects only the periderm (skin) of the potato tuber. Tubers are infected during the growing season, and lesions become visible in 3–5 weeks. Symptoms usually appear at the stolon end of the tuber as small pale-brown spots (Fig. 8.3). Lesions may be difficult to detect at harvest, particularly if the tubers are not washed. Tubers that appear to be disease-free at harvest may develop symptoms in storage.

8.1.3.2 Biomanagement

P. putida isolated from a soil selected for its important suppressive effect against the causal agent of potato silver scurf reduced the disease severity by 70 % after 30 days at 15 °C and by 22 % after 18 days at 24 °C (Martinez et al. 2002).

8.1.4 Bacterial Wilt or Brown Rot, *Ralstonia solanacearum*

Bacterial wilt of potato disease is more severe in tropical, subtropical and warm climates. In Ranchi district of Jharkhand (India), cultivation of rainfed potato was abandoned due to this disease, and similar instances are recorded in Indonesia and in Peru. In India, the losses reported are up to 55 % in Kumaon hills, 33–40 % in Maharashtra, 20–25 % in Hyderabad and over 75 % in some localities of Karnataka.

8.1.4.1 Symptoms

The disease affects both above- and underground parts of the plant. Infected plants show sudden wilting and complete collapse. Vascular browning has been reported to be the characteristic symptom of the disease and the bacterial mass ooze out (Fig. 8.4).

8.1.4.2 Biomanagement

Promising biocontrol agents include *Bacillus* spp., *B. polymyxa* and *P. fluorescens* which have been found to reduce the wilt development and incidence. Root colonization by *P. fluorescens* has been reported. *Bacillus* spp. reduced bacterial wilt incidence up to 79 %, while *P. fluorescens* by 43–75 %. Dipping of seed tubers in 0.25 % solution of the formulation of *B. cereus* (strain B4) and *B. subtilis* (strain B5) for 20 min was effective in reducing the wilt by 59–72 % and enhancement of tuber yield by 16.7–25.6 %, respectively (Sunaina et al. 2001). Preemptive colonization by *B. polymyxa* strain FU-6 and *P. fluorescens* was responsible for biological control of bacterial wilt of potato (Aspiras and de la Cruz 1986). Khan et al. (2000) found that tuber treatment followed by two soil applications of *P. fluorescens* and *B. subtilis* reduced the disease incidence by 76.31 and 53.15 % and increased the tuber yield by 220.17 and 129.12 %, respectively. Further, the population of the pathogen in the soil and rhizosphere was reduced by 10 and 100 times by these two antagonists (Table 8.3).

Fig. 8.4 *Left* – Bacterial wilt on potato plant. *Right* – Transverse section of potato tubers infected by bacterial wilt showing browning and necrosis of the vascular ring

Table 8.3 Effect of bioagents on bacterial wilt under wilt-infested conditions

Bioagent	Reduction in wilt (%)	Increase in yield (%)
Avirulent *Ralstonia solanacearum*	59.4	391.2
Enterobacter cloacae	66.3	391.7
Bacillus subtilis	72.4	465.2
Bacillus cereus	57.6	299.4

8.1.5 Common Scab, *Streptomyces scabies*

8.1.5.1 Symptoms

Scab spots are more or less circular brown, roughened areas, with irregular margins. Sometimes the ridged portions are in broken concentric rings. The spots may be raised or warty in appearance, level with the surface or somewhat sunken. Superficial russeting to deep pitting may occur (Fig. 8.5). White grubs, wire worms and millipedes are attracted to this affected tissue so that scab spots may be further enlarged or deepened. Scab begins when tubers start forming. Initially the spots may be so small that they are not noticed. As the tuber continues to grow, the areas of these reddish-brown spots also enlarge. An older tuber has too thick a protective layer on its surface to be invaded readily.

8.1.5.2 Biomanagement

Disease-suppressive strains of *Streptomyces scabies* control the scab disease in potato which is caused by pathogenic S. *scabies* strains. Soil treatment with *B. subtilis* gave effective control of common scab on potato.

Tanii et al. (1988) reported that strains of *P. fluorescens*, *Enterobacter agglomerans*, *Acinetobacter* sp. and nonfluorescent pseudomonads each significantly reduced the severity of common scab by 61, 51, 35 and 46 %, respectively, in field trials. The effect was enhanced by a combination of reducing the soil pH, beginning irrigation at the time of tuber formation and the application of urea to the soil

Seed treatment of potatoes with *B. subtilis* successfully reduced the incidence of S. *scabies* in greenhouse and field trials. Reduction of the disease incidence in field trials reached to a level of up to 67 % (Schmiedeknecht et al. 1998).

Soil treatment with *B. subtilis* gave effective control of common scab on potato.

8.1.5.3 Integrated Management

(a) *Bioagents and botanicals*: Compete Plus (six species of *Bacillus*, *Streptomyces griseoviridis* and *Trichoderma harzianum* plus organic nutrients) significantly reduced incidence of potato tubers with both black scurf (*Rhizoctonia solani*) and common scab (*Streptomyces scabies*).

8.1.6 Soft Rot, *Erwinia carotovora* pv. *carotovora*

8.1.6.1 Symptoms

Fruits develop subepidermal soft rot, which spreads rapidly, reducing the tissue to a pulp in

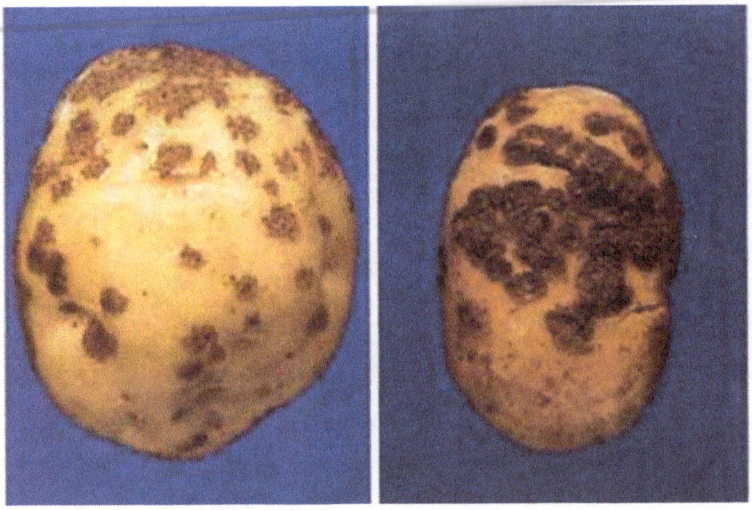

Fig. 8.5 Common scab symptoms on potato tubers

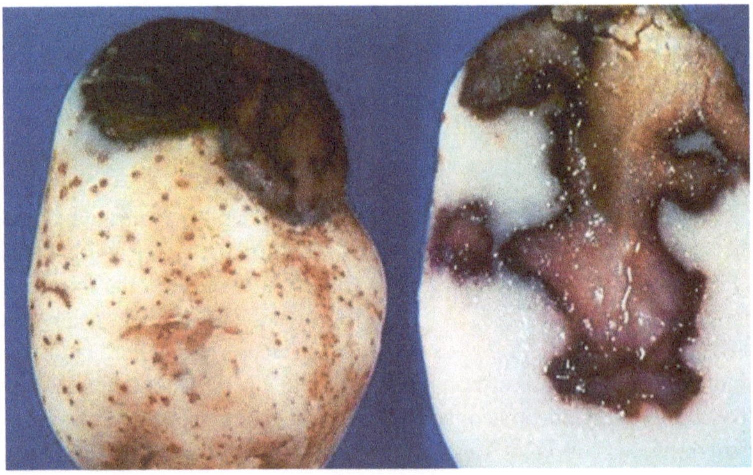

Fig. 8.6 Potato soft rot symptoms

4–6 days. Rotting takes place in tubers or in cut seed pieces. Flesh becomes whitish-creamy and soft. Rot occurs mostly in moist and warm conditions during harvesting, transit, storage and in soil (Fig. 8.6). Up to 100 % tubers are rotten.

8.1.6.2 Biomanagement

Strain W4P63 of *P. fluorescens* and *P. putida* effectively controlled *E. carotovora* infection and promoted the growth and yield increase effects of this rhizobacteria on potato (Kloepper 1983; Xu and Gross 1986). Disease control was attributed to production of siderophores that complex iron such that it is unavailable to the pathogen (Kloepper 1983). *P. putida* strain M17 was shown to control postharvest soft rot when applied as a seed piece treatment or as a postharvest treatment (Costa and Loper 1994).

Sharga and Lyon (1998) identified a *B. subtilis* isolate BS 107 that controlled *E. carotovora*. Control was attributed to antibiosis.

The strains of *Enterobacter* and *Acinetobacter* were antagonistic to *E. carotovora* subsp. *carotovora* and protected seed pieces from soft rot.

Fig. 8.7 Potato plants and roots infected with cyst nematode

Table 8.4 Effect of *Pseudomonas fluorescens* on cyst nematode and tuber yield of potato

Treatment	No. of J_2/g root at 54 DAP	No. of cysts/100 g soil	No. of eggs/g soil	Tuber yield (MT/ha)
P. fluorescens – 10 kg/ha	60	130	58	14.89
P. fluorescens – 5 kg/ha	114	179	72	11.43
P. fluorescens – 2.5 kg/ha	138	195	85	9.46
Control	164	241	102	8.42
CD (*P* = 0.05)	**11.24**	**18.47**	**5.42**	**1.12**

In some cases, antibiotic substance(s) produced by the bacteria was considered to be important in disease suppression.

8.1.7 Cyst Nematodes, *Globodera rostochiensis*, *G. pallida*

The avoidable yield loss in susceptible potato due to cyst nematodes was 99.5–99.8 % in summer and autumn crops. Total failure of the crop has been reported under severe infestation conditions. An average loss of about 9 % of global potato is accounted to the cyst nematodes amounting to about 45 million tons.

8.1.7.1 Symptoms

The disease usually occurs in patches. The infested plants exhibit typical symptoms of patchy growth of weak and stunted plants. The patches increase in time with continuous potato cropping. Under conditions of severe infestation, the plant growth is stunted and wilting occurs during the hot part of the day. Plants show tufting of leaves at the top as the outer leaves turn yellow and die. Root system is infected with adults and cysts (Fig. 8.7). Root system is smothered, secondary roots are induced at the collar region, and the plants can be easily pulled out. Tubers formed are less in number and reduced in size. In extreme cases, tuber formation is arrested. The total photosynthesis per plant is also significantly reduced as a result of reduced leaf area and this is reflected in reduced potato yield.

8.1.7.2 Biomanagement

Application of *P. fluorescens* at 10 g/m^2 gave good control of cyst nematodes (Mani et al. 1998).

Application of talc-based formulation of *P. fluorescens* (15×10^8 cfu/g) at 10 kg/ha reduced the penetration of roots by the juveniles of potato cyst nematodes at 45 days after planting by 63.4 %. Significant reduction in cyst count and egg production was observed with the above bio-agent. The above treatment also increased potato tuber yield significantly (76.2 % over control) (Table 8.4) (Seenivasan et al. 2007).

Plant health-promoting rhizobacteria, namely, *Agrobacterium radiobacter* and *Bacillus sphaericus* have been found to reduce toxic metabolites affecting the penetration of *G. pallida* by 41 % in potato roots and improving yield (Racke and Sikora 1992). The root penetration of *G. pallida* is reduced in the presence of *B. sphaericus* or *A. radiobacter* which may be due to resistance induced by bacteria (Hallman et al. 1998).

Reitz et al. (2000) showed that the lipopolysaccharides of *Rhizobium etli* G12 induce the systemic resistance to infection by the cyst nematode *G. pallida* in potato roots.

Fig. 8.8 Damping off of tomato seedlings

8.2 Tomato, *Lycopersicon esculentum*

8.2.1 Damping Off, *Pythium aphanidermatum*, *Rhizoctonia solani*

8.2.1.1 Symptoms

In the pre-emergence phase of the disease, the young seedlings are killed even before they emerge out of the soil surface. However, this disease mostly occurs at post-emergence stage which is characterized by the toppling over of the infected seedlings anytime after they emerge from the soil until the stem has hardened sufficiently to resist invasion by the pathogen. Infestation usually occurs at ground level or through roots. The infected tissue appears soft and water soaked. As the disease advances, the stem becomes constricted at the base of the plants that collapse later. Seedlings that are healthy looking 1 day may have collapsed the next morning (Fig. 8.8).

8.2.1.2 Biomanagement

Streptomyces diastaticus was effective in suppressing damping off of tomato (Singh and Reddy 1979).

Two strains of *B. subtilis* (FZB 13 and 44) tested on tomato against *P. aphanidermatum* in greenhouse trials partially compensated for the damage caused by the pathogen (Grosch et al. 1999).

B. subtilis strain RB14 produces the cyclic lipopeptide antibiotics iturin A and surfactin active against several phytopathogens. This strain has a very good potential to be used for the biological control of damping off of tomato caused by *R. solani* (Asaka and Shoda 1996).

P. fluorescens strain Pf1 induces resistance against *P. aphanidermatum* in tomato (Ramamoorthy et al. 1999).

Bacillus strains (*B. amyloliquefaciens* MB101 and *B. subtilis* MB14) were selected as potential antagonists from tomato rhizosphere that combat *R. solani* that causes damping off of tomato plant (Solankia et al. 2012).

8.2.1.3 Integrated Management

(a) *Bioagents and physical methods*: Combination of the seed/root application of *P. fluorescens* with soil solarization was very effective in management of damping off of tomato in nursery at the farmers' field.
(b) *Bioagents and botanicals*: Seed treatment with *P. fluorescens* and addition of neem cake to the nursery beds enhanced seed germination and seedling stand.

Seed treatment with *P. fluorescens* at 10 g/kg seed followed by soil application of *P. fluorescens* at 2.5 kg multiplied in 50 kg FYM per hectare, and seed treatment with *A. chroococcum* at 16 g/kg seed followed by soil application of *Azotobacter chroococcum* at 500 g multiplied in 50 kg FYM per hectare gave effective control of damping off in tomato (Rahman et al. 2002) (Table 8.5).

8.2.2 Early Blight, *Alternaria solani*

8.2.2.1 Symptoms

The pathogen not only affects leaves but also the fruits. Symptoms appear as small, isolated scattered pale-brown spots on leaves. They enlarge and develop into concentric rings in the necrotic tissue and induce yellowing leading to leaf blight (Fig. 8.9). Girdling of the stem at the base of the plant is also observed. The stem lesions are usually restricted to one side of a stem and become elongated and sunken. Similar symptoms also appear on the fruit causing fruit rot or stem end rot or blossom end rot (Fig. 8.9). Under favourable conditions to the pathogen, losses may go up to 70 %. It is more severe in June to July sown crop. In severe attacks, the entire plant may be defoliated, exposing the fruits to sun scald.

Table 8.5 Comparative efficacy of bioagents and organic amendments against damping off of tomato

Treatment	% disease index
Seed treatment with *Pseudomonas fluorescens* at 10 g/kg + soil application of 50 kg FYM enriched with 2.5 kg of *P. fluorescens*	7.93 (16.06)
Seed treatment with *Azotobacter croococcum* at 16 g/kg + soil application of 50 kg FYM enriched with 500 g of *A. croococcum*	8.46 (16.68)
Control (check)	30.64 (33.55)
CD at 5 %	**(3.23)**

Figures in parentheses indicate the arc sin transformed values

8.2.2.2 Biomanagement

B. subtilis was very effective in controlling the early blight disease since lowest per cent disease index (PDI) of 8.93 was recorded as compared to 19.37 PDI in control. The PDI recorded in the *B. subtilis* sprayed plants was on par with the PDI recorded in 0.2 % mancozeb fungicide (6.82 PDI). The highest fruit yield was recorded in plants sprayed with *B. subtilis* (3.24 kg/plant and 119.86 MT/ha) which was at par with the mancozeb sprayed plants (3.54 kg/plant and 131.32 MT/ha) (Amarananjundeswara et al. 2009).

Jagadish (2006) reported that the combination of all the five methods (seed treatment + soil application + nursery bed treatment + foliar spray + root dip method) was the best method of inoculation of *Pseudomonas* B-25. This method increased the plant height by 57 %, number of leaves by 53 %, stem girth by 46 % and number of branches by 71 % (Fig. 8.10). Combination of different methods of application of *Pseudomonas* strain B-25 controlled the early blight disease in tomato by 60.2 %.

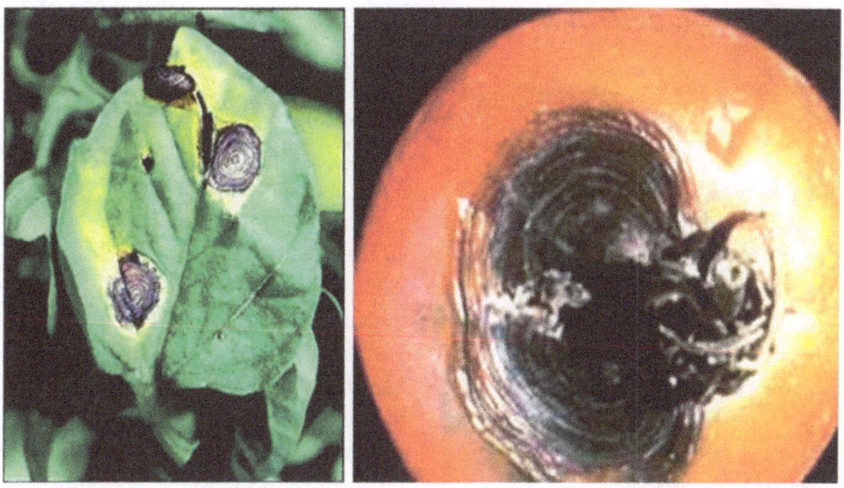

Fig. 8.9 Early blight on tomato leaf and fruit

Fig. 8.10 Influence of different methods of inoculation of *Pseudomonas* sp. B-25 on growth of tomato plants

Fig. 8.11 Buckeye rot on tomato fruits

Fig. 8.12 Tomato plants infected by *Fusarium* wilt

8.2.3 Buckeye Rot, *Phytophthora nicotianae* var. *parasitica*

This is a very serious disease and causes losses up to 92 %.

8.2.3.1 Symptoms

Spots with pale and dark-brown concentric rings appear on fruits. These may be small or may cover a major portion of the fruit surface causing fruit rot (Fig. 8.11), thereby lowering the yield considerably. One infected fruit with latent infection rots the entire basket overnight.

8.2.3.2 Biomanagement

Two strains of *B. subtilis* (FZB 13 and 44) tested on tomato against *P. nicotianae* in greenhouse trials partially compensated for the damage caused by the pathogen (Grosch et al. 1999).

8.2.4 Wilt, *Fusarium oxysporum* f. sp. *lycopersici*

8.2.4.1 Symptoms

The wilt disease is becoming a major disease of concern in many parts of the country. The disease is characterized by yellowing and wilting of leaves and finally the entire plant wilts (Fig. 8.12) and dies prematurely. Stem tissue often is discoloured throughout the plant. Vascular browning takes place in the root. The fungus can persist in the soil for many years and is virulent at moderate temperature (26–28 °C).

8.2.4.2 Biomanagement

A binary system of *Serratia marcescens* and *Streptomyces anulatus* was effective in controlling tomato wilt disease (Toyoda et al. 1993).

Pseudomonas strains have been isolated that reduce incidence of *Fusarium* wilt in tomato. Tomato plants develop increased resistance to *Fusarium* wilt if they have been treated with the endophytic strain 63-28 of *P. fluorescens*.

Use of *Bacillus* at transplanting can reduce disease incidence (diseases caused by soil fungi including *F. oxysporum*, *R. solani* and *P. capsici*) in contrast to traditional treatments (fungicide application) as folpat, captan and mancozeb compared to untreated control with the most efficient strain. Furthermore, the suppressive effect was maintained over time or among harvest times (Table 8.6) (Castillo et al. 2013).

Table 8.6 Disease incidence and severity at harvest time of tomato plants cv. Floradade subjected to different treatments with microcapsules containing strains of *Bacillus subtilis*

Treatments	Incidence (%)	Severity
Bacillus B1	0.0 d	0.0 c
Bacillus J1	0.0 d	0.0 c
Bacillus M2	12.0.c	1.5 c
B1 + J1 + M2 mix	0.0 b	0.0 c
Chemical treatment	27.0 b	3.5 b
Control	75.0 a	5.0 a
CV (%)	**10.4**	**1.2**

Values with same letters area not statistically different ($p<0.01$)

8.2.4.3 Integrated Management

(a) *Two bioagents*: Mao et al. (1998) found that combined inoculation of *Trichoderma virens* and *Burkholderia cepacia* resulted in increased plant stand and greater yield than those obtained with either biocontrol agent alone.

8.2.5 Southern Blight/Collar Rot, *Sclerotium rolfsii*

8.2.5.1 Symptoms

The first symptom of the disease is observed as soft tissue necrosis of bark of the stem near soil line. White growth of cottony mycelium is clearly visible on the affected portion just below the soil surface. Later, dense silvery fungus growth along with white- to light-brown mustard-like sclerotia is observed on the same portion. Progressive drooping, yellowing or wilting of the entire plant is observed (Fig. 8.13). Sometimes the plant is collapsed soon after infection. The disease is soilborne where the pathogen survives in the form of sclerotia.

8.2.5.2 Biomanagement

The disease was effectively managed under greenhouse and field conditions by seed + soil treatment with *B. subtilis.*

Under field conditions in Thailand, Jetiyanon et al. (2003) observed that a PGPR mixture containing *B. amyloliquefaciens* strain IN937a and *B. pumilus* strain IN937b induced systemic resistance against southern blight of tomato.

Fig. 8.13 Sothern stem blight on tomato

8.2.6 Stem Rot, *Botrytis cinerea*

8.2.6.1 Symptoms

The most serious symptom of *B. cinerea* in tomatoes is stem rotting. Lesions on the stem may result from direct infection or from progression of the rot along the infected leaves, until infection approaches the stem (Fig. 8.14).

8.2.6.2 Biomanagement

Mari et al. (1996) found that *Bacillus amyloliquefaciens* had a fungistatic effect on *B. cinerea* on mature green tomatoes stored at low temperature (10 °C), and significantly reduced pathogen growth during the first 7 days of storage.

Fig. 8.14 Stem rot on tomato

The soil + root dip treatments with isolate *Pseudomonas* C7 (French isolated) gave maximum inhibition of stem rot of tomato disease (Singh et al. 2003b).

8.2.7 Fusarium Crown and Root Rot, *Fusarium oxysporum* f. sp. *radicis-lycopersici*

8.2.7.1 Symptoms

The fungus invades susceptible plants through wounds and natural openings created by newly emerging roots. Early symptoms caused by the pathogen in tomato seedlings include stunting, yellowing and premature loss of cotyledons and lower leaves. A pronounced brown lesion that girdles the hypocotyl (root/shoot junction), root rot, wilting and death are advanced symptoms (Fig. 8.15).

8.2.7.2 Biomanagement

All PGPR strains (*B. subtilis* GBO3 and FZB24, *B. amyloliquefaciens* IN937a and *B. pumilus* SE34) tested promoted plant growth characteristics significantly. A higher promoting effect was provided by SE34. Experiments on population dynamics of PGPR strains revealed that, after 28 days of incubation, populations of strain SE34 remained stable, while the remaining bacterial

Fig. 8.15 Crown and root rot of tomato

Fig. 8.16 Symptoms of tomato foot and root rot

strains showed a slight decline in their population densities. The GBO3 and FZB24 strains provided higher disease suppression when applied individually. However, application of IN937a in a mixture with GBO3 provided a higher control efficacy. Treatment of tomato plants with acibenzolar-S-methyl (ASM) resulted in a small reduction in disease index, while application of hymexazol provided significantly higher control efficacy (Myresiotis et al. 2012).

Application of some microbes, such as *P. fluorescens* or *Serratia plymuthica*, can efficiently reduce the disease. The microbes can be applied by coating on the seeds or by drenching. Using studies with auto fluorescent proteins, it was observed that the pathogen and the biocontrol bacteria occupy the same niches on the root surface. It was also observed that the *Pseudomonas* biocontrol strain, applied on the seed, reaches these niches earlier than the pathogen (Lugtenberg 2011).

Seed treatment of tomato with endophytic bacterium *Bacillus pumilus* strain SE 34 prevented the entry of vascular wilt fungus *F. oxysporum* f. sp. *radicis-lycopersici* into the vascular stele and the mycelial growth was restricted to the epidermis and outer root cortex (Benhamou et al. 1998). Similarly, application of *P. fluorescens* strain 63-28 restricted the growth of *F. oxysporum* f. sp. *radicis-lycopersici* in tomato (M'Piga et al. 1997).

8.2.7.3 Integrated Management

(a) *Bioagents and chemicals*: Combined applications of the four PGPR strains (*B. subtilis* GBO3 and FZB24, *B. amyloliquefaciens* IN937a and *B. pumilus* SE34) with either ASM or hymexazol were significantly more effective (Myresiotis et al. 2012).

8.2.8 Foot and Root Rot, *Fusarium solani*

8.2.8.1 Symptoms

Fusarium foot rot causes varying degrees of interveinal chlorosis and necrotic spotting on young foliage. Above-ground symptoms may be restricted to single branches. In severe cases, plants die. A dark-brown lesion, about 1.25–2.50 cm long, is visible on the tap root or the main lateral root. Often the lesion completely girdles the root. The lesion usually occurs on the roots within the top 30 cm of soil. Internally, a brown discolouration of the vascular system extends 2.5–10.0 cm from the lesion (Fig. 8.16).

8.2.8.2 Biomanagement

Priming of seedlings with the eight selected bacterial strains reduced the proportion of diseased plants to as low as 19 %. In addition, in the absence of an added pathogen, three strains, namely, *P. putida* 1 T1, *P. extremorientalis*

Table 8.7 Control of tomato foot and root rot and stimulation of plant growth in salinated soil by selected bacterial isolates

Treatments[a]	Diseased plants (% ± SD)	Shoot length (cm)	Root length (cm)	Dry weight (g/plant)
Positive control	21±4.8	100	100	100
Negative control	46±10.8	(7.65)	(5.64)	(0.14)
P. fluorescens PCL1751	21±4.8*	105	122*	101
P. trivialis 3Re27	33±9.6	109	120	110
P. putida 1T1	20±4.2*	125*	134*	124*
P. chlororaphis RRj228	25±6.8*	95	121*	100
P. fluorescens SPB2145	33±6.8	112	105	101
P. extremorientalis TSAU20	25±9.6*	128*	130*	127*
S. rhizophila e-p10	19±4.2*	119*	139*	120*
S. plymuthica RR2-5-10	33±6.8	106	116	106

[a]Plants were grown under open natural conditions in pots containing salinated soil infested with *F. solani* spores, except for the positive control in which no spores were added to the soil
*Significantly different from the control at $P<0.05$

Table 8.8 Effect of bacterial strains on tomato (cv. Belle) germination, height and fruit yield in greenhouse experiments

Treatment	Plant height (cm)	% increase	Fruit yield (kg/m^2)	% increase
None	125±3.7	–	13.9±0.9	–
P. putida 1T1	159*±3.4	27	16.6*±1.2	19
P. extremorientalis TSAU20	157*±4.2	26	17.0*±0.7	22
S. rhizophila EP10	150*±4.7	20	15.9*±1.0	14

*Significantly different from the control at $P<0.05$

TSAU20 and *Stenotrophomonas rhizophila* EP10, showed a significant stimulatory effect on tomato plant growth, increasing the dry weight of whole tomato plants up to 27 % in comparison with the non-bacterized control. The strains also increased tomato fruit yield in a greenhouse varying from 14 to 22 % (Tables 8.7 and 8.8) (Egamberdieva et al. 2011).

8.2.9 Bacterial Wilt, *Ralstonia solanacearum*

8.2.9.1 Symptoms

The disease causes rapid wilting and death of the entire plant without any yellowing or spotting of leaves. Total collapse of the plant usually occurs when temperature reaches 32 °C and above. Plant wilts while still green (Fig. 8.17). The lower leaves may droop before wilting occurs. The vascular system becomes brown, root and rootlet rots leaving behind only the corky portion of the main root. When the stem of a wilting plant is cut across, the pith has a darkened water-soaked appearance, and on squeezing the cut stem, a white, yellow or greyish, slimy exudate may appear (Fig. 8.17). In later stages of the disease, decay of the pith may cause extensive hollowing of the stem.

8.2.9.2 Biomanagement

Bacillus megaterium, *Enterobacter cloacae*, *Pichia guilliermondii* and *Candida ethanolica* show antagonistic effects against *R. solanacearum* that causes bacterial wilt disease in tomato (Nguyen et al. 2011).

8.2.9.3 Integrated Management

(a) *Bioagents, AMF and botanicals*: Combined use of AMF *Gigaspora margarita* and 3–15 % charcoal compost (which contained antagonistic microorganisms such as *B. subtilis*, *Thermomonospora* sp. and *Thermoactinomyces* sp.) drastically reduced bacterial wilt in

Fig. 8.17 Bacterial wilt affected tomato plants and bacterial ooze from cut surface of stem

tomato. Moreover, AMF and charcoal compost stimulated rooting and increased the root volume and hence plant growth (Kabayashi 1989).

(b) *Bioagents and botanicals*: Integration of growing and incorporation of sunn hemp (green manure crop) into the soil and seed treatment with *P. fluorescens* (10^{10} cfu/ml) were highly effective in reducing the bacterial wilt incidence to 6.75 % and significantly increasing the fruit yield to 25.5 MT/ha as compared to high wilt incidence (55.6 %) and low yield (7.8 MT/ha) in control. Thus, the efficacy of *P. fluorescens* is increased under green manure-incorporated soil against bacterial wilt of tomato (Gopalakrishnan and Ajit Kumar 2006).

Bora and Talukdar (2003) reported that application of paste formulation of *P. fluorescens* in vermicompost exhibited lowest bacterial wilt incidence in tomato crop with maximum reduction of wilt pathogen in the rhizosphere. The treatment also resulted in significant effect on yield enhancement.

8.2.10 Bacterial Leaf Spot, *Xanthomonas campestris* pv. *vesicatoria*

8.2.10.1 Symptoms

It is a serious disease, occurring during the rainy season, and continues up to fruit initiation stage. It is most noticeable on fruits, but also causes damage to the foliage and stems. On green fruits, the initial spot is very small and water soaked (Fig. 8.18). It eventually enlarges to about 6 mm. As the bacterial spot matures, it becomes brown and scabby without extending deep into the fruit. On foliage, irregular greasy, dark-green spots, 2–3 mm wide, are observed (Fig. 8.18). The spots eventually dry and the tissue often tears. The disease is transmitted through the seed.

8.2.10.2 Biomanagement

In greenhouse trials, Companion applied alone significantly ($P < 0.05$) reduced bacterial spot severity or AUDPC values compared to the non-treated control. Data from two repeated field trials revealed significant disease reduction in the development of bacterial spot by treating tomato plants with Actinovate (*Streptomyces lydicus*) AG, BU EXP 1216C and BU EXP 1216S alone. Actinovate AG and BU EXP 1216C alone significantly increased the marketable yields when compared with the control in one of the field trials.

Highly effective products include Cease/Millstop, Citrex, Companion, Harpin, HMO 736, KPX-B2, KleenGrow, Kocide, K-Phite, Regalia, MBI-106020, SC 27, Seacide, MBI-10605, NAI 4201, Taegro, Actigard and Kasumin. The mean disease severities for these chemicals ranged from 1.0 to 2.0 compared to the pathogen control (3.7, disease severity). Correspondingly, the number of spots/leaf for these chemicals was in

Fig. 8.18 Bacterial spot lesions on tomato leaves and fruits

the range of 7.7–24.0. The moderately effective products against tomato bacterial leaf spot include KleenGrow, SC 27 and SUMM-IR4 with a mean disease severity of 2.3 (Vijay Krishna Kumar et al. 2011).

8.2.10.3 Integrated Management

(a) *Bioagents and chemicals*: In both greenhouse and field trials, tomato plants treated with all microbial products [Actinovate (*Streptomyces lydicus*) AG, BU EXP 1216C and BU EXP 1216S] in combination with chemical product acibenzolar-S-methyl (ASM, Actigard 50WG) (ASM is a SAR inducer labelled on tomato for control of bacterial spot) had significantly lower values of AUDPC than the control, and levels of disease reduction in combined treatments were generally greater than the standard chemicals Kocide 3000 and Manzate. Results from this study indicate a considerable potential for these microbial products with ASM to be incorporated into integrated management programmes for control of bacterial spot on tomato in Florida (Zhang et al. 2011).

8.2.11 Bacterial Speck, *Pseudomonas syringae* pv. *tomato*

8.2.11.1 Symptoms

The disease starts as small, round, water-soaked spots, which gradually become yellowish brown and finally blackish brown with sunken ashy centre and pale yellowish-green halo. Later on, spots become irregular to circular, measuring 1 mm in diameter, distributed over the entire leaf lamina and in severe cases forming large blotches (Fig. 8.19). Spots on petioles are more or less oval.

8.2.11.2 Biomanagement

Reduction of bacterial speck occurred in greenhouse tests in response to application of a nonpathogenic transpoon (Tn5) mutant of *P. syringae* pv. *tomato*. *P. syringae* pv. *tomato* can colonize the surface of tomato plants and survive for extended periods (49–51 days). This trait makes it possible to use nonpathogenic bacteria to compete with the pathogen for colonization and infection sites. Fluorescent *Pseudomonas* spp. partially controlled bacterial speck on young and mature plants with slightly better residual activity than copper compounds (Colin and Chafik 1986).

8.2.12 Cucumber Mosaic Virus

A severe epidemic of CMV in fresh market tomato in North Alabama resulted in an estimated 25 % yield loss in that region. CMV is transmitted by more than 65 aphid species.

8.2.12.1 Symptoms

Tomato plants are usually stunted and have poorly shaped leaves, or 'fern leaf', when infected by CMV (Fig. 8.20) Also certain strains of CMV can cause partial or total loss.

Fig. 8.19 Bacterial speck on tomato leaves and fruit

Fig. 8.20 CMV symptoms on tomato

8.2.12.2 Biomanagement

Seed treatment with *P. fluorescens* strain 89B-27 and *S. marcescens* strain 90–166 has consistently reduced the number of cucumber mosaic virus-infected plants (CMV) and delayed the development of symptoms in tomato (Raupach et al. 1996). PGPR have been reported to act as inducers of systemic resistance towards cucumber mosaic virus in tomato (Raupach et al. 1996). PGPR strains reduced cucumber mosaic virus disease severity in tomato. The percentage of plants with CMV symptoms in the four PGPR treatments ranged from 14.7 % to 22.1 %, compared with 75.8 % of plants with symptoms in the plants without PGPR treatment.

In the 1996 field experiment, area under the disease progress curve (AUDPC) values, indicating disease symptom progression over time, was significantly lower in all PGPR treatments compared with the disease control (Fig. 8.21). Similarly, ELISA values in all PGPR treatments, and the percentage of infected plants (Fig. 8.22) (based on ELISA) in three PGPR treatments, were significantly lower than in the disease control (Zehnder et al. 1999).

The percentage of infected plants in the disease control treatment was over threefold greater than in the IN937a and IN937b treatments. Plant height measurements taken 30 days after transplanting indicated that plant growth in the PGPR treatments was greater than in the disease control. The increased growth may have resulted from PGPR-induced resistance against CMV, PGPR-induced growth promotion or both factors

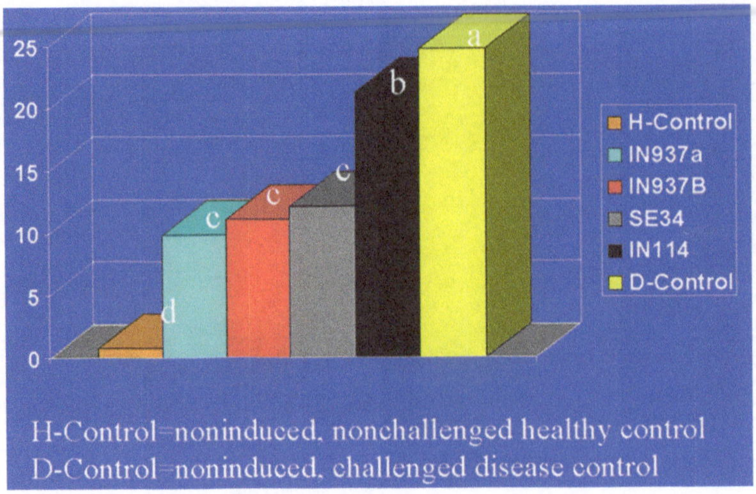

Fig. 8.21 PGPR-ISR against CMV on field tomato: AUDPC values

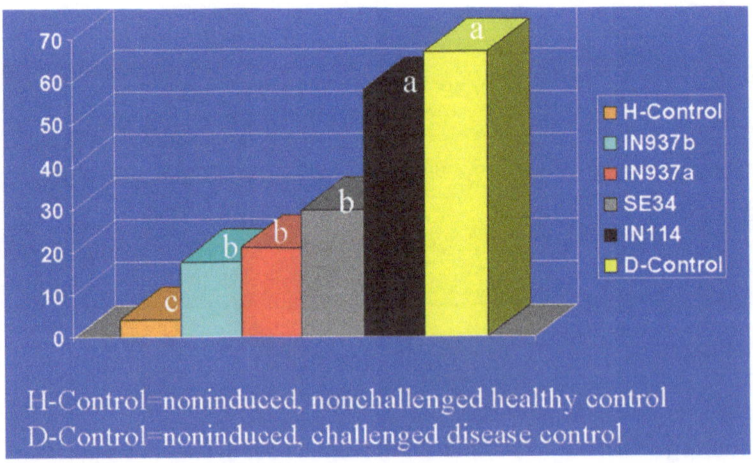

Fig. 8.22 PGPR-ISR against CMV on field tomato: % infected plants

combined. Importantly, yields in the SE34, IN937a and IN937b treatments were significantly greater than in the disease control (Fig. 8.23) (Zehnder et al. 1999).

Zehnder et al. (2000) reported that plant growth-promoting rhizobacteria reduced the level of aphids transmitting cucumber mosaic virus in tomato plants. The percentage of symptomatic plants in the most effective PGPR treatments ranged from 32 to 58 %, compared with 88–98 % in the non-rhizobacterized challenged disease control treatment.

Sikora and Murphy (2003) demonstrated that PGPR were significantly effective in controlling cucumber mosaic virus by reducing 80 % population of aphids, the vectors of CMV in tomato.

8.2.13 Tomato Mottle Virus

Tomato mottle, caused by the tomato mottle virus (ToMoV), poses a major threat for both transplant and field production of tomato in West Central and Southwest Florida. Tomato mottle

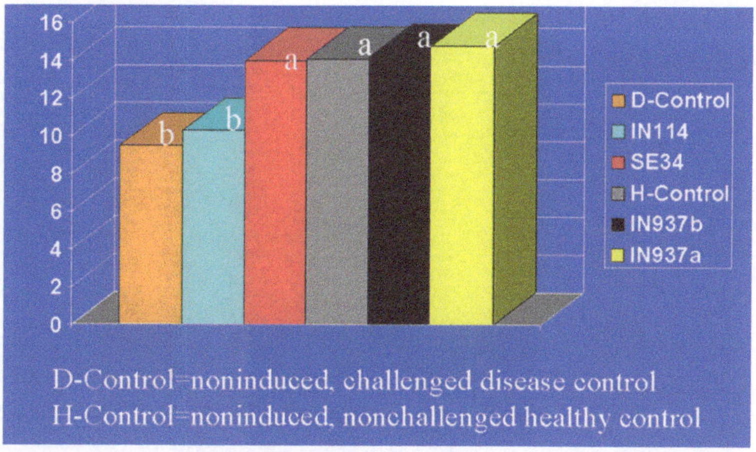

Fig. 8.23 PGPR-ISR against CMV on field tomato: yields (kgs/plot)

Fig. 8.24 Symptoms of tomato mottle virus (marked yellowing and distortion)

virus belongs to Geminiviridae family and is transmitted by adult sweet potato whiteflies, *Bemisia tabaci*, biotype B (also known as the silverleaf whitefly, *B. argentifolii*).

8.2.13.1 Symptoms

Symptoms of ToMoV in field-grown tomatoes include chlorotic mottling and upward curling of leaflets and an overall reduction in plant height and in the number and size of fruits (Fig. 8.24).

8.2.13.2 Biomanagement

Treatment of tomato seeds/soil application with powder formulation of PGPR (*Bacillus amyloliquefaciens*, *B. subtilis*, *B. pumilus*) reduced symptom severity of ToMoV and increased fruit yield (Murphy et al. 2000). Tomato plants treated with PGPR, applied as seed treatment, a spore preparation mixed with potting medium (referred to as 'powder') or a combined seed and powder treatment showed significantly lesser disease severity rating in all PGPR powder-based treatments than in either of seeds or control treatments. The use of PGPR could become a component of an integrated programme for management of this virus in tomato (Murphy et al. 2000)

Experiments conducted over 5 years in Alabama and Florida evaluating strains of PGPR for induction of resistance against whitefly-transmitted tomato mottle virus demonstrated a reduction in the incidence of viral infection and an increase in tomato yield in PGPR-treated plants (Zehnder et al. 1999).

The IN937b and SE34 PGPR strains both provided protection against ToMoV in the Florida tomato field experiment. Visual symptom ratings and dot blot analysis of leaves at 40 days after transplanting indicated that the PGPR powder and powder + seed formulations were more effective for ISR against ToMoV than the PGPR seed formulations (Fig. 8.25). The severity of virus symptoms was significantly lower in all PGPR powder and powder + seed treatments than in the nontreated control. Contrast analysis of the percentage of plants testing positive for ToMoV DNA from leaf samples collected 40 days after planting generated similar results (Fig. 8.26). By 80 days after planting, most plants in all treatments were

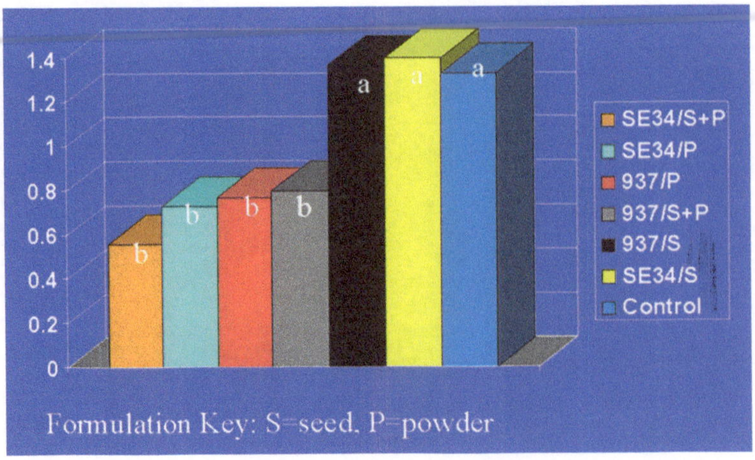

Fig. 8.25 PGPR-ISR against ToMV on field tomato: symptom ratings

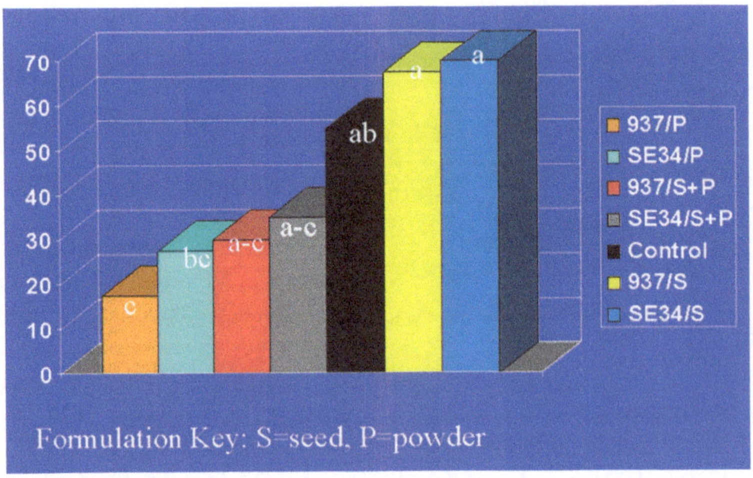

Fig. 8.26 PGPR-ISR against ToMV on field tomato: % infected plants

infected with ToMoV, and differences in the percentages of infected plants among treatments were not significant.

At the first harvest date at Bradenton (80 days after transplanting), tomato yields were higher in PGPR powder or seed + powder treatments than in the control or seed-only treatments; however yield differences were statistically significant only between the 937b powder treatment and the control.

8.2.14 Tomato Spotted Wilt Virus

8.2.14.1 Symptoms

The most characteristic symptoms are sudden browning of leaves followed by cessation of growth of the plants. Later on, the leaves become distorted and areas of necrosis are common. Leaves exhibit bronze colouring. The intensity of bronzing varies from inconspicuous green-brown glaze to a distinct dark-brown to an almost black

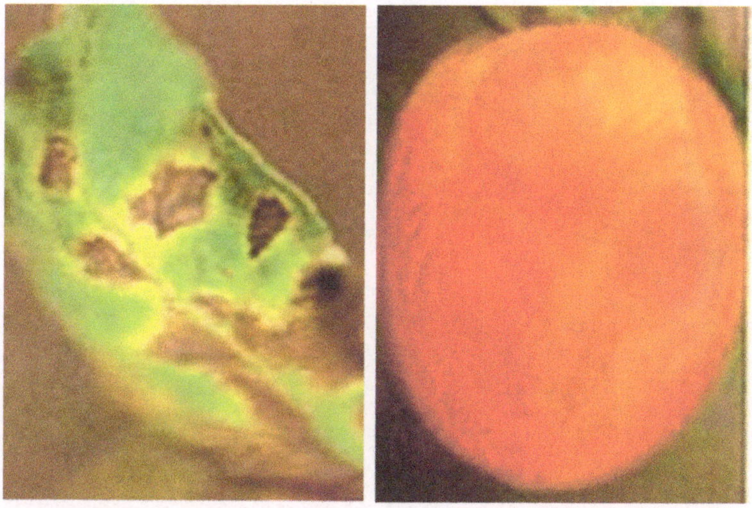

Fig. 8.27 Spotted wilt virus on leaf and fruit

Fig. 8.28 Tomato leaves and fruit infected with tobacco mosaic virus

glazed area. Bronzed areas usually roll inward and affected areas often become necrotic. The fruits on infected plants show numerous spots of about 1.2 cm in diameter with concentric circular markings. On ripe fruits, these markings consist of alternate bands of red and yellow which are diagnostic for the virus (Fig. 8.27).

8.2.14.2 Biomanagement

Kandan et al. (2005) demonstrated that *P. fluorescens* strains applied to seed significantly reduced tomato spotted wilt virus, with a concomitant increase in growth promotion in both the glasshouse and field experiments. In *P. fluorescens-treated* tomato plants, increased activity of polyphenol oxidases, β-1, 3-glucanase and chitinase was observed.

8.2.15 Tobacco Mosaic Virus

8.2.15.1 Symptoms

The foliage shows mosaic (mottled) areas with alternating yellowish- and dark-green areas (Fig. 8.28). Leaves are sometimes fernlike in appearance and sharply pointed. Infections of young plants reduce fruit set and occasionally cause blemishes and distortions of the fruit (Fig. 8.28). The dark-green areas of the mottle

Fig. 8.29 Plant growth promotion by the selected PGPR strains

often appear thicker and somewhat elevated giving the leaves a blister-like appearance. Symptoms on other plant hosts include various degrees of chlorosis, curling, mottling mosaic, dwarfing, distortion and blistering of the leaves. Many times the entire plant is dwarfed and flowers are discoloured.

8.2.15.2 Biomanagement

Kirankumar (2007) demonstrated that seed treatment plus soil application was the best method of inoculation of *Pseudomonas* B-25. It controlled the tobacco mosaic virus disease in tomato plants by 100 % and showed increased plant height by 49.66 %, number of leaves by 88.64 %, biomass by 66.55 % and fruit yield by 102.00 % when compared to pathogen-inoculated control (Fig. 8.29). He also showed that plants inoculated with *Pseudomonas* B-25 exhibited maximum peroxidase, PALase and chitinase activity.

8.2.16 Root-Knot Nematode, *Meloidogyne incognita, M. javanica*

M. incognita was responsible for 30.57–46.92 % loss in fruit yield of tomato, while *M. javanica* caused 77.5 % loss in yield.

8.2.16.1 Symptoms

Above-ground symptoms are stunting, yellowing, wilting, reduced yield and premature death of plants. Belowground symptoms are swollen or knotted roots (root galls) or a stubby root system (Fig. 8.30). Root galls vary in size and shape depending on the nematode population levels, and species of root-knot nematode present in the soil.

8.2.16.2 Biomanagement

The inoculants Equity (contains 47 strains of bacilli in a liquid formulation), BioYield (contains *B. subtilis* strain GBO3 and *B. amyloliquefaciens* strain GB99 in a chitosan carrier) and FZB42 (a strain of *B. amyloliquefaciens*) induced significant reductions in nematode galls per plant on tomato (Cadenaa et al. 2008). AgBlend, containing microbial metabolites, reduced the number of galls. Treatment with each of the inoculants also increased root weight (Table 8.9).

Mixtures of two PGPR strains plus a formulation carrier chitosan, particularly LS213 (containing *B. subtilis* strain GBO3 and *B. amyloliquefaciens* strain GB99 in a chitosan carrier), exhibited a more consistent and significant level of growth promotion on tomato seedlings. The transplant mix delivery system for biological preparations has demonstrated potential to enhance transplant vigour (Fig. 8.31), provide protection against root-knot nematodes (Fig. 8.32) and other foliar diseases in the field and enhance yield. These results indicate a synergy in plant growth promotion and induced resistance by the combination of chitosan and the mixtures of two bacterial strains (Reddy et al. 2001).

Fig. 8.30 Heavy galling of tomato roots with *Meloidogyne incognita*

Table 8.9 Effect of PGPR on root weight and root galling due to *Meloidogyne incognita*

Treatment	Root-knot index	Root weight (g)/plant
Control	4.2	15.08
AgBlend	3.3	27.18
BioYield	2.1	18.18
Equity	3.3	22.64
FZB42	2.7	16.77

The plants treated with *P. fluorescens* gave maximum root and shoot length and significantly ($P=0.05$) suppressed females per root system (40.52 %), J2/g of root (39.80 %), galls per root system (41.50 %) and egg masses per root system (43.23 %) resulting in improved growth over control plants.

The effects of six isolates of PGPR *Pseudomonas putida*, *P. fluorescens*, *Serratia marcescens*, *Bacillus amyloliquefaciens*, *B. subtilis* and *B. cereus* were studied on tomato plant growth and root-knot nematode reproduction after 45 days from nematode infection. The highest shoot dry weight/g (43.00 g) was observed in plants treated with *S. marcescens* followed by *P. putida* (34.33 g), *B. amyloliquefaciens* (31.66 g), *P. fluorescens* (30.0 g), *B. subtilis* (29.0 g), *B. cereus* (27.0 g) and nematode alone (untreated) 20 g/plant. The maximum plant height was observed when the plant was treated with *S. marcescens*, followed by *P. fluorescens*, *P. putida*, *B. amyloliquefaciens* and *P. putida* at 52.66, 50.66, 48 and 48 cm, respectively (Almaghrabi et al. 2013) (Fig. 8.33).

The highest fruit yield was observed with *S. marcescens* (319.6 g/plant). The highest number of fruits/plant was observed when the plants were treated with *S. marcescens* (10.66), followed by *B. amyloliquefaciens* (8.66), *P. putida* (8),

Fig. 8.31 Growth promotion of tomato seedling transplants by PGPR

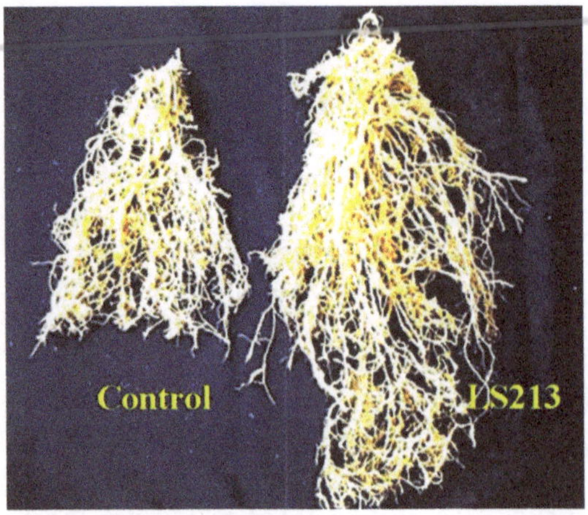

Fig. 8.32 Effect of PGPR on root-knot nematode of tomato seedlings

P. fluorescens (8) and *B. cereus* (7.66) (Fig. 8.33) (Almaghrabi et al. 2013).

On the other hand, the lowest numbers of J_2/10 g of soil (78), galls/root (24.33), egg masses/root (12.66) and egg/egg masses (280.66) were observed in the plants treated with *S. marcescens* (Fig. 8.34) (Almaghrabi et al. 2013).

Application of *P. fluorescens* at 10 g/m^2 in nursery beds gave good control of root-knot nematodes. The level of infestation of root-knot nematode *M. incognita* on tomato was reduced with fewer galls and egg masses in the soil following root dipping with *P. fluorescens* strain Pf1 (Santhi and Sivakumar 1995). Similarly, application of the bacterium *P. chitinolytica* reduced the

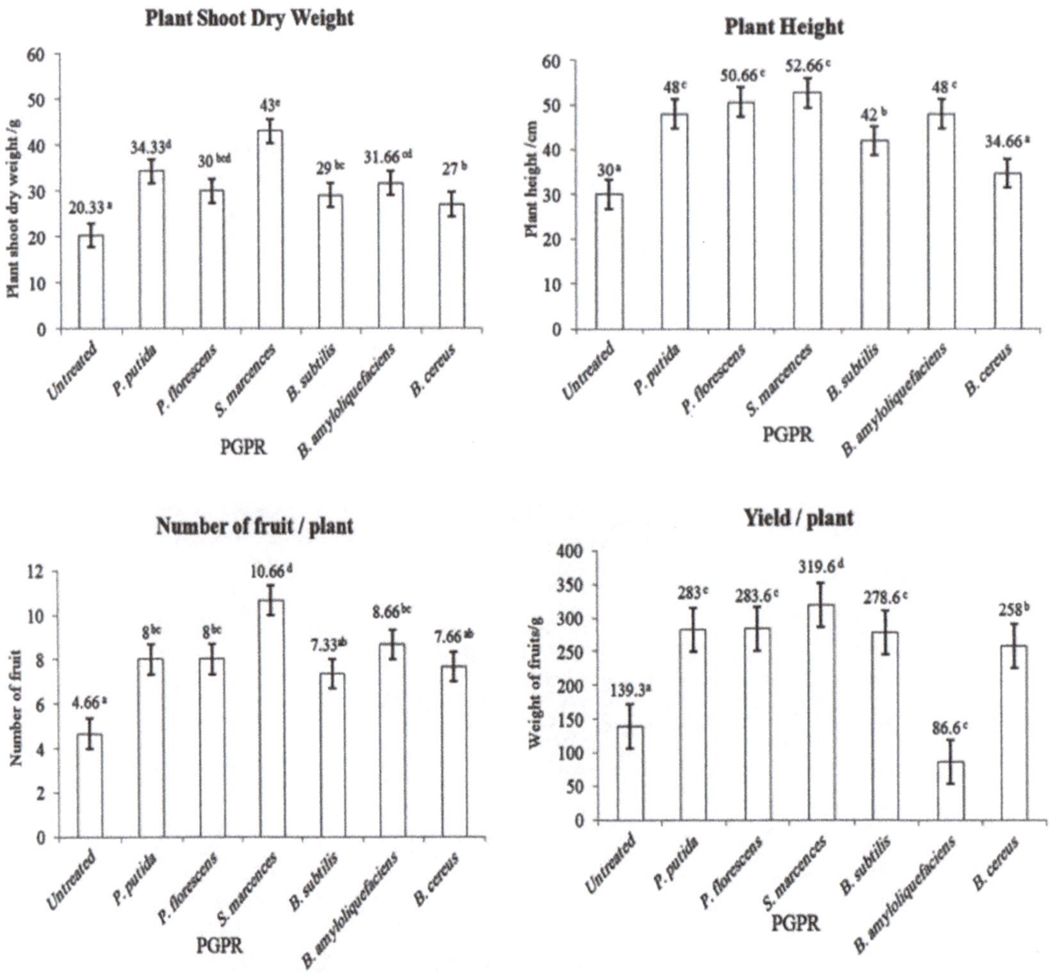

Fig. 8.33 Influence of six isolates of plant growth-promoting rhizobacteria (*PGPR*) on tomato plant performance

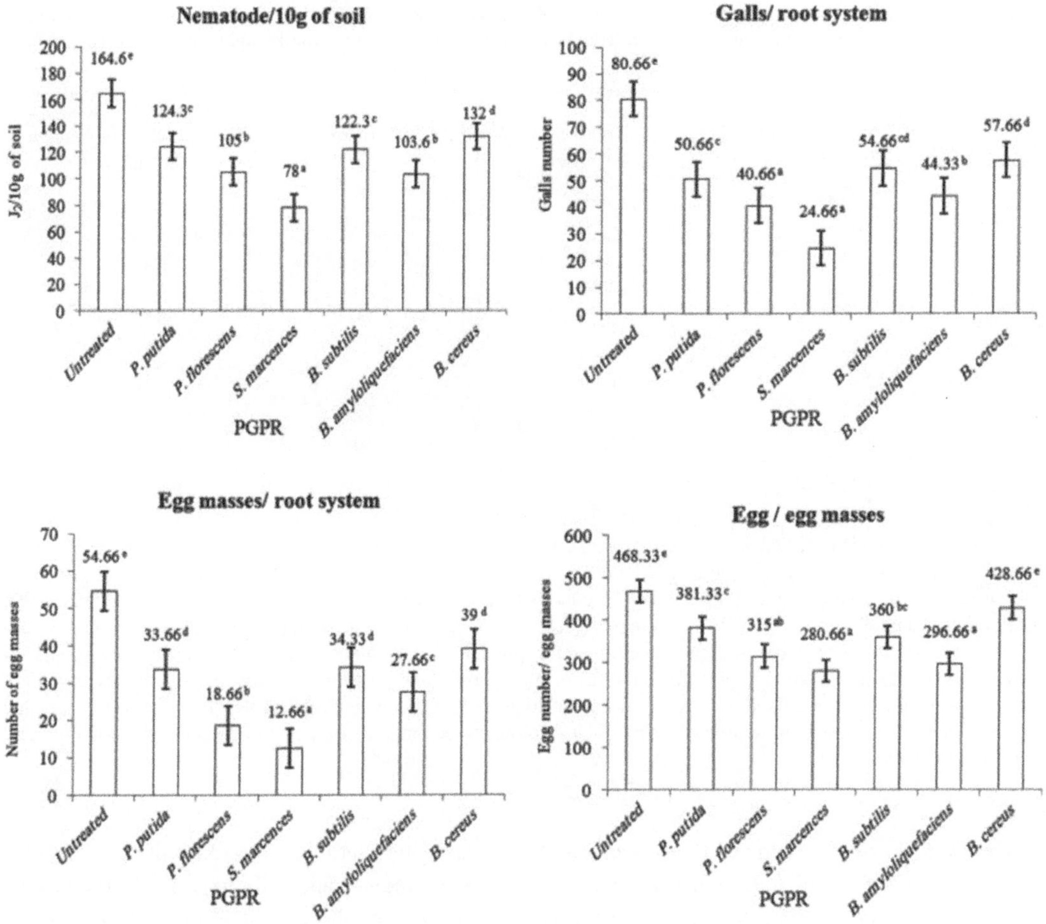

Fig. 8.34 Influence of six isolates of plant growth-promoting rhizobacteria (*PGPR*) on root-knot nematode reproduction

root-knot nematode infection in tomato crop (Spiegel et al. 1991).

8.2.16.3 Integrated Management

(a) *Bioagents and botanicals*: Nursery bed treatment with *P. fluorescens* (with 1×10^9 spores/g) and *Pochonia chlamydosporia* (10 ml/pot containing 4×10^6 spores/ml) each at 20 g/m^2 and field application of 5 tons of FYM enriched with the above bioagents each at 5 kg significantly reduced root-knot nematodes in tomato by 76 % over untreated check. The yield increase was up to 21.7 % with benefit/cost ratio of 4.9.

Application of PGPR (*Azotobacter chroococcum*, *B. subtilis* and *Pseudomonas putida*) alone and in combination with cattle manure resulted in a significant increase in the growth of nematode-inoculated plants. The highest increase (79 %) in the growth of nematode-inoculated plants was observed when *P. putida* was used with cattle manure. *P. putida* resulted in maximum reduction in galling and nematode multiplication, followed by *B. subtilis* and *A. chroococcum*. *P. putida* with cattle manure was useful to achieve greater biocontrol of *M. incognita* on tomato (Siddiqui and Futai 2009).

The efficacy of neem cake along with *P. fluorescens* against root-knot and reniform nematodes (*M. incognita* and *Rotylenchulus reniformis*) was evaluated on tomato. Tomato

Table 8.10 Effect of application of *P. fluorescens* enriched with neem cake or vermicompost on root population densities of *M. incognita* and *R. reniformis*, fruit yield and root colonization by the bioagent in tomato (mean of 5 replicates)

Treatments[a]	No. of *M. incognita*/ 10 g root	No. of *R. reniformis*/ 10 g root	Increase in fruit yield (%)	Root colonization of *P. fluorescens* (cfu/g $\times 10^6$)
T1	24.03	21.23	8	3.0
T2	15.58	14.62	10	6.3
T3	17.80	17.05	15	4.2
T4	12.46	21.52	26	8.3
T5	16.47	16.10	21	5.3
T6	23.35	21.23	6	0.0
T7	29.37	26.10	4	0.0
T8	20.47	18.79	12	3.6
T9	21.36	19.84	10	4.5
T10	44.50	34.80	–	0.0
CD – 5 %	**3.46**	**2.84**	**–**	**0.76**

[a]T1 – seedlings of tomato treated with *P. fluorescens* (cfu 2×10^8/ml); T2 – untreated seedlings sown in the plots mixed with neem cake enriched with *P. fluorescens* 20 g/m²; T3 – untreated seedlings sown in the plots mixed with vermicompost enriched with *P. fluorescens* 100 g/m²; T4 – seedlings treated with *P. fluorescens* sown in plots mixed with neem cake enriched with *P. fluorescens*; T5 – seedlings treated with *P. fluorescens* sown in plots mixed with vermicompost enriched with *P. fluorescens*; T6 – untreated seedlings transplanted in plots mixed with neem cake alone 20 g/m²; T7 – untreated seedlings transplanted in plots mixed with vermicompost alone 20 g/m²; T8 – seedlings treated with *P. fluorescens* transplanted in plots mixed with neem cake alone 20 g/m²; T9 – seedlings treated with *P. fluorescens* transplanted in plots mixed with vermicompost alone 20 g/m²; T10 – untreated seedlings transplanted in the plots without any treatment, served as control

seedling treatment with *P. fluorescens* and application of neem cake enriched with *P. fluorescens* was found to be significantly effective in reducing the population of nematodes. This treatment reduced nematode populations of *R. reniformis* by 64 % and *M. incognita* by 72 %. There was significant increase in the yield of tomato by 26 % (Table 8.10). In general the growth of tomato plants treated with *P. fluorescens* was better (Rao 2011).

Application of *P. fluorescens* with neem cake helped in the increased root colonization of *P. fluorescens* in tomato. Root colonization of *P. fluorescens* was more when the PGPR waere applied along with neem cake in comparison to its application along with vermicompost, in tomato (Table 8.10). Both neem cake and vermicompost when enriched with *P. fluorescens* are able to support the growth and multiplication of this PGPR. However, when it is applied along with neem cake, it had an increased bio-efficacy. Neem cake application is changing the root surface phenomenon and making it more favourable for higher colonization of these PGPR. It was observed that the higher the root colonization of *P. fluorescens*, the higher was its bio-efficacy (Rao 2011).

(b) *Two bioagents*: Gautam et al. (1995) reported that combined application of *B. subtilis* and *Paecilomyces lilacinus* suppressed root-knot nematode population beyond application of bioagents individually. Biocontrol agents suppressed the number of root galls, females, eggs and second-stage juveniles.

Two bacteria *Bacillus licheniformis* and *Pseudomonas mendocina* and two fungi *Acrophialophora fusispora* and *Aspergillus flavus* were evaluated for their efficacy as biocontrol agents individually and in various combinations against *M. incognita* infecting tomato. Individually, *B. licheniformis* gave best control although when all four agents were used, nematode multiplication was reduced by 82.06 %.

Consortial formulation of biocontrol agents, viz. *P. fluorescens* Pf 128 and *B. subtilis* Bbv 57, recorded the highest defence enzymatic activity (peroxidase, polyphenol oxidase and phenylalanine ammonia lyase) and lowest nematode population in tomato roots compared to other strains either alone or in combination (Sankarimeena et al. 2012).

The greatest increase in growth of nematode-inoculated tomato plants and reduction in nematode galling due to *M. incognita* was observed when *Paenibacillus polymyxa* was used with *P. lilacinus* or *P. chlamydosporia.*

(c) *Bioagents and AMF*: PGPR (*Bacillus* sp. and *B. polymyxa*) increased colonization of roots by AM fungi (*Glomus versiforme* and *G. mosseae*), and AM fungi increased the numbers of PGPR in the rhizosphere. Dual inoculations of AM fungi plus PGPR provided greater control of *M. incognita* and greater promotion of plant growth than single inoculations, and the best combination was *G. mosseae* plus *Bacillus* sp. The results indicate that specific AM fungi and PGPR can stimulate each other and can interact to suppress root-knot disease development (Liu et al. 2012).

(d) *Bioagents, AMF and botanicals*: The efficacy of *P. fluorescens*, *Glomus mosseae* and organic amendments against root-knot nematodes (*M. incognita*) was evaluated to produce nematode-free seedlings of tomato.

Seed treatment with talc-based formulation of *P. fluorescens* was able to increase the growth of the seedlings of tomato. Seed treatment with talc-based formulation of *P. fluorescens* and talc- or neem-based formulation of *P. fluorescens* along with coco-peat increased control of *M. incognita.* This resulted in the production of increased number of nematode-free seedlings of tomato. There was significant increase in the production of nematode-free tomato seedlings in *G. mosseae* with coco-peat treatment + seed treatment with talc-based formulation of *P. fluorescens* + neem-based formulation of *P. fluorescens* with coco-peat treatment. There appears to be synergistic interaction among *P. fluorescens*, *G. mosseae* and neem cake, and they are complementing each other resulting in additive effect on the shoot and root growth.. Neem cake helped in the increased root colonization of *P. fluorescens* and *G. mosseae* in tomato. When these seedlings were transplanted in the main field, there was significant reduction in the infestation of *M. incognita*, *M. javanica* (root-knot nematodes) and *Rotylenchulus reniformis* (reniform nematode) on tomato (Table 8.11) (Gavaskar et al. 2011).

8.2.17 Reniform Nematode, *Rotylenchulus reniformis*

R. reniformis was responsible for 42.25–49.02 % loss in fruit yield of tomato.

8.2.17.1 Symptoms

General symptoms include reduced root system, leaf chlorosis, overall stunting of host plants, reduced yields and plant longevity. Female nematodes and their eggs are often visible when plant roots are viewed under a dissecting microscope (Fig. 8.35).

8.2.17.2 Biomanagement

Multiplication rate of *R. reniformis* was reduced significantly (44.5 %) when soil drenching with *B. subtilis* (isolate Bs_1) was done a week before nematode inoculation. But, it was not so in the case of treatments that received simultaneous inoculations of the bacterium and the nematode. These results indicated that *B. subtilis* could induce systemic resistance in tomato against *R. reniformis* (Niknam and Dhawan 2001).

8.2.17.3 Integrated Management

Bioagents and botanicals: Nursery bed treatment with *P. fluorescens* (with 1×10^9 spores/g) and *P. chlamydosporia* (with 1×10^6 spores/g) each at 20 g/m^2 and field application of 5 tons of FYM enriched with the above bioagents each at 5 kg significantly reduced reniform nematodes in tomato by 72 % over control. The yield increase was up to 21.7 % with benefit/cost ratio of 4.9.

Table 8.11 Effect of application of *P. fluorescens*, *G. mossease* and coco peat on growth of tomato seedlings, infestation of *M. incognita* and root colonization with bioagents (mean of 5 replicates)

Treatments[a]	Seedling height (cm)	% nematode-free seedlings	Root colonization of *P. fluorescens* (cfu/g × 10^6)	Root colonization of *G. mossease* (%)
T1	19.04	87	1.56	–
T2	20.52	90	3.42	–
T3	21.60	91	4.82	–
T4	21.28	85	0.00	27
T5	23.85	92	2.25	–
T6	21.35	90	4.94	–
T7	22.86	92	6.39	–
T8	24.27	94	5.68	34
T9	25.16	97	8.26	42
T10	22.37	93	4.66	37
T11	24.46	94	23.76	40
T12	17.26	86	15.20	–
CD – 5 %	**2.66**	**–**	**1.72**	**–**

[a]T1 – seed treatment with talc-based formulation of *P. fluorescens*, T2 – talc-based formulation of *P. fluorescens* with coco-peat treatment, T3 – neem-based formulation of *P. fluorescens* with coco-peat treatment, T4 – *G. mossease* with coco-peat treatment, T5 – seed treatment with talc-based formulation of *P. fluorescens* + *G. mossease* with coco-peat treatment, T6 – seed treatment with talc-based formulation of *P. fluorescens* + talc-based formulation of *P. fluorescens* with coco-peat treatment, T7 – seed treatment with talc-based formulation of *P. fluorescens* + neem-based formulation of *P. fluorescens* with coco-peat treatment, T8 – seed treatment with talc-based formulation of *P. fluorescens* + talc-based formulation of *P. fluorescens* with coco-peat treatment + *G. mossease* with coco-peat treatment, T9 – seed treatment with talc-based formulation of *P. fluorescens* + neem-based formulation of *P. fluorescens* with coco-peat treatment + *G. mossease* with coco-peat treatment, T10 – talc-based formulation of *P. fluorescens* with coco-peat treatment + *G. mossease* with coco-peat treatment, T11 – neem-based formulation of *P. fluorescens* with coco-peat treatment + *G. mossease* with coco-peat treatment, T12 – control without seed and coco-peat treatment

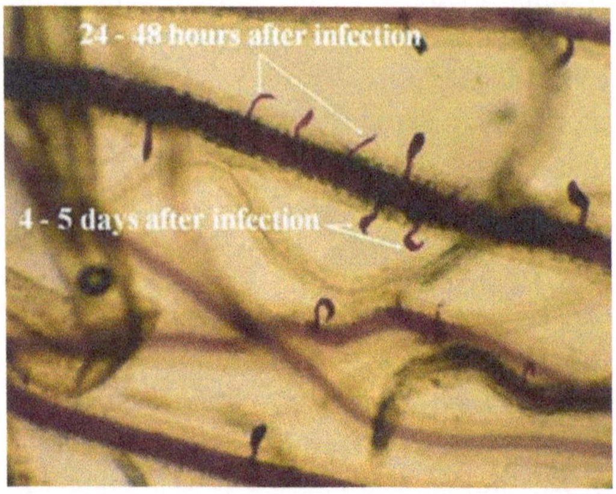

Fig. 8.35 Tomato root infected with *Rotylenchulus reniformis*

8.2.18 Root-Knot Nematode, *Meloidogyne* spp., and Wilt, *Fusarium oxysporum* f. sp. *lycopersici*, Disease Complex

8.2.18.1 Symptoms

The interaction among soilborne pathogens is well documented in tomatoes on old production land when *Fusarium* wilt disease and root-knot nematode (*Meloidogyne* spp.) are both present (Fig. 8.36). Young plants are very susceptible to the combination of pests, collapsing prior to harvest. The nematode, by impairing water and nutrient availability, disrupts root function and plant growth processes. These effects combined

Fig. 8.36 The simultaneous occurrence of both root-knot nematode (*Meloidogyne* spp.) and *Fusarium* wilt disease (*Fusarium oxysporum*) causing enhanced disease development and tomato yield loss

with vascular blocking due to the wilt fungi can be particularly severe and, if widespread, result in total crop failure. The root-knot nematode, by causing the development of root galls, provides a nutrient-rich food source which the fungi colonize rapidly. Root-knot nematodes can thus significantly enhance disease development and yield loss, elevating primary or secondary pathogens to major pest status even though population levels or pathogenic potential of the fungi was initially very low and yield losses would have been minimal in the absence of the nematode.

8.2.18.2 Biomanagement

Paenibacillus polymyxa GBR-462 and GBR-508 and *P. lentimorbus* GBR-158 showed the strongest antifungal and nematicidal activities. These three strains used in pot experiment reduced the symptom development of the disease complex (wilting and plant death) and increased plant growth. The control effects were estimated to be 90–98 % and also reduced root gall formation by 64–88 % compared to the untreated control. The protective properties of selected *Paenibacillus* strains make them as potential tool to reduce deleterious impact of disease complex plants (Table 8.12) (Son et al. 2009).

Table 8.12 Control effect of *Paenibacillus* strains on the disease complex in tomato caused by *Fusarium oxysporum* f. sp. *lycopersici* (*FO*) and *Meloidogyne incognita* (*MI*)

Treatments	Wilt severity index[a]	CE (%)	Gall severity index[b]	CE (%)
FO + MI	4.1a	–	5.0a	–
FO + MI + GBR-508	0.1b	97.6	0.6c	88.0
FO + MI + GBR-462	0.3b	92.7	1.0b	78.0
FO + MI + GBR-158	0.4b	90.2	1.8b	64.0
Untreated	0.0	–	0.0	–

In a column, means followed by a common letter are not significantly different by Duncan's multiple range test at $P=0.05$

CE control effect

[a]Wilt severity was graded using a 0–5 scale, where 0 = no visible symptoms; 1 = epinasty and chlorosis/wilting of primary leaves; 2 = chlorosis/wilting of second and third leaves, primary leaves maybe lost; 3 = wilting above third leaves, second and third leaves maybe lost; 4 = wilting of whole plant; and 5 = plant died

[b]Severity of root galling was assessed on a 0–5 rating scale, in which 0 = 0–10 % of galled roots; 1 = 11–20 %; 2 = 21–50 %; 3 = 51–80 %;4 = 81–90 %; and 5 = 91–100 %

8.2.19 Root-Knot Nematode, *Meloidogyne* spp., and Root Rot, *Rhizoctonia solani*, Disease Complex

Damping off of tomato was more severe in soil infested with both *M. javanica* and *R. solani* than with the fungus alone. *M. javanica* increased the extent of damage by pre- and post-emergence phases of damping off caused initially by *R. solani* in tomato.

8.2.19.1 Symptoms

Plants in untreated field soil or in sterilized soil inoculated with both organisms developed a root rot in about 42 days. If the nematode preceded the fungus by 3 weeks, the root rot was more severe and appeared within 14–21 days. The fungus penetrated either directly or through ruptures in the root created by the mature female nematode. *R. solani* colonized nematode giant cells and root xylem cells. Vascular discolouration occurred in both roots and stem; however, no fungus was isolated from stems.

8.2.19.2 Integrated Management

(a) *Two bioagents*: Combined application of *Pseudomonas aeruginosa* and *Paecilomyces lilacinus* significantly suppressed soilborne root-infecting pathogens such as *R. solani* and *M. javanica* (Siddiqui et al. 2000).

P. aeruginosa–B. subtilis treatment was the most effective in the suppression of root rot disease complex with enhancement of plant growth (Siddiqui and Ehteshamul-Haque 2001).

8.2.20 Root-Knot Nematode, *Meloidogyne incognita*, and Bacterial Wilt, *Ralstonia solanacearum*, Disease Complex

8.2.20.1 Symptoms

Significant reduction in the root-knot index and larval development of *M. incognita* was reported in soil where *R. solanacearum* was present. *R. solanacearum* and *M. incognita* alone as well as in different combinations reduced plant growth and yield significantly with the nematode followed by the bacterium combination showing the maximum reduction in growth.

Wilt disease development occurred earlier and with a higher mortality rate in both wilt-resistant and susceptible tomato cultivars grown in *R. solanacearum*- and *M. incognita*-infested soil.

8.2.20.2 Integrated Management

(a) *Bioagents and botanicals*: Treatment of nursery bed with *P. fluorescens* (10^9 cfu/g) and *Trichoderma harzianum* (10^6 cfu/g) each at the rate of 20 g/m^2 and subsequent application of 5 MT of farmyard manure enriched with 5 kg each of *P. fluorescens* (10^9 cfu/g) and *P. lilacinus* (10^6 cfu/g) per hectare significantly reduced *M. incognita* in tomato roots by 70 %, reduced the incidence of bacterial wilt and increased the yield by 24.2 %. Benefit/cost ratio (calculated for the additional cost of biopesticides and additional returns accrued by the application of biopesticide) was 4.4 (Rao et al. 2009).

8.2.21 Whitefly, *Bemisia tabaci*

Whitefly is a well known vector which transmits tomato leaf curl virus and causing severe loss in yield. Loss in yield ranging from 38 % to 93 % during different months of the year has been reported. The greenhouse whitefly is widely distributed in the Western Hemisphere including Bermuda, Hawaii, Europe and Australia where it can become an important pest of tomatoes in greenhouses.

8.2.21.1 Damage

Both nymphs and adults feed on the lower surface of the leaves, causing deformation of young leaves (Fig. 8.37). Severe cases lead to chlorotic leaves accompanied by wilting. Whiteflies also excrete honeydew, causing sooty mould.

8.2.21.2 Biomanagement

Tomato plants treated with *Bacillus amyloliquefaciens* strain 937a, *B. subtilis* strain 937b and

Fig. 8.37 Whiteflies on tomato leaf

B. pumilus strain SE 35, as seed treatment and as powder formulation applied to planting medium, had reduced the crawlers, nymphs and pupae of whitefly. Among the three, *B. amyloliquefaciens* strain 937a was the most effective as powder formulation (Murphy et al. 2000).

Lower whitefly populations were noticed on *B. subtilis*-treated tomato plants grown under greenhouse conditions (Hanafi et al. 2007). The mechanism of this resistance was not identified, but a later study showed that treatment of tomato plants with PGPR reduced adult whitefly emergence and even restored whitefly population suppression normally achieved by jasmonic acid (JA)-pathway plant responses (Valenzuela-Soto et al. 2010).

8.3 Brinjal, *Solanum melongena*

8.3.1 Damping Off, *Pythium aphanidermatum*

8.3.1.1 Symptoms

The pathogen causes pre- and post-emergence damping off in the nursery beds. In Pre-emergence phase of disease, the young seedlings are killed before they emerge through the soil surface. In fact, the seeds may rot or the seedlings may be killed before the hypocotyl has broken the seed coat. The radical and the plumule, when they come out of the seed, undergo complete rotting. Since this happens under the soil surface, the disease is often not recognized at all by the farmer, who attributes the failure of emergence to poor quality of the seed.

The post-emergence mortality of seedlings is generally very conspicuous. This phase of the disease is characterized by the toppling over of infected seedlings any time after they emerge from the soil (Fig. 8.38) until the stem has hardened sufficiently to resist invasion. Infection usually occurs through the roots or at the ground level. The infected tissue appears soft, stained and water soaked. As the disease advances, the stem becomes constricted at the base and the plant collapsed. Seedlings that are apparently healthy 1 day may have collapsed by the following morning. Generally, the cotyledons and leaves wilt slightly before the seedlings are prostrated, although sometimes they remain green and turgid until collapse occurs. In fields and nurseries, the disease usually radiates from initial infection points, causing large spots or areas in which nearly all the seedlings are killed.

8.3.1.2 Biomanagement

Ramesh and Korikanthimath (2003) evaluated two isolates of *P. fluorescens* against *Pythium aphanidermatum* (damping off agent) in brinjal. Application of these biocontrol agents as seed and soil treatments reduced both pre and post-emergence damping off, ranging from 18 to 42 % and 23 to 55 %, respectively. The growth parameters and vigour index recorded in nursery were higher in *P. fluorescens* treatments. The least population of the pathogen was observed after 20 days of treatment.

8.3.1.3 Integrated Management

(a) *Bioagents and botanicals*: Seed treatment with *P. fluorescens* at 10 g/kg seed followed by soil application of *P. fluorescens* at 2.5 kg multiplied in 50 kg FYM per hectare and seed treatment with *Azotobacter chroococcum* at 16 g/kg seed followed by soil application of *A. chroococcum* at 500 g multiplied in 50 kg FYM per hectare gave effective control of damping off in brinjal (Rahman et al. 2002).

Fig. 8.38 Damping-off symptoms on brinjal seedlings

Table 8.13 Comparative efficacy of soil solarization, bioagents and organic amendment against damping off of brinjal

Treatment	% disease index
Soil solarization with transparent polythene sheets for 30 days	1.50 (6.93)
Seed treatment with *Pseudomonas fluorescens* at 10 g/kg + soil application of 50 kg FYM enriched with 2.5 kg of *P. fluorescens*	7.20 (15.37)
Seed treatment with *Azotobacter croococcum* at 16 g/kg + soil application of 50 kg FYM enriched with 500 g of *A. croococcum*	10.88 (19.17)
Control (check)	42.82 (29.74)
CD at 5 %	**(3.13)**

Figures in parentheses indicate the arc sin transformed values

(b) *Bioagents and physical methods*: Combination of the seed/root application of *P. fluorescens* with soil solarization was very effective in management of damping off of brinjal in nursery at the farmers' field.

(c) *Bioagents, botanicals and physical methods*: Soil solarization gave least damping off followed by seed treatment with *Pseudomonas fluorescens/Azotobacter croococcum* + soil application of FYM enriched with *T. viride* (Table 8.13) (Rahman et al. 2002).

8.3.2 Collar Rot, *Sclerotinia sclorotiorum*, *Sclerotium rolfsii*

8.3.2.1 Symptoms

Infection takes place on leaves, twigs, flowers and fruits. Water-soaked lesions develop and the infected tissue is macerated by the pathogen. White mass of mycelium grow on the surface and later embedded sclerotia are formed (Fig. 8.39). These sclerotia become black after drying. The pathogen is soilborne.

8.3.2.2 Biomanagement

Treatment of brinjal seeds with *B. subtilis* was found to be effective for managing collar rot (Chaliha 1995). Das and Phookan (2002) reported that treatment of brinjal seed with *B. subtilis* and *P. fluorescens* could significantly manage the collar rot infection with yield enhancement.

8.3.3 Phomopsis Blight, *Phomopsis vexans*

8.3.3.1 Symptoms

In the seedbeds, the disease appears as damping off. After transplanting, the leaves show clearly defined circular, grey to brown spots with light-coloured centre. The affected leaves turn yellow and die. The lesions on stem are dark brown becoming grey in the centre. Mostly the stem

Fig. 8.39 Collar rot on brinjal

Fig. 8.40 *Phomopsis* fruit rot on brinjal

base is attacked and is characterized by constriction of the base or a grey dry rot. Pale, sunken spots develop on the fruit and may progress to cover the entire fruit surface (Fig. 8.40). The internal portion of the fruit rots. If the fungus enters through the calyx, the whole fruit is mummified due to dry rot.

8.3.3.2 Biomanagement

The soil + root dip treatments with isolate *Pseudomonas* C7 (French isolated) gave maximum inhibition of *Phomopsis* blight disease (Singh et al. 2003a, b).

8.3.4 Bacterial Wilt, *Ralstonia solanacearum*

8.3.4.1 Symptoms

Characteristic symptoms of bacterial wilt are rapid and complete wilting of normal grown-up plants (Fig. 8.41). Pathogen is mostly confined to vascular

Fig. 8.41 Bacterial wilt affected plants of brinjal

region. In advanced cases, it may invade the cortex and pith and cause yellowish-brown discolouration of tissues. Infected plant parts when cut and immersed in clear water, a white streak of bacterial ooze coming out from cut ends is visible.

8.3.4.2 Biomanagement

The bacterial wilt of brinjal was successfully controlled by incorporation of *P. fluorescens*.

8.3.5 Root-Knot Nematodes, *Meloidogyne* spp.

M. incognita was responsible for 27.30–48.55 % loss in fruit yield of brinjal.

8.3.5.1 Symptoms

Plants infected by root-knot nematodes are generally less vigorous than healthy plants. In foliage, symptoms of nutrient deficiency and diurnal wilting are visible due to reduced function of root system. Affected plants are normally stunted and eventually wilt and die. The most characteristic symptom is formation of root galls (knots) and these can be seen with the naked eye (Fig. 8.42). The infested roots eventually rot and affected plants die.

8.3.5.2 Biomanagement

Nursery bed treatment with *Bacillus macerans* at 25 g/m^2 + soil drench (2 % solution) 10 days after sowing gave maximum reduction in nematode population and maximum yield of 3.825 MT/ha (Sheela and Nisha 2004).

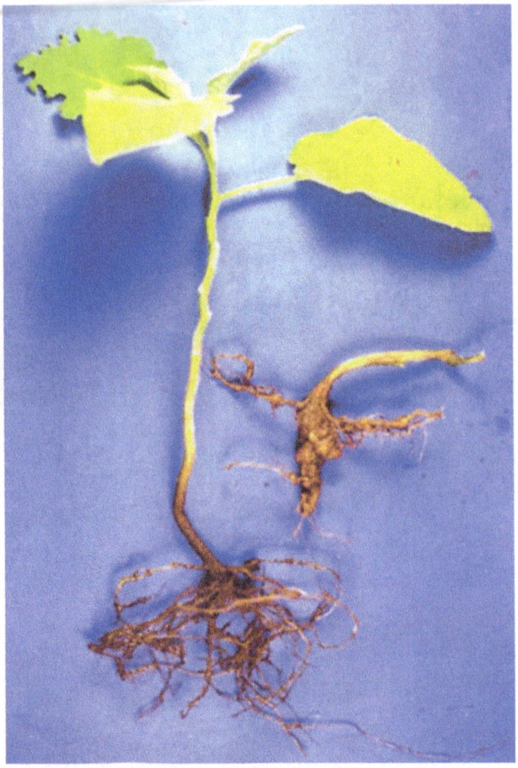

Fig. 8.42 Root-knot nematode on brinjal. *Left* – Healthy. *Right* – Infected

Chahal and Chahal (1986) reported that *Azotobacter chroococcum* caused significant inhibition of hatching of egg masses of *M. incognita* and did not allow the larvae to penetrate into the roots of brinjal to form galls. *A. chroococcum* was also responsible for reduction in soil and root population of root-knot nematodes and root galling. Among different strains of *A. chroococcum* tested, strain 23 proved better than other strains (Chahal and Chahal 1988, 1999).

8.3.5.3 Integrated Management

(a) *Bioagents and botanicals*: The treatment of *P. fluorescens* at 10 kg/ha + neem cake at 1 t/ha was found effective in reducing *M. incognita* population (49.28 %), number of root galls/egg masses (68.45 %) and gall index (58.5 %) and increasing the yield (30.66 %) of brinjal with benefit/cost ratio of 4.64 (Table 8.14) (Zore et al. 2013)

(b) *Two bioagents*: The nursery seedling stand index was good with seed treatment with *P. fluorescens* (50 g/kg) + soil application (10 g/m^2 seed bed) followed by soil application of neem-based of *T. harzianum* + *P. fluorescens* and neem-based formulation of *P. fluorescens*, where the crop stand index was 4.6 and 4.0, respectively.

8.3.6 Root-Knot Nematode, *Meloidogyne incognita*, and Bacterial Wilt, *Ralstonia solanacearum*, Disease Complex

Eggplant is prone to many soilborne diseases, among which the bacterial wilt (*R. solanacearum*) in combination with root-knot nematode (*M. incognita*) takes heavy toll every year all over the world (Fig. 8.43).

Table 8.14 Effect of *Pseudomonas fluorescens* on root-knot nematode and fruit yield of brinjal

Treatment	Gall index	Yield (t/ha)	Benefit/cost ratio
Soil application of *P. fluorescens* at 20 kg/ha at transplanting	4.06	18.444	4.67
Soil application of *P. fluorescens* at 20 kg/ha + neem cake at 1 t/ha	3.33	20.741	4.64
Untreated control	8.85	15.873	–
CD (P=0.05)	**0.26**	**13.12**	–

8.3.6.1 Biomanagement

P. fluorescens at 5 g/kg soil significantly reduced the final soil nematode population and wilt disease incidence in brinjal. The highest reduction in bacterial population in soil was observed in the treatment *P. fluorescens* (Barua and Bora 2008) (Table 8.15). Further, combined application of wheat bran formulations of *P. fluorescens* significantly reduced root galling, egg mass production and final population of *M. incognita* and *R. solanacearum* in soil and increased population of the antagonists in soil (Barua and Bora 2009).

Root dip of brinjal seedlings in *P. fluorescens* (5 g/l of water for 30 min) gave maximum suppression of root-knot nematode population in soil followed by *P. fluorescens* (soil application at 10 g/plant + root dip at 5 g/l of water for 30 min), *P. fluorescens* as soil application (10 g/plant) (Dhawan et al. 2008).

8.3.6.2 Integrated Management

(a) *Two bioagents*: The combination formulation of *P. fluorescens* + *Pochonia chlamydosporia* performed well (yield was 1.782 kg/3 m^2) followed by combination of *Trichoderma harzianum* + *P. fluorescens* (1.680 kg). Root gall index (RGI) and wilt

Fig. 8.43 Root-knot nematode and bacterial wilt disease complex in brinjal

Table 8.15 Effect of different treatments on root galling, egg mass production and population of nematodes and bacteria in brinjal

Treatment/dose	No. of galls/ plant	Egg masses/ plant	Nema popn. in soil	Nema popn. in root	Bact. popn. (10^8 cfu/g soil)
P. fluorescens – 5 g/kg soil	64	40	309	54	13
P. fluorescens – 10 g/kg soil	56	34	290	49	7
P. fluorescens – 15 g/kg soil	44	28	235	40	4
Carbofuran – 1.5 kg a.i./ha	26	16	176	24	21
Streptocycline – 500 ppm	83	38	452	70	11
Neem cake – 15 g/kg soil	39	24	226	96.8	11
Control	92	53	484	416	42
CD (P=0.05)	**0.093**	**0.129**	**80.00**	**11.1**	**2.75**

disease incidence due to *R. solanacearum* (WDI) was minimum in combination treatment of *P. fluorescens* + *P. chlamydosporia* followed by *T. harzianum* + *P. fluorescens* where the RGI was 1.81 and 1.93, respectively, and that of WDI was 25.13 % and 27.28 %, respectively (Naik 2004).

8.4 Chilli and Bell Pepper, *Capsicum annuum*

8.4.1 Damping Off, *Pythium aphanidermatum*

Fig. 8.44 Damping off on chillies

8.4.1.1 Symptoms

Symptoms include leaf chlorosis, vascular discolouration and wilting of chilli pepper plants. High temperature and high moisture are conducive to symptom development. The shrinking of the cortical tissue of the hypocotyl and toppling over of the infected seedlings takes place. It affects germination and stand of seedlings in nursery beds due to pre-emergence damping off. The pathogen is responsible for the death of seedlings in nursery beds (Fig. 8.44).

8.4.1.2 Biomanagement

P. fluorescens showed highest induction of systemic resistance on chilli plants with increased seed germination and vigour index when challenge inoculated with *F. solani* (Devika Rani et al. 2009).

Manoranjitham et al. (1999) reported that seed treatment of chilli with *P. fluorescens* (5 g/kg) increased germination percentage, shoot length, root length and biomatter production and reduced the incidence of damping-off disease. Vigour index of the seedlings was also increased compared to the individual applications of the antagonist and control.

Application of PGPR formulations of *B. subtilis* (strain GBO3) and *B. amyloliquefaciens* (strain IN937a) to transplant plugs at seeding has resulted in increased plant vigour (Fig. 8.45), reduction in root diseases and increased yield in capsicum.

8.4.1.3 Integrated Management

(a) *Bioagents and botanicals*: Seed treatment with *P. fluorescens* at 10 g/kg seed followed by soil application of *P. fluorescens* at 2.5 kg

Fig. 8.45 Increased growth and plant vigour seen with the application of PGPR. Pepper plants on the *right* were treated with *Bacillus subtilis* strain GBO3 and *B. amyloliquefaciens* strain IN937a compared to the untreated check on the *left*

multiplied in 50 kg FYM per hectare, and seed treatment with *A. chroococcum* at 16 g/kg seed followed by soil application of *A. chroococcum* at 500 g multiplied in 50 kg FYM per hectare gave effective control of damping off in chilli (Rahman et al. 2002).

(b) *Bioagents, botanicals and physical methods*: Soil solarization gave least damping off followed by seed treatment with *Pseudomonas fluorescens/Azotobacter croococcum* + soil application of FYM enriched with *T. viride* (Table 8.16) (Rahman et al. 2002).

(c) *Bioagents and physical methods*: Combination of the seed/root application of *P. fluorescens* with soil solarization was very effective in management of damping off of capsicum in nursery at the farmers' field.

(d) *Two bioagents*: Combined seedling root dip treatment and soil application of nonpathogenic *Fusarium* (Fo 52) and *P. fluorescens* gave maximum inhibition of wilt (Singh et al. 2002).

Table 8.16 Comparative efficacy of soil solarization, bioagents and organic amendment against damping off of chilli

Treatment	% disease index
Soil solarization with transparent polythene sheets for 30 days	1.28 (6.45)
Seed treatment with *Pseudomonas fluorescens* at 10 g/kg + soil application of 50 kg FYM enriched with 2.5 kg of *P. fluorescens*	6.75 (14.82)
Seed treatment with *Azotobacter croococcum* at 16 g/kg + soil application of 50 kg FYM enriched with 500 g of *A. croococcum*	6.68 (14.80)
Seed treatment with captan at 2.5 g/kg + soil drenching with captan at 0.25 % at 6 L/m^2	4.75 (12.29)
Control (check)	20.13 (26.22)
CD at 5 %	**(3.09)**

Figures in parentheses indicate the arc sin transformed values

8.4.2 *Phytophthora* Root Rot, Fruit Rot and Leaf Blight, *Phytophthora capsici*

8.4.2.1 Symptoms

This is a disease of rainy season, characterized by small, water-soaked spots appearing on fruits leading to complete rot of fruits. On leaves, water-soaked bleached spots appear, resulting in blighting. It causes damage to plants and affects fruit yield. Infection of the root and lower portion of the stem leads to plant wilting (Fig. 8.46), which is the most conspicuous symptom of the disease. In furrow-irrigated fields, there is a row-delimited pattern of wilted plants (Fig. 8.46). Over time, wilting is displayed over the entire field. *P. capsici* produces specialized swimming cells, known as zoospores, which enable spread of the pathogen from plant to plant.

Fig. 8.46 Wilting of chilli plant and a field severely affected by *Phytophthora* root rot

8.4.2.2 Biomanagement

Pure culture of *P. fluorescens* was effective in reducing chilli dieback (*P. capsici*). Formulations of *P. fluorescens* were also effective in reducing *P. capsici* infection and in increasing seed germination, vigour index and field emergence (Srinivas et al. 2006).

Bacillus megaterium KL39, a biocontrol agent of red-pepper *Phytophthora* blight disease, produces an antifungal antibiotic active against a broad range of plant pathogenic fungi (Jung and Kim 2003).

8.4.2.3 Integrated Management

(a) *Bioagents and chemicals*: Seed treatment with carbendazim (0.2 %) + seedling dip with *P. fluorescens* (10 g/l) +2 sprays of *P. fluorescens* at 45 and 60 days after transplanting +2 sprays of hexaconazole at 75 and 90 days after transplanting recorded the least disease incidence of fruit rot (16.0 PDI). This treatment also gave the highest yield (0.85 t/ha) and highest net returns (Rs 25,940/ha) (Mesta et al. 2009).

Fig. 8.47 Bell pepper *Fusarium* wilt

8.4.3 Wilt, *Fusarium* spp., *Rhizoctonia solani* and *Phytophthora capsici*

8.4.3.1 Symptoms

Symptoms include leaf chlorosis, vascular discolouration and wilting of chilli pepper plants (Fig. 8.47). High temperature and high moisture are conducive to symptom development.

8.4.3.2 Biomanagement

Use of *Bacillus* at transplanting can reduce disease incidence in contrast to traditional treatments (fungicide application) such as folpat, captan and mancozeb as much as 64 % and 72 % compared to untreated control with the most efficient strain. Furthermore, the suppressive effect

Table 8.17 Effect of four strains of *Bacillus* and commercial products on incidence and severity of wilt and root rot, plant height and fruit yield on pepper

Treatments	Disease incidence (%)	Disease severity (%)	Plant height (cm)	Yield (kg)	% increase/decrease over fungicide
Strain *Bacillus* B1	2.10 c	2.35 b	61.20 ab	15.10 a	(+) 74
Strain *Bacillus* B3	3.05 ac	3.10 ab	60.05 ab	10.39 b	(+) 20
Strain *Bacillus* B9	3.00 ac	3.05 ab	59.40 ab	9.17 b	(+) 6
Strain *Bacillus* B13	2.75 bc	2.85 b	66.40 a	11.16 ab	(+) 29
Strains *Bacillus* mix	2.90 ac	3.00 ab	63.55 ab	10.25 b	(+) 18
Fungicides	3.50 ab	3.25 ab	58.19 ab	8.67 b	–
Control	3.85 a	3.85 a	55.55 b	4.08 c	(–) 53

Fig. 8.48 Effect of *Bacillus subtilis* in the biomass production of pepper plants (*root*). Treatments with *Bacillus* (*left*) and without *Bacillus* (*right*)

was maintained over time or among harvest times (Table 8.17) (Castillo et al. 2013).

The effects obtained by applying *Bacillus* as biofungicide were positive for crop development, because *Bacillus* increased plant height compared to traditional crop management through the use of synthetic agrochemicals (Fig. 8.48). In a similar way, positive effects of *Bacillus* application were observed on crop yield, by preventing soil pathogen attack. Application of *Bacillus* increased pepper yields in contrast to the traditional crop management by up to 74 % cumulative assessment in three harvest times (Castillo et al. 2013).

8.4.3.3 Integrated Management

(a) *Two bioagents*: Combined seedling root dip treatment and soil application of nonpathogenic *Fusarium* (Fo 52) and *P. fluorescens* gave maximum inhibition of wilt (Singh et al. 2002).

8.4.4 Anthracnose or Dieback, *Colletotrichum dematium*

8.4.4.1 Symptoms

The disease is characterized by small grey spots on leaves. It causes necrosis of twigs and lower

Fig. 8.49 Anthracnose on chilli leaves and fruit

branches or the entire top of the plant may wither away. Black dots (acervuli) are formed all over the necrotic surface of the affected twig. Later dark-brown to black sunken spots having pink spore masses appear on fruits (Fig. 8.49). It may cause losses up to 25 % under suitable weather conditions. Fruit rot extends to the seed cavity making it internally seed-borne.

8.4.4.2 Biomanagement

Ramamoorthy and Samiyappan (2001) reported that isolates Pf1 and ATB of *P. fluorescens* gave increased growth of chilli plants, produced maximum amount of indole acetic acid in vitro and decreased the incidence of *Colletotrichum capsici* fruit rot under greenhouse conditions.

Under field conditions, Jetiyanon et al. (2003) observed that a PGPR mixture containing *B. amyloliquefaciens* strain IN937a and *B. pumilus* strain IN937b induced systemic resistance against anthracnose of long cayenne pepper (*Capsicum annuum* var. *acuminatum*) caused by *Colletotrichum gloeosporioides*.

Antibiotic-producing *B. subtilis* KP07 was tested for elicitation of induced systemic resistance (ISR) in control of *Colletotrichum acutatum* on red pepper (*Capsicum annum*) in greenhouse and field conditions based on its antibiosis. The elicitation of ISR by KP07 against *C. acutatum* revealed minimum disease severity (7.9 %) when compared to chemical control (17.0 %) under field conditions. Additionally, the study also revealed the significant increase in fruit yield by soil drenching of KP07 strain (10^8 cfu/ml) under field conditions. The maximum fruit yield (102.4 kg/plot) was recorded with KP07 followed by 93.8 kg/plot with chemical control. It has also been observed that KP07-treated plants showed higher chlorophyll content in the fresh leaves compared to untreated control. Hence the results of this study identify *B. subtilis* KP07 as a promising biocontrol agent against anthracnose disease and also growth promotion on red-pepper plants (Jang Sun Suh et al. 2011).

8.4.5 Powdery Mildew, *Leveillula taurica*

8.4.5.1 Symptoms

Powdery mildew infection causes whitish powdery patches both on the upper and lower leaf surfaces (Fig. 8.50).

8.4.5.2 Integrated Management

(a) *Bioagents and chemicals*: Seed treatment with carbendazim (0.2 %) + seedling dip with *P. fluorescens* (10 g/l) +2 sprays of *P. fluorescens* at 45 and 60 days after transplanting +2 sprays of hexaconazole at 75 and 90 days after transplanting recorded the least disease incidence of powdery mildew (26.7 PDI). This treatment also gave highest yield (0.85 t/ha) and highest net returns (Rs 25, 940/ha) (Mesta et al. 2009).

Fig. 8.50 Powdery mildew on chilli leaves and fruit

Fig. 8.51 Chilli plant infected by *Rhizoctonia solani*, with a reddish-brown discolouration at the soil line level of the stem and root (Courtesy: P. Bosland)

8.4.6 Root Rot, *Rhizoctonia solani*

8.4.6.1 Symptoms

Root rot caused by *R. solani* generally affects seedlings. But, *R. solani* can also infect mature plants and induce root rot (Fig. 8.51), which leads to wilting and death of chilli plants.

8.4.6.2 Biomanagement

The soil + root dip treatments with isolate *Pseudomonas* C7 (French isolated) gave maximum inhibition of root rot disease (Singh et al. 2003a, b).

8.4.7 Charcoal Rot, *Macrophomina phaseolina*

8.4.7.1 Symptoms

The disease gets its name from the dusty, black appearance of stems and root tissue, caused by the tiny black microsclerotia of the fungus (Fig. 8.52). Above-ground symptoms of charcoal rot are yellowing of the foliage and wilting.

8.4.7.2 Biomanagement

The soil + root dip treatments with isolate *Pseudomonas* C7 (French isolated) gave maximum inhibition of charcoal rot disease (Singh et al. 2003a, b).

8.4.8 Bacterial Wilt, *Ralstonia solanacearum*

8.4.8.1 Symptoms

The disease occurs in scattered plants or groups of plants in the field. Characteristic symptom is wilting of the entire plant with no leaf yellowing (Fig. 8.53). Cross sections cut from roots and lower stems of diseased plants exude milky streams of bacteria from the vascular system when suspended in water.

8.4.8.2 Integrated Management

(a) *Bioagents and botanicals*: Seed treatment + seedling treatment + soil application of *P. fluorescens*, along with vermicompost as substrate, was most effective on bacterial wilt disease reduction in bell pepper (Bora and Bora 2009).

8.4.9 Bacterial Leaf Spot of Sweet Pepper, *Xanthomonas campestris* pv. *vesicatoria*

8.4.9.1 Symptoms

Leaves, fruit and stems can be attacked by the disease. Leaf spots begin as circular, water

Fig. 8.52 Charcoal rot symptoms on chilli stem

Fig. 8.53 Bacterial wilt of bell pepper

soaked that becomes necrotic with brown centres with yellowish borders. The spots are sunken on the upper leaf surface and slightly raised on the lower surface. On stems, spots are elongated. Affected leaves turn yellow and drop. Affected fruits have raised brown spots that are wartlike in appearance (Fig. 8.54).

8.4.9.2 Biomanagement

Highly effective products include Citrex, Harpin, KPX-B2, KleenGrow, Kocide, K-Phite, Regalia, Seacide, MBI 10605, SUMM-IR4, NAI-4201, Taegro, Actigard, Kasumin and AMV4024 Sum.

Fig. 8.54 Bacterial spot on capsicum leaves and fruit

The disease severities for these products were in the range of 1.0–1.6 with no significant differences as compared to the pathogen control that recorded a disease severity of 3.7. The other products that were found to be moderately effective include Cease/Millstop, Companion, Harpin, HMO 736, KleenGrow, Kocide, MBI 106020, SC 27 and Taegro (Vijay Krishna Kumar et al. 2011).

8.4.10 Bell Pepper Root-Knot Nematode, *Meloidogyne incognita*

M. incognita was responsible for 24.54–28.00 % loss in fruit yield of bell pepper.

Fig. 8.55 Root-knot nematode on bell pepper

8.4.10.1 Symptoms

Nematode infestations damage the plant roots, and therefore symptoms reflect poorly functioning root systems. Above-ground symptoms of severe root-knot infestations include patches of chlorotic, stunted, necrotic or wilted plants. Nematode-infested plants are more susceptible to moisture or temperature stress and exhibit stress symptoms earlier than other plants. Furthermore, root systems that have been damaged by nematodes are often more susceptible to infection by soil-inhabiting fungi such as *Fusarium* and *Verticillium* species. Feeding by root-knot nematodes result in characteristic galls on roots. Severely galled roots may appear malformed and the root system shortened and thickened (Fig. 8.55). Root galls caused by root-knot nematodes on sweet pepper are frequently small.

8.4.10.2 Biomanagement

Seed treatment with *P. fluorescens* alone was effective against *M. incognita* infecting capsicum under field conditions (Naik 2004).

Burkholderia cepacia decreased egg hatch of *M. incognita* and mobility of second-stage juveniles (Meyer et al. 2000).

8.4.10.3 Integrated Management

(a) *Bioagents and botanicals*: The efficacy of neem cake along with *P. fluorescens*, against root-knot and reniform nematodes (*Meloidogyne incognita* and *Rotylenchulus reniformis*), was evaluated on capsicum (Rao 2011).

Capsicum seedling treatment with *P. fluorescens* and soil application of neem cake enriched with *P. fluorescens* was found to be significantly effective in reducing the population of nematodes. This treatment reduced root populations of *R. reniformis* by 69 % and *M. incognita* by 68 %. There was significant increase in the yield of capsicum by 19 % (Table 8.18). In general the growth of capsicum plants treated with *P. fluorescens* was better (Rao 2011).

(b) *Two bioagents*: Integrated management of *M. incognita* infecting capsicum was achieved by seed treatment with *P. fluorescens* at 50 g/kg seed combined with nursery bed treatment with *Pochonia chlamydosporia* at 50 g/m^2. The above treatment was significantly effective in increasing the seedling growth and root colonization with bioagents and in reducing root galling (Table 8.19). These seedlings, when transplanted in field, significantly reduced nematode population both in soil and roots, root galling and increased root colonization by bioagents, propagule density in soil, parasitization of eggs and fruit yield (Rao et al. 2004) (Table 8.20).

Table 8.18 Effect of neem cake enriched with *P. fluorescens* and vermicompost on root population densities of *M. incognita* and *R. reniformis*, PGPR root colonization and yield of capsicum (mean of 5 replicates)

Treatments	No. of *M. incognita*/ 10 g root	No. of *R. reniformis*/ 10 g root	Root colonization of *P. fluorescens* (CFU/g $\times 10^6$)	Increase in yield of capsicum (%)
T1	22.54	19.83	1.7	6
T2	15.28	13.65	4.0	14
T3	16.43	15.93	2.3	12
T4	12.32	10.07	6.4	19
T5	14.13	14.95	3.6	16
T6	22.17	19.83	–	5
T7	26.74	24.38	–	4
T8	18.72	17.55	2.5	8
T9	19.86	18.53	2.0	5
T10	38.20	32.50	–	–
CD – 5 %	**3.24**	**2.68**	**0.45**	–

T1 – seedlings of capsicum treated with *P. fluorescens* (cfu 2×10^8/ml); T2 – untreated seedlings sown in the plots mixed with neem cake enriched with *P. fluorescens* 20 g/m^2; T3 – untreated seedlings sown in the plots mixed with vermicompost enriched with *P. fluorescens* 100 g/m^2; T4 – seedlings treated with *P. fluorescens* sown in plots mixed with neem cake enriched with *P. fluorescens*; T5 – seedlings treated with *P. fluorescens* sown in plots mixed with vermicompost enriched with *P. fluorescens*; T6 – untreated seedlings transplanted in plots mixed with neem cake alone 20 g/m^2; T7 – untreated seedlings transplanted in plots mixed with vermicompost alone 20 g/m^2; T8 – seedlings treated with *P. fluorescens* transplanted in plots mixed with neem cake alone 20 g/m^2; T9 – seedlings treated with *P. fluorescens* transplanted in plots mixed with vermicompost alone 20 g/m^2; T10 – untreated seedlings transplanted in the plots without any treatment, served as control

Table 8.19 Effect of integration of bioagents on plant growth and management of *Meloidogyne incognita* infecting capsicum in nursery

Treatment/dose	Seedling weight (g)	No. of galls/ 10 seedlings	Root colonization of *P. chlamydosporia* (cfu/g)	Root colonization of *P. fluorescens* (cfu/g)
P. chlamydosporia – nursery treatment (25 g/m^2)	3.6	62	15,896	–
P. chlamydosporia – nursery treatment (50 g/m^2)	3.8	59	17,459	–
P. fluorescens – seed treatment (50 g/kg)	4.0	55	–	12,563
P. fluorescens – seed treatment (50 g/kg) + *P. chlamydosporia* – nursery treatment (25 g/m^2)	4.5	51	16,256	12,249
P. fluorescens – seed treatment (50 g/kg) + *P. chlamydosporia* – nursery treatment (50 g/m^2)	4.4	53	16,789	11,897
Control	3.2	82	–	–
CD (P=0.05)	**0.34**	**7.96**	**959.72**	**729.65**

Table 8.20 Effect of integration of bioagents on plant growth and management of *Meloidogyne incognita* infecting capsicum under field conditions

Treatment[a]	Root-knot index (1–10 scale)	Yield (kg/6 m^2)	% eggs parasitized by *P. chlamydosporia*	Root colonization (cfu/g)		Propagule density (cfu/g soil)	
				Pc	*Pf*	*Pc*	*Pf*
T1	6.3	4.4	41.00	22,456	–	19,569	–
T2	5.6	4.7	45.00	25,789	–	22,845	–
T3	6.0	5.1	–	–	22,568	–	15,987
T4	4.8	5.3	40.00	21,679	21,789	19,234	16,843
T5	4.4	5.5	42.67	24,587	21,567	23,221	15,359
Control	8.1	4.0	–	–	–	–	–
CD 5 %	**0.45**	**0.28**	**3.78**	**1246.76**	**1089.86**	**1178.32**	**873.65**

[a]T1 – *P. chlamydosporia* – nursery treatment (25 g/m^2), T2 – *P. chlamydosporia* – nursery treatment (50 g/m^2), T3 – *P. fluorescens* – seed treatment (50 g/kg), T4 – *P. fluorescens* – seed treatment (50 g/kg) + *P. chlamydosporia* – nursery treatment (25 g/m^2), T5 – *P. fluorescens* – seed treatment (50 g/kg) + *P. chlamydosporia* – nursery treatment (50 g/m^2)

8.4.11 Root-Knot Nematode, *Meloidogyne incognita*, and Foot and Root Rot, *Phytophthora capsici*, Disease Complex

The disease incidence was raised to 54 % when hot pepper was inoculated with *Meloidogyne* sp. and *Phytophthora* sp.

8.4.11.1 Biomanagement

Capsicum seed treatment with consortia of *P. fluorescens* (three strains) at the rate of 10 g/kg and subsequent application of consortia formulation of *P. fluorescens* at the rate of 15 g/m^2 significantly reduced *M. incognita* in capsicum roots by 74 %, reduced the incidence of *Phytophthora* foot and root rot and increased the yield by 26 %. Benefit/cost ratio (calculated for additional cost of biopesticides and additional returns accrued by the application of biopesticide) was 3.8 (Rao et al. 2009).

8.4.12 Root-Knot Nematode, *Meloidogyne incognita*, and Bacterial Wilt, *Ralstonia solanacearum*, Disease Complex

8.4.12.1 Symptoms

Capsicum is prone to many soilborne diseases, among which the bacterial wilt (*R. solanacearum*) in combination with root-knot nematode (*M. incognita*) takes heavy toll every year all over the world. The disease ratings of bacterial wilt in plants inoculated with *M. incognita* and *R. solanacearum* in different combinations (inoculation of *M. incognita* 2 weeks/1 month later than *R. solanacearum* or inoculation of both pathogens simultaneously) were higher than the ratings in plants inoculated with *R. solanacearum* alone. It is concluded that a complex infection of root-knot nematodes and *R. solanacearum* increased disease severity and reduced the resistance of bacterial wilt-resistant chilli plants.

8.4.12.2 Integrated Management

(a) *Two bioagents*: Combined application of neem-based formulations of *P. fluorescens* and *Pochonia chlamydosporia/Trichoderma harzianum* at 40 g/m^2 in nursery beds and transplanting these seedlings in the main field resulted in significant reduction in disease index and root-knot index in capsicum to the tune of 70 % and increased the crop yield by 37 % (Rao et al. 2002) (Table 8.21). Combinations of the bioagents did not affect the colonization of individual bioagents on the roots, and hence the transplants carried the bioagents to the main field. This has resulted in the effective management of the pathogens involved in the disease complex.

Naik (2004) reported that combination of *T. harzianum* along with *P. fluorescens* increased the fruit yield (3.320 kg/3 m^2

Table 8.21 Effect of integration of neem-based bioagents on the growth of transplants and management of disease complex and yield of capsicum

Treatment	Seedling weight (g)	Root-knot index (1–10)	Disease index (1–9)	Yield in kg/4 m²
Seed treatment with *P. fluorescens*	421	5.6	6.4	4.3
Seed treatment with neem-based formulation of *P. fluorescens*	428	5.2	6.7	4.7
Nursery treatment with *P. fluorescens*	435	4.6	5.4	4.8
Nursery treatment with *P. fluorescens* + *P. chlamydosporia*	463	4.1	5.2	4.0
Nursery treatment with *P. fluorescens* + *T. harzianum*	493	3.8	3.5	5.1
Control	340	8.7	8.2	2.6
CD at 5 %	**27.20**	**0.49**	**0.38**	**0.25**

Table 8.22 Effect of integration of bioagents on plant growth and management of disease complex in bell pepper in nursery beds

Treatment[a]	Seedling length (cm)	Seedling weight (g)	No. of galls/10 plants	Root colonization (cfu/g)	
				P. lilacinus	*B. pumilus*
T1	12.63	3.8	52	13,342	–
T2	15.53	4.0	50	16,539	–
T3	15.17	4.2	55	–	14,431
T4	18.43	4.4	44	12,984	13,964
T5	19.25	4.6	40	16,839	14,291
T6	10.53	3.3	76	–	–
CD (*P*=0.05)	**1.29**	**0.37**	**6.28**	**1087.95**	**976.75**

[a]T1 – nursery bed treatment with *P. lilacinus* (25 g/m²), T2 – nursery bed treatment with *P. lilacinus* (50 g/m²), T3 – nursery bed treatment with *B. pumilus*, T4 – T1+ T3, T5 – T2 + T3, T6 – control

plot). The least gall index was present where the combination of *T. harzianum* and *P. fluorescens* was used (1.5) and *P. fluorescens* + *P. chlamydosporia* (1.7). But among combination treatments of bioagents, all the treatments were on par (*P. fluorescens* + *T. harzianum*, *P. fluorescens* + *P. chlamydosporia*). Results related to the percent bacterial wilt disease (*R. solanacearum*) incidence also showed similar trend in which *T. harzianum* + *P. fluorescens* performed very well and the percent incidence was 16.90.

Naik et al. (2003) reported that soil application of organically developed *Paecilomyces lilacinus* (50 g/m² of formulated product containing 10^6 spores/g) along with pure culture of *B. pumilus* (10^8 cfu/ml) to the nursery beds of bell pepper increased plant growth and root colonization and reduced root galling by the bioagents. The above seedlings when transplanted in field significantly reduced root galling and wilt incidence, and increased root colonization, propagule density in soil, egg parasitization and fruit yield in the main field (Tables 8.22 and 8.23).

(b) *Bioagents, botanicals and AMF***:** The efficacy of *P. fluorescens* and *Glomus mosseae*, against root-knot and reniform nematodes (*Meloidogyne incognita* and *Rotylenchulus reniformis*), was evaluated to produce nematode-free seedlings of capsicum.

Seed treatment with talc-based formulation of *P. fluorescens* was able to increase the growth of the seedlings of capsicum. Seed treatment with talc-based formulation of *P. fluorescens* + neem-based formulation of *P. fluorescens* along with coco-peat reduced root galling due to *M. incognita*. This

Table 8.23 Effect of integration of bioagents on root galling, wilt incidence, root colonization, propagule density and yield of bell pepper under field conditions

Treatment[a]	Root-knot index	Mortality %	Yield kg/6 m^2	Root colonization (cfu/g)		Propagule density (cfu/g soil)	
				P. lilacinus	*B. pumilus*	*P. lilacinus*	*B. pumilus*
T1	6.5	68.90	4.8	30,653	–	21,346	–
T2	5.3	52.45	5.4	34,289	–	23,467	–
T3	7.0	34.62	7.5	–	25,875	–	17,321
T4	4.3	25.86	8.2	31,764	21,459	20,457	15,347
T5	4.1	17.69	8.5	34,762	20,764	22,689	14,569
T6	8.4	91.45	0.5	–	–	–	–
CD (P=0.05)	**0.58**	**8.94**	**0.35**	**1176.52**	**1087.54**	**1234.87**	**897.34**

[a]T1 – nursery bed treatment with *P. lilacinus* (25 g/m^2), T2 – nursery bed treatment with *P. lilacinus* (50 g/m^2), T3 – nursery bed treatment with *B. pumilus*, T4 – T1+ T3, T5 – T2 + T3, T6 – control

Table 8.24 Effect of application of *P. fluorescens*, *G. mossease* and coco peat on growth of tomato seedlings, infestation of *M. incognita* and root colonization with bioagents (mean of 5 replicates)

Treatments	Seedling height (cm)	% nematode-free seedlings	Root colonization of *P. fluorescens* (cfu/g × 10^6)	Root colonization of *G. mossease* (%)
T1	19.04	87	1.56	–
T2	20.52	90	3.42	–
T3	21.60	91	4.82	–
T4	21.28	85	0.00	27
T5	23.85	92	2.25	–
T6	21.35	90	4.94	–
T7	22.86	92	6.39	–
T8	24.27	94	5.68	34
T9	25.16	97	8.26	42
T10	22.37	93	4.66	37
T11	24.46	94	23.76	40
T12	17.26	86	15.20	–
CD – 5 %	**2.66**	**–**	**1.72**	**–**

T1 – seed treatment with talc-based formulation of *P. fluorescens*, T2 – talc-based formulation of *P. fluorescens* with coco-peat treatment, T3 – neem-based formulation of *P. fluorescens* with coco-peat treatment, T4 – *G. mossease* with coco-peat treatment, T5 – seed treatment with talc-based formulation of *P. fluorescens* + *G. mossease* with coco-peat treatment, T6 – seed treatment with talc-based formulation of *P. fluorescens* + talc-based formulation of *P. fluorescens* with coco-peat treatment, T7 – seed treatment with talc-based formulation of *P. fluorescens* + neem-based formulation of *P. fluorescens* with coco-peat treatment, T8 – seed treatment with talc-based formulation of *P. fluorescens* + talc-based formulation of *P. fluorescens* with coco-peat treatment + *G. mossease* with coco-peat treatment, T9 – seed treatment with talc-based formulation of *P. fluorescens* + neem-based formulation of *P. fluorescens* with coco-peat treatment + *G. mossease* with coco-peat treatment, T10 – talc-based formulation of *P. fluorescens* with coco-peat treatment + *G. mossease* with coco-peat treatment, T11 – neem-based formulation of *P. fluorescens* with coco-peat treatment + *G. mossease* with coco-peat treatment, T12 – control without seed and coco-peat treatment

resulted in production of increased number of nematode-free seedlings of capsicum. There was significant increase in the production of nematode-free capsicum seedlings in *G. mosseae* with coco-peat treatment + seed treatment with talc-based formulation of *P. fluorescens* + neem-based formulation of *P. fluorescens* with coco-peat treatment. There appears to be synergistic interaction among *P. fluorescens*, *G. mosseae* and neem cake, and they are complementing each other resulting in additive effect on plant height. Neem cake helped in the increased root colonization of *P. fluorescens* and *G. mossease* in capsicum (Table 8.24). When these seedlings were transplanted in the main field, there was significant reduction in the infestation of *M. incognita*, *M. javanica* (root-knot nematodes) and *Rotylenchulus reniformis* (reniform nematode) on capsicum (Gavaskar et al. 2011).

8.4.13 Root-Knot Nematode, *Meloidogyne javanica*, and Wilt, *Fusarium oxysporum*, Disease Complex

8.4.13.1 Symptoms

The disease complex involving *M. javanica* and *F. oxysporum/F. solani* inflicts severe losses to chilli crop. The simultaneous occurrence of both root-knot nematode and Fusarium wilt disease caused enhanced disease development and chilli yield loss.

8.4.13.2 Integrated Management

(a) *Bioagents and botanicals*: Integration of *P. fluorescens* at 50 kg/ha with neem seed powder at 250 kg/ha in nursery beds was found to be effective in increasing the number of germinated seedlings (81/0.5 m^2 compared to 39/0.5 m^2 in control) and fresh weight of seedlings (136.7 g compared to 61 g in control). The above treatment was also effective in reducing root galling due to *M. incognita* (0.7 RKI compared to 3.5 RKI in control) and percent root infection by *F. solani* (5.5 % compared to 41.5 % in control) in nursery beds (Table 8.25) (Kumar et al. 2009).

(b) *Two bioagents*: *Pseudomonas aeruginosa* and *Paecilomyces lilacinus* when used together significantly reduced infection of the disease complex on chilli (Perveen et al. 1998). Use of *P. aeruginosa* and *P. lilacinus* used alone or together significantly ($P<0.05$) increased plant height and reduced infection of *M. javanica* and root-infecting fungi, viz. *Macrophomina phaseolina*, *Rhizoctonia solani*, *F. solani* and *F. oxysporum* on chilli. *P. aeruginosa* was more effective than *P. lilacinus* in reducing *M. javanica* infection (Table 8.26).

Table 8.25 Effect of biocontrol agents and organic amendments against *M. incognita–F. solani* disease complex in chilli

Treatment	No. of seedlings emerged/bed	Fresh total wt. of seedlings/bed (g)	Root-knot index	% root infection
Untreated control	39	61.0	3.5	41.5
Pseudomonas fluorescens	65	105.7	1.9	17.3
Neem seed powder	64	105.0	1.3	15.5
Farmyard manure	45	71.9	3.3	35.5
P. fluorescens + neem seed powder	81	136.7	0.7	5.5
P. fluorescens + farmyard manure	75	124.9	1.5	13.0
Carbofuran	72	117.0	0.6	25.5
CD ($P=0.05$)	**3.53**	**7.91**	**0.06**	**0.65**

Table 8.26 Effect of *Pseudomonas aeruginosa* and *Paecilomyces lilacinus* on plant height and control of root rot disease complex in chilli

			Infection %			
Treatment	Plant height (cm)	Root- knot index	*M. phaseolina*	*R. solani*	*F. solani*	*F. oxysporum*
Control	10.5	3.3	31	19	75	37
P. lilacinus (*PL*)	14.5	2.9	19	0	56	44
P. aeruginosa (*PA*)	16.5	2.5	12	6	75	21
PL + *PA*	14.7	2.1	12	25	69	12
CD ($P=0.05$)	**2.2**	**0.34**	**6.1**	**6.1**	**6.1**	**6.1**

Bare root dip treatment with *P. aeruginosa* along with or without *T. harzianum*, *T. koningii* and *T. hamatum* significantly controlled infection of roots by *F. solani* and *M. javanica* on chilli. Combined use of *T. harzianum* along with *P. aeruginosa* caused the greatest reduction in root galling by *M. javanica* (Siddiqui et al. 1999).

8.4.14 Chilli Diseases and Pests

8.4.14.1 Integrated Methods

(a) *Bioagents, botanicals and chemicals*: The incidence of powdery mildew, leaf spot, wilt, dieback/anthracnose as well as sucking pest and fruit borer was significantly reduced in IPM plot treated with PGPR (*Pseudomonas fluorescens*) + pesticides + botanicals compared to control. The average per cent incidence of leaf spot, powdery mildew and fruit rot/wilt and also aflatoxin contamination came down from 26.16 to 8.95, 17.25 to 15.95, 14.32 to 9.65 and 17.91 to 11.89, respectively (Table 8.27). The damage caused by insect pests such as sucking pests, defoliators and fruit borer also came down significantly (Table 8.28). As a result, the number of pesticides sprays and the cost of protection is reduced significantly by 25–30 % in IPM plots as compared to non-IPM plot in adjacent region. Overall, the average yield in IPM plot was 3.146 t/ha compared to 2.460 t/ha in non-IPM plot. In terms of monetary value, Rs. 2,03,243 per ha was recorded in IPM plot as against Rs. 1,48,581 in non-IPM plot, which is a net gain of Rs. 54,662 per ha (Table 8.29). In addition, untold ecological damage which is inevitable in non-IPM practice, as a result of indiscriminate use of pesticides, was minimized in IPM practice (Naik et al. 2011).

Table 8.27 Incidence of various diseases of chilli in IPM plots compared to non-IPM plots in 2008–2011

	Range of incidence	
Diseases	IPM (%)	Non-IPM (%)
Leaf spot	08.95–26.16	10.62–36.73
Powdery mildew	15.95–17.25	18.58–28.00
Fruit rot	09.65–14.32	16.69–17.48
Wilt	11.89–17.91	13.40–26.33

Table 8.28 Incidence of insect pests of chilli in IPM plots compared to non-IPM plots in 2008–2011

	Range of incidence	
Insect pests	IPM	Non-IPM
Thrips/leaf	1.77–2.20	2.59–3.18
Mites/plant	1.16–1.61	1.79–2.06
Aphids/plant	0.00–0.00	0.41–1.81
Spodoptera litura	0.35–0.73	0.62–1.50
Helicoverpa armigera	0.29–0.41	0.63–0.93
Foliage damage (%)	2.29–2.66	5.89–5.73
Fruit damage (%)	1.63–1.92	5.17–7.65

Table 8.29 Economics of chilli production in IPM and non-IPM plots in 2008–2011 at Nelahal

			Cost of cultivation (Rs/ha)			
Treatment	Dry chilli yield* (qt/ha)	Total income (Rs/ha)	Production cost	Protection cost	Total cost	Net profit (Rs/ha)
IPM plots	31.26	236,797/–	20,950/–	12,604/–	33,554/–	203,243/–
Non-IPM plots	24.6	186,941/–	18,950/–	16,410/–	38,360/–	148,581/–

References

Almaghrabi OA, Massoud SI, Abdelmoneim TS (2013) Influence of inoculation with plant growth promoting rhizobacteria (PGPR) on tomato plant growth and nematode reproduction under greenhouse conditions. Saudi J Biol Sci 20:57–61

Al-Mughrabi KI (2010) Biological control of the potato dry rot caused by *Fusarium* species using PGPR strains. Biol Control 53:280–284

Amarananjundeswara H, Chikkasubbanna V, Shiva Reddy N (2009) Sustainable management of early blight in tomato caused by *Alternaria solani* on hybrid Sun-7611 grown under greenhouse. International conference on horticulture, Bangalore

Asaka O, Shoda M (1996) Biocontrol of *Rhizoctonia solani* damping-off of tomato with *Bacillus subtilis* RB14. Appl Environ Microbiol 62:4081–4085

Aspiras RB, de la Cruz (1986) Biocontrol of bacterial wilt in tomato and potato through preemptive colonization using *Bacillus polymyxa* and *Pseudomonas fluorescens*. Philipp J Crop Sci 11:1–4

Barua L, Bora BC (2008) Comparative efficacy of *Trichoderma harzianum* and *Pseudomonas fluorescens* against *Meloidogyne incognita* and *Ralstonia solanacearum* complex in brinjal. Indian J Nematol 38:86–89

Barua L, Bora BC (2009) Compatibility of *Trichoderma harzianum* and *Pseudomonas fluorescens* against *Meloidogyne incognita* and *Ralstonia solanacearum* complex on brinjal. Indian J Nematol 39:29–34

Benhamou N, Kloepper JW, Tuzun S (1998) Induction of resistance against Fusarium wilt of tomato by combination of chitosan with an endophytic bacterial strain: ultrastructural and cytochemistry of the host response. Planta 204:153–168

Bora LC, Bora P (2009) Biological approach for management of bacterial wilt of *bhoot chilli* in Assam. International conference on horticulture, Bangalore

Bora LC, Talukdar NC (2003) Final project report on biological management of wilt of tomato and brinjal. Assam Agricultural University, Jorhat

Brewer MT, Larkin RP (2005) Efficacy of several potential biocontrol organisms against *Rhizoctonia solani* on potato. Crop Prot 24:939–950

Cadenaa MB, Burellec N, Lawrencea KS, van Santenb E, Kathy S, Kloepper JW (2008) Suppressiveness of root-knot nematodes mediated by rhizobacteria. Biol Control 47:55–59

Castillo HFD, Castillo FR, Gallegos GM, Rodríguez RH, Aguilar C (2013) Biological control of root pathogens by plant-growth promoting Bacillus spp., Weed and pest control – conventional and new challenges. In: Soloneski S (ed), ISBN: 978-953-51-0984-6, InTech, doi:10.5772/54229

Chahal PPK, Chahal VPS (1986) Effect of *Azotobacter chroococcum* on the hatching of egg masses and eggs of *Meloidogyne incognita*. Plant Soil 95:289–292

Chahal PPK, Chahal VPS (1988) Biological control of root-knot nematode on brinjal (*Solanum melongena* L.) with *Azotobacter chroococcum*. In: Maqbool MA, Golden AM, Gaffar A, Krusberg LQ (eds) Advances in plant nematology. National Nematological Research Centre, University of Karachi, Karachi, pp 257–263

Chahal VPS, Chahal PPK (1999) Final technical report of PL – 480 project on microbial control of plant parasitic nematodes submitted to USDA

Chaliha K (1995) Biological control of collar rot of brinjal caused by Sclerotinia sclerotiorum. M. Sc. thesis, Assam Agricultural University, Jorhat

Colin JE, Chafik Z (1986) Comparison of biological and chemical treatments for control of bacterial speck of tomato under field conditions in Morocco. Plant Dis 70:1048–1050

Costa JM, Loper J (1994) Derivation of mutants of *Erwinia carotovora betavasculorum* deficient in export of pectolytic enzymes with potential for biological control of potato soft rot. Appl Environ Microbiol 60:2278–2285

Das BC, Phookan AK (2002) Final project report on biological control of Sclerotinia sclerotiorum under agro-ecological conditions of Assam. Assam Agricultural University, Jorhat

Devika Rani GS, Naik MK, Patil MB, Prasad PS (2009) Biological control of *Fusarium solani* causing wilt of chilli. Indian Phytopathol 62:190–198

Dhawan SC, Singh S, Kamra A (2008) Biomanagement of root-knot nematode, *Meloidogyne incognita* by *Pochonia chlamydosporia*. Indian J Nematol 38:119–121

Egamberdieva D, Berg G, Chebotar V, Tikhonovich I, Kamilova F, Validov SZ, Lugtenberg B (2011) Evaluation of rhizosphere bacteria for biological control of Fusarium foot and root rot of tomato (TFRR) in salinated soil. In: Proceedings of the 2nd Asian PGPR conference, Beijing, pp 75–79

Gautam A, Siddiqui ZA, Mahmood I (1995) Integrated management of *Meloidogyne incognita* on tomato. Nematol Medit 23:245–247

Gavaskar J, Priti K, Rao MS (2011) Production of nematode free, PGPR and mycorrhiza colonized seedlings of capsicum and tomato. In: Proceedings of the 2nd Asian PGPR conference, Beijing, pp 423–430

Gopalakrishnan C, Ajith Kumar K (2006) Increasing the field efficacy of *Pseudomonas fluorescens* in the management of bacterial wilt in tomato, *Lycopersicon esculentum* Mill. National symposium on improving input use efficiency in Horticulture, Indian Institute of Horticultural Research, Bangalore, pp 197–198

Grosch R, Junge H, Kerbs B, Bochow H (1999) Use of *Bacillus subtilis* as a biocontrol agent. III. Influence of *Bacillus subtilis* on fungal root diseases and on yield in soilless culture. Z Pflanzenkr Pflanzenschutz 106:568–580

Hallman JK, Hasky Grinther K, Hoffman HS, Reitz M, Sikora RA (1998) Similarities and differences in the mode of action of two rhizosphere bacteria antagonistic to *Globodera pallida* on potato. Bull Control Fungal Bact Plant Pathol 21:41–43

Hanafi A, Traore M, Schnitzler WH, Woitke M (2007) Induced resistance of tomato to whiteflies and *Pythium* with the PGPR *Bacillus subtilis* in a soilless crop grown under greenhouse conditions. In: Hanafi A, Schnitzler WH (eds) Proceedings of the VIIIth IS on protected cultivation in mild winter climates., pp 315–322. Acta Hort 747

Jagadish DR (2006) Evaluation of different methods of application of Pseudomonas B-25 strain for biological control of early blight of tomato caused by Alternaria solani Mill. M.Sc. (Agri.) thesis, University of Agricultural Sciences, Dharwad (India)

Jang Sun Suh, Kotnala Balaraju, Young Eun Cho, Jin Woo Park, Se Won Lee, Kyungseok Park (2011) Biological control of *Phytophthora capsici* and *Colletotrichum acutatum* by a PGPR strain *Bacillus subtilis* KP07 on red-pepper. In: Proceedings of the 2nd Asian PGPR conference, Beijing, p 351

Jetiyanon K, Fowler WD, Kloepper JW (2003) Broad-spectrum protection against several pathogens by PGPR mixtures under field conditions in Thailand. Plant Dis 87:1390–1394

Jung HK, Kim SD (2003) Purification and characterization of an antifungal antibiotic from *Bacillus megaterium* KL 39, a biocontrol agent of red-pepper *Phytophthora* blight disease. Korean J Microbiol Biotechnol 31:235–241

Kabayashi N (1989) Charcoal compost fixed antagonistic microorganisms and its effect to soil-borne diseases (Abstr.). Ann Phytopathol Soc Jpn 55

Kandan A, Ramiah M, Vasanthi VJ, Radjacommare R, Nandakumar R, Ramanathan A, Samiyappan R (2005) Use of Pseudomonas fluorescens based formulations for management of tomato spotted wilt virus (TSWV) and enhanced yield in tomato. Biocontrol Sci Technol 15:553–569

Khan ANA, Amarananjundeswara H, Khan MM, Nagaraj MS (2000) Recent advances in the management of bacterial wilt, *Ralstonia solanacearum* of potato through biological agents. National seminar on hitech horticulture, Bangalore

Kirankumar R (2007) Evaluation of plant growth promoting rhizobacterial strains against TMV on tomato. M. Sc (Agri.) thesis, University of Agricultural Sciences, Dharwad (India)

Kloepper JW (1983) Effect of seed piece inoculation with plant growth-promoting rhizobacteria on population of *Erwinia carotovora* on potato roots and daughter tubers. Phytopathology 73:217–219

Kumar V, Haseeb A, Sharma A (2009) Integrated management of *Meloidogyne incognita* and *Fusarium solani* disease complex of chilli. Indian Phytopathol 62:324–327

Liu R, Dai M, Wu X, Li M, Liu X (2012) Suppression of the root-knot nematode [*Meloidogyne incognita* (Kofoid & White) Chitwood] on tomato by dual inoculation with arbuscular mycorrhizal fungi and plant growth-promoting rhizobacteria. Mycorrhiza 22: 289–296

Lugtenberg B (2011) Bioactive metabolites involved in the microbiological control of tomato foot and root rot. In: Proceedings of the 2nd Asian PGPR conference, Beijing, pp 34–38

Mani MP, Rajeswari S, Sivakumar CV (1998) Management of the potato cyst nematodes, *Globodera* spp. through plant rhizosphere bacterium *Pseudomonas fluorescens* Migula. J Biol Control 12:131–134

Manoranjitham SK, Prakasam V, Rajappan K (1999) Effect of antagonists on *Pythium aphanidermatum* and the growth of chilli seedlings. J Biol Control 13:101–106

Mao W, Lewis JA, Lumsden RD, Hebber KP (1998) Biocontrol of selected soilborne disease of tomato and pepper plants. Crop Prot 17:535–542

Mari M, Guizzardi M, Brunelli M, Folchi A (1996) Postharvest biological control of grey mould (*Botrytis cinerea*) on fresh market tomatoes with *Bacillus amyloliquefaciens*. Crop Prot 15:699–705

Martinez C, Michaud M, Bélanger RR, Tweddell RJ (2002) Identification of soils suppressive against *Helminthosporium solani*, the causal agent of potato silver scurf. Soil Biol Biochem 34:1861–1868

Mesta RK, Mohankumar HD, Shivaprasad M, Tatagar MH, Rao MS (2009) Management of fruit rot and powdery mildew of chilli through *Pseudomonas fluorescens*. International conference on horticulture, Bangalore

Meyer SLF, Massoud SI, Chitwood DJ, Roberts DP (2000) Evaluation of *Trichoderma virens* and *Burkholderia cepacia* for antagonistic activity against root-knot nematode *Meloidogyne incognita* on bell pepper. Nematologica 2:871–879

M'Piga P, Belanger RR, Paulitz TC, Benhamou N (1997) Increased resistance to *Fusarium oxysporum* f. sp. *radicis-lycopersici* in tomato plants treated with the endophytic bacterium *Pseudomonas fluorescens* strain 63-28. Physiol Mol Plant Pathol 50:301–320

Murphy JF, Zehnder GW, Schuster DJ, Sikora EJ, Polston JE, Kloepper JW (2000) Plant growth promoting rhizobacterial mediated protection in tomato against tomato mottle virus. Plant Dis 84:779–784

Myresiotis CK, Karaoglanidis GS, Vryzas Z, Papadopoulou-Mourkidou E (2012) Evaluation of plant-growth-promoting rhizobacteria, acibenzolar-S-methyl and hymexazol for integrated control of Fusarium crown and root rot on tomato. Pest Manag Sci 68:404–411

Naik D (2004) Biotechnological approaches for the management of wilt disease complex in capsicum (*Capsicum annum* L.) and egg plant (*Solanum melongena*) with special emphasis on biological control. Ph. D. thesis, Kuvempu University, Shimoga

Naik D, Rao MS, Shylaja M, Reddy MS, Rahiman BA (2003) Management of wilt disease complex in capsicum using *Bacillus pumilus* and *Paecilomyces lilacinus*. In: Proceedings of international plant growth promoting rhizobacteria workshop, Calicut, pp 276–281

Naik MK, Manjunatha H, Amaresh YS, Hosmani AK, Bheemanna M, Reddy S, Sudha MS, Chennappa G, Sreenivas AG (2011) Interactions between PGPRs and crops with special reference to chilli (*Capsicum annuum* L.). In: Proceedings of the 2nd Asian PGPR conference, Beijing, pp 159–163

Nguyen MT, Ranamukhaarachchi SL, Hannaway DB (2011) Efficacy of antagonistic strain of *Bacillus megaterium*, *Enterobacter cloacae*, *Pichia guilliermondii* and *Candida ethanolica* against bacterial wilt disease of tomato. J Phytol 3(2):1–10

Niknam GR, Dhawan SC (2001) Induction of systemic resistance by *Bacillus subtilis* isolate Bs1 against *Rotylenchulus reniformis* in tomato. National Congress on Centenary of Nematology in India – appraisal & future plans, Indian Agricultural Research Institute, New Delhi, pp 143–144

Perveen S, Ehteshamul-Haque S, Ghaffar A (1998) Efficacy of *Pseudomonas aeruginosa* and *Paecilomyces lilacinus* in the control of root rot-root knot disease complex on some vegetables. Nematol Mediterr 26:209–212

Racke J, Sikora RA (1992) Isolation, formulation and antagonistic activity of rhizosphere bacteria toward the potato cyst nematode, *Globodera pallida*. Soil Biol Biochem 24:521–526

Rahman MA, Vijaya M, Chiranjeevi CH, Sai Reddy C, Reddy IP (2002) Disease management of solanaceous vegetable crops at nursery stage with bio-control agents. In: Proceedings of international conference on vegetables, Bangalore, pp 352–355

Ramamoorthy V, Raguchander T, Samiyappan R (1999) Isolation, characterization and screening fluorescent pseudomonads for managing damping-off disease of major vegetable crops. Symposium on plant disease management for sustainable agriculture. Indian Phytopathological Society, Southern Zone held at C.P.C.R.I. Kayankulam, Kerala, p 26

Ramamoorthy V, Samiyappan R (2001) Induction of defense-related genes in *Pseudomonas fluorescens* treated chilli plants in response to infection by *Colletotrichum capsici*. J Mycol Plant Pathol 31:146–155

Ramesh R, Korikanthimath (2003) Use of rhizosphere bacteria and soil fungi for the management of damping off and wilt in brinjal. In: 6th international PGPR workshop, Calicut

Rao MS (2011) Potential of PGPR in the management of nematodes and as bionematicides–research initiatives – national and international. In: Proceedings of the 2nd Asian PGPR conference, Beijing, pp 216–224

Rao MS, Naik D, Shylaja M (2004) Bio-intensive management of root-knot nematodes on bell pepper using *Pochonia chlamydosporia* and *Pseudomonas fluorescens*. Nematol Medit 32:159–163

Rao MS, Naik D, Shylaja M, Parvatha Reddy P (2002) Prospects for the management of nematode disease complex in capsicum using biological control agents. In: Proceedings of international conference on vegetables, Bangalore, pp 347–351

Rao MS, Ramachandran N, Sowmya DS, Ratnamma K, Chaya MK, Manoj Kumar R (2009) Biological control of nematode induced disease complex in certain vegetable crops. In: International conference on horticulture, Bangalore

Raupach GS, Liu L, Murphy JF, Tuzun S, Kloepper JW (1996) Induced systemic resistance in cucumber and tomato against cucumber mosaic cucumovirus using plant growth promoting rhizobacteria PGPR. Plant Dis 80:891–894

Reddy MS, Ryu CM, Rodríguez-Kábana R, Dawkins T, Kloepper JW (2001) Can beneficial bacteria from Auburn boost tomato growth and yield? Online Highlights 48(2)

Reitz M, Rudolph K, Schröder I, Hoffmann-Hergarten S, Hallmann J, Sikora RA (2000) Lipopolysaccharides of *Rhizobium etli* strain G12 act in potato roots as an inducing agent of systemic resistance to infection by the cyst nematode *Globodera pallida*. Appl Environ Microbiol 66:3515–3518

Sadfi N, Chérif M, Fliss I, Boudabbous A, Antoun H (2001) Evaluation of bacterial isolates from salty soils and *Bacillus thuringiensis* strains for the biocontrol of Fusarium dry rot of potato tubers. J Plant Pathol 83:101–118

Sadfi N, Chérif M, Hajlaoui MR, Boudabbous A (2002) Biological control of the potato tubers dry rot caused by *Fusarium roseum* var. *sambucinum* under greenhouse, field and storage conditions using *Bacillus* spp. isolates. J Phytopathol 150:640–648

Sankarimeena K, Jonathan EI, Devarajan K, Raghuchander T (2012) *Pseudomonas fluorescens* induced systemic resistance in tomato against *Meloidogyne incognita*. Indian J Nematol 42:5–10

Santhi A, Sivakumar CV (1995) Biocontrol potential of *Pseudomonas fluorescens* (Migula) against root-knot nematode, *Meloidogyne incognita* (Kofoid and White, 1919) Chitwood, 1949 on tomato. J Biol Control 9:113–115

Schmiedeknecht G, Bochow H, Junge H (1998) Use of *Bacillus subtilis* as biocontrol agent. II. Biological control of potato diseases. J Plant Dis Prot 105:376–386

Seenivasan N, Devarajan K, Selvaraj N (2007) Management of potato cyst nematodes, *Globodera* spp. through biological control. Indian J Nematol 37:27–29

Sharga BM, Lyon GD (1998) *Bacillus subtilis* BS 107 as an antagonist of potato black leg and soft rot bacteria. Can J Microbiol 44:777–783

Sheela MS, Nisha MS (2004) Impact of biological control agents for the management of root-knot nematode in brinjal. National symposium on paradigms in nematological research for biodynamic farming, University of Agricultural Sciences, Bangalore, pp 63–64

Siddiqui IA, Ehteshamul-Haque S (2001) Suppression of the root rot–root knot disease complex by *Pseudomonas aeruginosa* in tomato: the influence of inoculum density, nematode populations, moisture and other plant-associated bacteria. Plant Soil 237:81–89

Siddiqui IA, Ehteshamul-Haque S, Ghaffar A (1999) Root-dip treatment with *Pseudomonas aeruginosa* and *Trichoderma* spp. in the control of root rot-root knot disease complex in chilli (*Capsicum annuum* L.). Pak J Nematol 17:67–75

Siddiqui IA, Qureshi SA, Sultana V, Ehteshamul-Haque S, Ghaffar A (2000) Biological control of root rot-root knot disease complex of tomato. Plant Soil 227:163–169

Siddiqui ZA, Futai K (2009) Biocontrol of *Meloidogyne incognita* on tomato using antagonistic fungi, plant-growth-promoting rhizobacteria and cattle manure. Pest Manag Sci 65:943–948

Sikora EJ, Murphy JF (2003) Identification and management of cucumber mosaic virus in Alabama. In: 6th international PGPR workshop, Calicut

Singh RS, Reddy CS (1979) Suppression of damping-off of tomato and seedling blight of chickpea and sugar beet by strains of *Streptomyces diastaticus*. Indian Phytopathol 32:374–377

Singh RS, Singh PP, Bedi JS (2002) Final report of DBT scheme on biocontrol of seed and soil-borne diseases of vegetables. Punjab Agricultural University, Ludhiana, 90 pp

Singh SK, Nene YL, Reddy MV (2003a) Influence of crop system on *Macrophomina phaseolina* populations in soil. Plant Dis 74:812–814

Singh US, Mishra DS, Varshney S, Zaidi NW, Khan T, Joshi D, Sharma R, Singh S, Singh N (2003b) Potential and effectiveness of fungi and bacteria as biocontrol agents for plant disease management. In: Integrated pest management: principles and application (In press)

Solankia MK, Kumara S, Pandeyb AK, Srivastavaa S, Singha RK, Kashyapa PL, Srivastavaa AK, Aroraa DK (2012) Diversity and antagonistic potential of *Bacillus* spp. associated to the rhizosphere of tomato for the management of *Rhizoctonia solani*. Biocontrol Sci Technol 22:203–217

Son SH, Khan Z, Kim SG, Kim YH (2009) Plant growth-promoting rhizobacteria, *Paenibacillus polymyxa* and *Paenibacillus lentimorbus* suppress disease complex caused by root-knot nematode and fusarium wilt fungus. J Appl Microbiol 107:524–532

Spiegel Y, Cohn E, Galper S, Sharon E, Chet I (1991) Evaluation of a newly isolated bacterium, *Pseudomonas chitinolytica* sp. nov., for controlling the root-knot nematode *Meloidogyne javanica*. Biocontrol Sci Technol 1:115–125

Srinivas C, Niranjana SR, Praveen Kumar L, Chandra Nayaka S, Shetty HS (2006) Effect of fungicides and bioagents against *Phytophthora capsici* on seed quality of chilli. Indian Phytopathol 59:62–67

Sunaina V, Kishore V, Shekhawat G (2001) Impact of plant growth promoting *Bacillus* bioagents on potato yields in field plot experiments. J Indian Potato Assoc 28:107–108

Tanii A, Horita H, Takeuchi T (1988) Biological control of scab and black scurf of potato plant with seed tuber bacterization. Plant Prot 42:235–240

Tiwari RK, Mistry NC, Singh B, Gandhi CP (2014) Indian horticulture database-2013. National Horticulture Boad, Guragaon, 289 pp

Toyoda H, Morimoto M, Kakutani K, Morikawa M, Fukamizo T, Goto S, Terada H, Ouchi S (1993) Binary microbe system for biological control of Fusarium wilt of tomato: enhanced root colonization of an antifungal rhizoplane bacterium supported by a chitin-degrading bacterium. Ann Phytopathol Soc Jpn 59:375–386

Valenzuela-Soto JH, Estrada-Herna'ndez MG, Ibarra-Laclette E, De'lanoFrier JP (2010) Inoculation of tomato plants (*Solanum lycopersicum*) with growth promoting *Bacillus subtilis* retards whitefly *Bemisia tabaci* development. Planta 231:397–410

Vijay Krishna Kumar KS, Yellareddygari KR, Reddy MS, Kloepper JW, Sudini H, Zhou XG, Surendranatha Reddy EC, Yuan Zhi-lin, Yang Zhi-ling, Qi Wang, Binghai Du Shouan Zhang, Ravindra Reddy B, Bhaskara Reddy BV (2011) Over-view of commercial potential of plant growth-promoting rhizobacteria (PGPR) in the USA. In: Proceedings of the 2nd Asian PGPR conference, Beijing, pp 539–546

Xu GW, Gross DC (1986) Selection of fluorescent pseudomonads antagonistic to *Erwinia carotovora* and suppression of potato seed piece decay. Phytopathology 76:414–422

Zehnder GW, Yao C, Murphy JF, Sikora ER, Kloepper W (2000) Induction of resistance in tomato against cucumber mosaic cucumovirus by plant growth promoting rhizobacteria. Biocontrol 45:127–137

Zehnder GW, Yao C, Murphy JF, Sikora ER, Kloepper JW, Schuster DJ, Polston JE (1999) Microbe induced resistance against pathogens and herbivores: evidence of effectiveness in agriculture. In: Agarwal AA, Tuzun S, Bent E (eds) Induced plant defenses against pathogens and herbivores. APS Press, St. Paul, pp 335–355

Zhang S, Mersha Z, Fu Y (2011) Efficacy of microbial products for managing bacterial spot of tomato. In: Proceedings of the 2nd Asian PGPR conference, Beijing, p 455

Zore MP, Pawar SA, Kadam DB, Chandlee AG (2013) Management of root-knot nematode, *Meloidogyne incognita*, Race-II (Kofoid and White, 1919) Chitwood, 1949 infesting brinjal (*Solanum melongena* L.). Indian J Nematol 43:90–92

Bulbous, Cruciferous and Malvaceous Vegetable Crops

9

9.1 Bulbous Vegetable Crops

9.1.1 Onion, *Allium cepa*, and Garlic, *Allium sativum*

9.1.1.1 Basal Rot, *Fusarium oxysporum* f. sp. *cepae*

Basal rot generally occurs when soil temperatures are very warm (optimum 29 °C).

i) Symptoms

The early symptoms in the field are yellowing of leaves and tip dieback. As the disease progresses, the whole plant may collapse and, if the plant is pulled, it often comes out without any roots attached since they have decayed. The basal plate of the onion becomes pinkish-brown and secondary bacterial rots may develop in the affected area (Fig. 9.1). If infection occurs late in the season, the symptoms may not show up until the onions are in storage.

ii) Biomanagement

Burkholderia cepacia was effective for the management of basal rot of onion when added on the seed (Kawamoto and Lorbeer 1976). Seedling dip in antagonists like *B. cepacia* gives protection to seedlings.

Bacillus amyloliquefaciens BL-3, *Paenibacillus polymyxa* BL-4 and *Pseudomonas putida* Cha94 were highly inhibitory to conidial germination of fungi such as *Fusarium oxysporum*, *Aspergillus* sp. and *Botrytis allii*, which were the basal- and neck rot-causing pathogens for onion during storage (Lee et al. 2001).

9.1.1.2 White Rot, *Sclerotium cepivorum*

White rot is a very destructive disease that begins in the field and can carry over into storage.

i) Symptoms

The first above-ground symptoms are yellowing and dieback of the leaf tips, followed by a collapse of the affected leaves. However, these symptoms alone could easily be confused with other types of damage (i.e. onion maggot). When the bulbs and roots are examined, a white, fluffy mould and soft rot will be observed. Masses of tiny black sclerotia can also be seen within this mould (Fig. 9.2). These sclerotia remain in the soil for many years. Infected bulbs can rot in storage boxes and stain other bulbs. White rot typically develops in patches in the field and is less of a problem when soils are warm (higher than 24 °C) and dry.

ii) Biomanagement

Seed treatment with the *B. subtilis* isolate BACT 2 significantly controlled onion white rot and BACT 4 and BACT 8 decreased bulb neck diameter, which is desirable to speedy maturity and curing and prolonging keeping quality (Utkhede and Rahe 1983).

P.P. Reddy, *Plant Growth Promoting Rhizobacteria for Horticultural Crop Protection*,
DOI 10.1007/978-81-322-1973-6_9, © Springer India 2014

iii) Integrated Management

(a) *Physical methods and bioagents*: Application of *B. subtilis* to the solarized soil effectively controlled the disease (Pereira et al. 1996).

9.1.1.3 Neck Rot, *Botrytis allii*

i) Symptoms

Bulbs are usually first affected at the neck (Fig. 9.3), although the decay can spread downwards to affect the whole bulb. Occasionally, symptoms develop at the side or the base of the bulb if there has been physical damage at this point. Scales of diseased bulbs become soft and brown, giving them a 'cooked' appearance. A dense grey mould growth develops, producing huge numbers of spores. This is often accompanied by hard, black, crust-like structures (the resting bodies or sclerotia of the fungus). Affected tissues gradually dry up so that the bulb suffers from a dry rot and eventually becomes sunken and mummified.

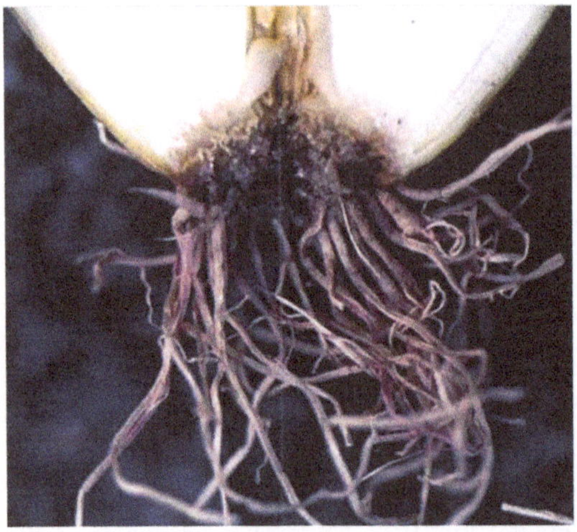

Fig. 9.1 Basal rot on onion bulbs

ii) Biomanagement

Bacillus amyloliquefaciens BL-3 and *Paenibacillus polymyxa* BL-4 were applied in the rhizoplane of onion at transplanting. BL-3 completely suppressed the neck rot of onion. Further, strain BL-3 produced a heat-stable antifungal protein that reduced decay regardless of the inoculum level or the isolate of pathogenic fungi tested and was effective at temperatures of 10–30 °C (Lee et al. 2001).

9.1.1.4 Root-Knot Nematodes, *Meloidogyne* spp.

i) Symptoms

Above-ground symptoms on onions heavily infected with M. *hapla* are general stunting, uneven growth; thicker necks and smaller bulbs; and also delayed maturity. The diagnostic symptoms are found on roots as galls or root thickenings of various sizes and shapes with protruding egg masses (Fig. 9.4). On onions, galls are usually small and barely noticeable, often no more than slight swellings. Bulb weight of onions was reduced by as much as 70 % in heavily infested sections of commercial fields. In field microplots, bulb weight of the

Fig. 9.2 White rot of onion

Fig. 9.3 Symptoms of onion neck rot

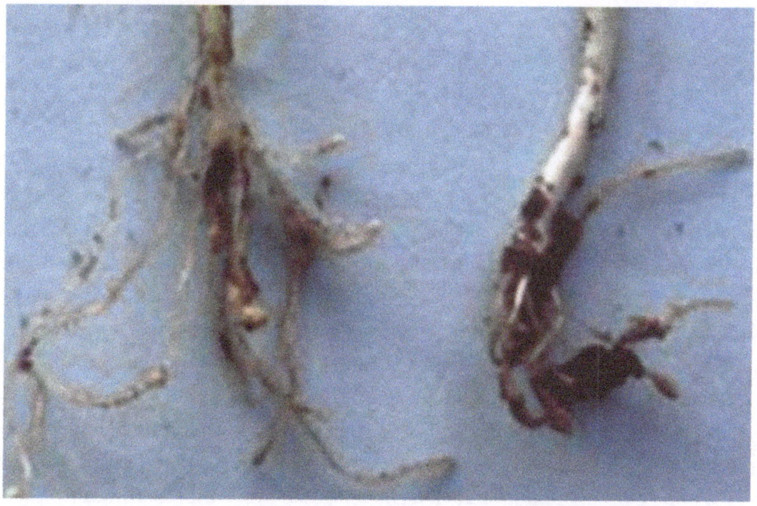

Fig. 9.4 Root-knot nematode on onion

onion cultivars Norstar and Paragon were reduced by about 50 % at an infestation level of 20 eggs/ml soil.

ii) Integrated Management

(a) *Bioagents and botanicals*: Seed treatment with *P. fluorescens* (10^9 cfu/g) at 10 g/kg and subsequent soil application of 5 tons of FYM enriched with 5 kg each of *P. fluorescens* (10^9 cfu/g) and *Pochonia chlamydosporia* (10^6 cfu/g) per ha significantly reduced *M. incognita* population in roots by 69 % and increased bulb yield by 21 % (Anon 2012).

9.2 Cruciferous Vegetable Crops

9.2.1 Cabbage, *Brassica oleracea* var. *capitata*, and Cauliflower, *Brassica oleracea* var. *botrytis*

9.2.1.1 Chinese Cabbage (*Brassica rapa*) Damping Off, *Rhizoctonia solani*

i) Symptoms

R. solani attacks the seedlings at the bottom of the young stem (hypocotyl) or at the soil line. The fungus attacks in the seedling stage, causing damping off as well as wire stem. The tissue

Fig. 9.5 Damping off of crucifer seedlings

Fig. 9.6 *Alternaria* leaf spot on cabbage

becomes water-soaked and rapidly collapse causing the seedlings to topple over and die (Fig. 9.5). At later stages, plants are brownish to black just above and below the soil line. The stems are usually somewhat smaller than normal but tough and woody. The pathogen is responsible for bottom rot, head rot and foot rot in older plants.

ii) Biomanagement

Pseudomonas CMR12a and mutants of this strain impaired in phenazine and/or biosurfactant production were applied at a concentration of 10^7 cfu g-1 soil to control damping off caused by *R. solani* AG 2-1 on Chinese cabbage. Data recorded 14 days after inoculation indicated that disease severity had decreased from 3.5 to 1.42 in CMR12a-treated seedlings. Both phenazines and biosurfactants were responsible for the profound effects of *Pseudomonas* CMR12a.

iii) Integrated Management

(a) *Bioagents and botanicals*: Soil incorporation of coconut fiber (CF) decreased the viability of *R. solani* sclerotia dramatically after 4 weeks of incubation from 75 to 20 % while the mean number of sclerotia killed by mycoparasitism was significantly raised from 4 to 66 %. The incorporation of 5 % CF into soil also resulted in a steady increase in population of fluorescent pseudomonads and *Trichoderma* spp. which might contribute to the death of sclerotia (Hua and Höfte 2011).

9.2.1.2 *Alternaria* Black Spot, *Alternaria brassicicola*

i) Symptoms

On leaves several small, dark brown zonate spots are produced, expanding rapidly to form circular lesions up to 1 cm in diameter. The enlargement of lesions may lead to formation of concentric circles which coalesce in the centre (Fig. 9.6). In humid weather, the fungus may cause bluish growth in the centre of the spots which hampers the photosynthetic activity of the plant, thereby hampering the overall productive potential of the plants. Sometimes cauliflower heads are also infected by the pathogen. The heads show browning, starting at the margin of the individual flowers or flower clusters.

ii) Biomanagement

Leiffert et al. (1993) reported the biocontrol of *A. brassicicola* on cabbage by *P. fluorescens* isolates CL42, CL66 and CL82 and *Serratia plymuthica* isolate CL43.

In greenhouse tests, *S. griseoviridis* controlled *A. brassicicola* on cauliflower and cabbage.

iii) Integrated Management

(a) *Bioagents and botanicals*: Seed treatment with *P. fluorescens* and soil treatment of nursery beds with neem cake enhanced seed germination and seedling stand.

9.2.1.3 Black Rot, *Xanthomonas campestris* pv. *campestris*

i) Symptoms

The first sign of the disease appears near the leaf margins, characterized by chlorosis, which progresses towards the centre of the leaf blade in 'V' shape (Fig. 9.7). In affected portions, veins and veinlets turn brown and finally black. In case of severe infection, the leaves wither. The vascular blackening may extend to the main stem causing systemic spread of the infection. Infected plants become stunted, often one side of cotyledons turn yellow to black, bend down and drop-off prematurely. Many lower leaves are also shed off early. Young plants may be killed in the seedbed itself. In cabbage and cauliflower, if the heads are already formed, soft rot of the head may set in.

ii) Biomanagement

B. subtilis was found effective in managing black rot of cabbage when applied as seed treatment, seedling root-dip treatment and soil drenching (Bhattacharya 1996).

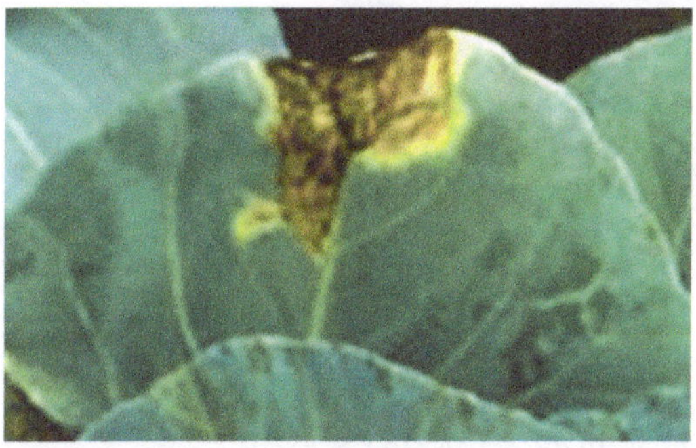

Fig. 9.7 Black rot on cabbage leaf

Jalali and Parashar (1995) reported that Hsb-19 strain of *Bacillus* was an effective and safe means to suppress the progression of black rot disease. They recorded higher protection than that obtained with the application of streptocycline when the antagonist was applied 24 to 74 h before inoculation with *X. campestris* pv. *campestris*. This provided 83.15 % disease control over the pathogen-inoculated check.

Bacillus spp. isolated from healthy cabbage, kale and radish have reduced black rot incidence in kale and cabbage in greenhouse and field experiments. Four of these *Bacillus* strains produced lipopeptides (antibiotics) active against black rot pathogen during its late growth phase. Recently, it was demonstrated that lipopeptides can stimulate ISR in plants, probably by interacting with plant cell membranes and inducing temporary alterations in the plasma membrane which could raise plant defences.

Pseudomonas aeruginosa suppress the black rot of crucifers by antagonizing *X. campestris* pv *campestris* (Xcc), which is a phytopathogen for black rot (Mishra and Arora 2012).

9.2.1.4 Root-Knot Nematode, *Meloidogyne incognita*, and Club Root, *Plasmodiophora brassicae* Disease Complex

i) Biomanagement

Among biocontrol agents *P. fluorescens* and *T. viride* gave effective control of club root and root-knot disease complex and increased the yield (Table 9.1).

Table 9.1 Efficacy of bioformulation mixtures against root-knot nematode – club root disease complex in cabbage under field conditions

Treatment	Club root index (PDI)[a]	% decrease in nema. popn. over initial	Root-knot index	Yield (MT/ha)
B. subtilis	43.0 (40.97)	29.0	3.33	41.33
P. fluorescens	23.0 (28.66)	46.2	1.33	59.10
Carbendazim	18.0 (25.10)	26.4	3.66	51.11
Carbofuran	46.0 (42.70)	30.2	2.00	45.77
Carbendazim + carbofuran	11.5 (19.81)	31.1	2.00	54.66
Control	60.0 (50.79)	+52.2	4.00	33.77

[a]Values in parenthesis are transformed values

Table 9.2 Efficacy of bioformulation mixtures against root-knot nematode – club root disease complex in cabbage under greenhouse conditions

Treatment	Club root index	Nematode incidence	
		Population	Root-knot index
Trichoderma viride	25.99 (30.65)	129	2.66
Pseudomonas fluorescens	28.20 (32.07)	112	2.33
T. viride + *P. fluorescens*	25.33 (30.22)	114	2.33
T. viride + chitin	25.44 (30.29)	108	2.33
P. fluorescens + chitin	25.66 (30.43)	111	2.00
T. viride + *P. fluorescens* + chitin	22.22 (28.12)	108	2.00
Chitin alone	31.70 (34.26)	139	3.00
Carbendazim	19.90 (26.49)	264	4.66
Carbofuran	39.90 (39.17)	106	1.66
Carbendazim + carbofuran	15.00 (22.79)	103	1.66
Plasmodiophora brassicae alone	48.90 (44.37)	0.033	0.133
Meloidogyne incognita alone	0.03 (0.60)	280	5.00

Fig. 9.8 DBM damage on cabbage

ii) Integrated Management

(a) *Bioagents and botanicals*: PGPR strains (*P. fluorescens*, *B. subtilis*) combined with fungal biocontrol agents (*Trichoderma viride*, *T. harzianum*) were found to be effective in reducing the nematode-fungal disease complex in cabbage (Loganathan et al. 2001). The bioformulation mixture of *P. fluorescens*, *T. viride* and chitin effectively reduced the disease complex in cabbage and cauliflower under greenhouse conditions (Table 9.2).

9.2.1.5 Diamondback Moth (DBM), *Plutella xylostella*

i) Damage

This is a major pest of cruciferous crops, particularly cabbage and cauliflower during the months of January to June and also during drought periods in monsoon. The first instar larvae mine the epidermal surface of the leaves. Second instar onwards the larvae feed externally by making holes in the leaves (Fig. 9.8).

ii) Integrated Management

(a) *Bioagents and chemicals*: Integration of gibberellic acid at 1,000 ppm + *P. fluorescens* at 5 kg/ha + *Bacillus thuringiensis* var. *kurstaki* at 1 kg/ha in alternation with *P. xylostella* granulosis virus at 1.5×10^{13} OB/ha recorded significantly lower incidence of *P. xylostella* (4.33 and 5.67 larvae/10 plants as compared to 43.00 in control) and increased curd yield of cauliflower (32.33 MT/ha and 35.33 MT/ha as compared to 21.33 MT/ha in control) (Mohana Sundaram and Dhandapani 2009).

9.3 Malvaceous Vegetable Crops

9.3.1 Okra, *Abelmoschus esculentus*

9.3.1.1 Root Rot, *Macrophomina phaseolina*, *Rhizoctonia solani* and *Fusarium solani*

i) Biomanagement

Ehteshamul-Haque and Ghaffar (1993) observed that *Sinorhizobium meliloti*, *Rhizobium leguminosarum* bv. *viciae* and *Bradyrhizobium japonicum* used either as seed dressing or as soil drench reduced infection of root rot in okra plants under field conditions.

9.3.1.2 Yellow Vein Mosaic Virus (YVMV)

It is estimated that if the plants are infected within 20 days after germination, the loss in yield was recorded up to 98 %. For the plants that infected at 35 and 50 days after germination, the loss in yield was estimated to be 83 and 49 %, respectively.

i) Symptoms

YVMV is caused by geminivirus and transmitted by whitefly. Characteristic symptoms appear as yellow vein and veinlets leaving green tissue in interveinal area (Fig. 9.9). Severely infected leaves sometimes become completely yellow. Fruits also change in colour to yellow and become hard in the early stage of development. The infected plants are stunted and bear very few yellow-coloured fruits.

ii) Biomanagement

The greenhouse experiment revealed that *Pseudomonas* 218(1) was the most promising isolate which controlled YVMV by 86.67 % (Table 9.3). All the isolates induced systemic

Fig. 9.9 YVMV infection on okra leaves

Table 9.3 Plant growth parameters of bhendi and YVMV control as influenced by inoculation with rhizobacteria (at 60 days after sowing)

Treatments	Plant height (cm)	Total biomass (g/plant)	% disease control	Fruit yield (kg/plant)
Pseudomonas B-40 + YVMV	43.00 (23.44)	21.46 (42.09)	80.00	0.400 (60)
Pseudomonas 212 (1) + YVMV	44.26 (33.10)	20.86 (37.79)	60.00	0.350 (40)
Pseudomonas 212 (4) + YVMV	48.20 (35.96)	22.42 (46.36)	60.00	0.520 (108)
Pseudomonas 205 (4) + YVMV	50.00 (43.44)	23.22 (53.40)	80.00	0.560 (124)
Fluorescent *Pseudomonas* 218 (1) + YVMV	51.22 (46.15)	24.58 (56.63)	86.67	0.580 (132)
P. fluorescens NCIM 2099 + YVMV	50.18 (45.09)	23.40 (54.32)	86.67	0.570 (128)
Diseased control (only YVMV)	35.10	15.54	0.00	0.250
Healthy control (no rhizobacteria, no YVMV)	41.64	20.10	0.00	0.390
CD (*P*=0.05)	**2.75**	**1.41**	–	**0.006**

Figures in parentheses indicate per cent increase in columns 2, 3 and 5 over diseased control resistance in okra plants. Plants inoculated with the lignite-based formulation of fluorescent *Pseudomonas* 218(1) recorded the highest phenol content, peroxidase activity and PALase activity (20.83 %, 79.35 % and 47.05 %, respectively, higher than the diseased control). The plants treated with this strain also showed the highest reduction in the insect population (83.33 % less than that of the diseased control). Semi quantitative PCR analysis revealed lower viral load accumulation in plants inoculated with these isolates. The fluorescent *Pseudomonas* 218(1) recorded maximum plant height, total biomass, chlorophyll content and fruit yield, which were increased by 46, 56, 62 and 132 %, respectively, over the diseased control (Jagadeesh et al. 2011).

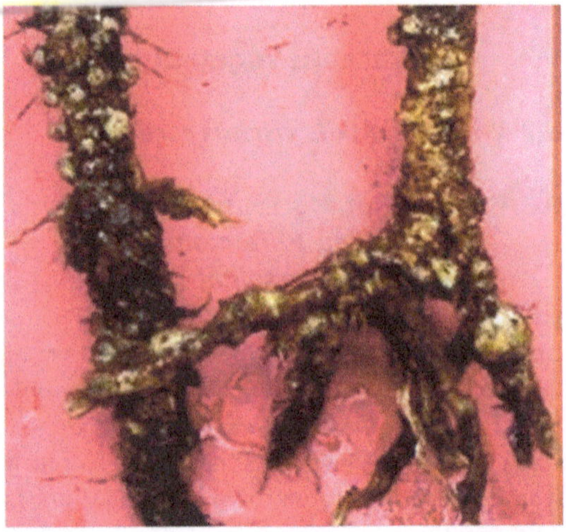

Fig. 9.10 Root-knot nematode on okra

9.3.1.3 Root-Knot Nematodes, *Meloidogyne* spp.

M. incognita was responsible for 28.08–90.90 % loss in fruit yield of okra, while *M. javanica* caused 20.20–41.20 % loss in yield.

i) Symptoms

Symptoms include leaf chlorosis (yellowing), stunting, general unthriftiness and premature wilting of plants. Plants exhibiting stunted or decline symptoms usually occur in patches of nonuniform growth rather than as an overall decline of plants within an entire field. Galls (knots) are produced on roots (Fig. 9.10), which stunt and weaken plants.

ii) Biomanagement

Root-knot nematodes in okra can be managed by the application of *Bacillus macerans* at 1.2×10^8 cells/m^2 before sowing. *Azotobacter* sp. significantly reduced root-knot nematode infestation in okra.

9.3.1.4 Root-Knot Nematode, *Meloidogyne incognita*, and Wilt, *Fusarium oxysporum* f. sp. *vasinfectum*, Disease Complex

Meloidogyne–Fusarium disease complex has been considered important on okra leading to reduction in its productivity.

i) Integrated Management

(a) *Bioagents and botanicals*: Soil application of 25 g/m^2 of deoiled neem cake enriched with *P. fluorescens* (2×10^6 cfu/g) has proved to be an effective treatment in combating the damage caused by *M. incognita* and *F. oxysporum* f. sp. *vasinfectum* to the tune of 68 %. This treatment also increased the yield of okra fruits by 24 % under field conditions (Chaaya et al. 2010).

9.3.1.5 Root-Knot Nematode, *Meloidogyne incognita*, and Root rot, *Rhizoctonia solani*, Disease Complex

i) Symptoms

M. incognita predisposed roots to *R. solani* resulted in severe root rot and subsequent plant death. Okra plants inoculated with either *R. solani* or *M. incognita* alone were free from root decay for the entire period of 6-week study. Three weeks after nematode and fungus inoculation, black sclerotia of *R. solani* were visible on nematode-induced galls, while on non-galled portions on the same root system were free of sclerotia. Prior to root rot development, *R. solani* demonstrated marked preference for root galls on nematode-infected roots. It is hypothesized that the leakage of nutrients from the root was responsible for attracting the fungus to the galls and for initiating sclerotial formation.

Table 9.4 Effect of bioagents and chemicals for the management of disease complex caused by *Meloidogyne incognita* and *Rhizoctonia solani*

Treatment[a]	Pre-emergence damping off (%)	Post-emergence damping off (%)	Plant height (cm)	No. of galls/root system	No. of egg masses/ root system
T_1	32.98	34.34	27.30	236.00	138.25
T_2	11.05	13.99	36.40	70.50	44.68
T_3	10.86	12.52	36.82	74.75	46.25
T_4	6.26	7.73	44.26	54.25	26.25
T_5	20.45	24.95	31.85	102.25	58.75
T_6	18.69	25.14	32.76	107.50	60.25
T_7	18.88	23.38	36.10	72.50	42.50
T_8	20.26	24.94	32.58	76.00	40.26
T_9	18.78	24.76	31.40	146.00	62.30
T_{10}	5.01	7.73	43.60	51.75	30.75
T_{11}	0.00	0.00	45.94	0.00	0.00
CD (*P*=0.05)	**3.94**	**4.25**	**3.26**	**0.71**	**0.67**

[a]T_1 – *M. incognita* + *R. solani* (inoculated and untreated control), T_2 – *T. harzianum* (seed treatment), T_3 – *P. fluorescens* (seed treatment), T_4 – *T. harzianum* + *P. fluorescens* (seed treatment), T_5 – *T. harzianum* (soil application), T_6 – *P. fluorescens* (soil application), T_7 – *T. harzianum* + *P. fluorescens* (soil application), T_8 – carbosulfan (seed treatment), T_9 – carbendazim (seed treatment), T_{10} – carbosulfan + carbendazim 50 (seed treatment), T_{11} – uninoculated and untreated control

M. incognita and *R. bataticola*, when inoculated simultaneously in soil, reduce the germination of seeds in okra. Combined attacks of both these pathogens cause significantly greater damage to the crop than that of the damage caused by either pathogen alone.

ii) Integrated Management

(a) *Two bioagents*: The lowest pre-emergence (5.01 %) and post-emergence (7.73 %) damping off were observed in the treatment where both carbosulfan 25 SD and carbendazim 50 WP were applied together as seed treatment (T_{10}) which was at par with the treatment with dual application of *T. harzianum* (6.26 %) and *P. fluorescens* (7.73 %) as seed treatment (T_4). As evidenced from the results, the seed treatment was found to be significantly superior to soil application of bioagents (Bhagawati et al. 2009) (Table 9.4).

The maximum plant height was recorded in the uninoculated and untreated control (T_{11}) followed by the treatment receiving *T. harzianum* + *P. fluorescens* as seed treatment (T_4) which were at par with the treatment receiving carbosulfan 25 SD + carbendazim 50 WP as seed treatment (T_{10}) (Bhagawati et al. 2009) (Table 9.4).

The minimum number of galls and egg masses in roots was recorded in the treatment receiving carbosulfan 25 SD + carbendazim 50 WP as seed treatment (T_{10}) which was on par with the treatment receiving *T. harzianum* + *P. fluorescens* as seed treatment (T_4). Further, the treatment with *T. harzianum* + *P. fluorescens* as seed treatment was found to be significantly better in reducing the host infection and nematode multiplication than soil application of both these bioagents (Bhagawati et al. 2009) (Table 9.4).

References

Anon (2012) Annual Report 2011–12, Indian Institute of Horticulture Research, Bangalore, pp 23–24

Bhagawati B, Choudhury BN, Sinha AK (2009) Management of *Meloidogyne incognita-Rhizoctonia solani* complex on okra through bioagents. Indian J Nematol 39:156–161

Bhattacharya AK (1996) Management of black rot of cabbage caused by Xanthomonas campestris *pv.* campestris (Pammel) Dowson. M. Sc. thesis, Assam Agricultural University, Jorhat

Chaaya MK, Rao MS, Ramachandran N (2010) Biomanagement of disease complex in okra (*Abelmoschus esculentus* L. Moench) using microbial antagonists. National Conf. on Innovations in Nematological Research for Agricultural Sustainability –

Challenges and A Roadmap Ahead, Tamil Nadu Agricultural University, Coimbatore, pp 96–97

Ehteshamul-Haque S, Ghaffar A (1993) Use of rhizobia in the control of root rot diseases of sunflower, okra, soybean and mungbean. J Phytopathol 138:157–163

Hua GKH, Höfte M (2011) The use of coconut fiber and antagonistic *Pseudomonas* CMR12a for biological control of *Rhizoctonia* damping-off on Chinese cabbage (*Brassica rapa*). In: Proceedings of the 2nd Asian PGPR conference, Beijing, P.R. China, pp 233–239

Jagadeesh KS, Patil NM, Krishnaraj PU, Patil MS, Vastrad AS 2011. Evaluation of rhizobacterial strains against bhendi yellow vein mosaic virus (BYVMV) disease in okra. In: Proceedings of the 2nd Asian PGPR conference, Beijing, P.R. China, pp 548–555

Jalali I, Parashar RD (1995) Biocontrol of *Xanthomonas campestris* pv. *campestris* in *Brassica juncea* with phylloplane antagonist. Plant Dis Res 10:145–147

Kawamoto SO, Lorbeer JW (1976) Protection of onion seedlings from *Fusarium oxysporum* f. sp. *cepae* by seed and soil infestation with *Pseudomonas cepacia*. Plant Dis Rep 60:189–191

Lee JT, Bae DW, Park SH, Shim CK, Kwak YS, Kim HK (2001) Occurrence and biological control of postharvest decay in onion caused by fungi. Plant Pathol J 17:141–148

Leiffert C, Sigee DC, Stanley R, Knight C, Epton HAS (1993) Biocontrol of *Botrytis cinerea* and *Alternaria brassicola* on Dutch white cabbage by bacterial antagonists at cold temperatures. Plant Pathol 42:270–279

Loganathan M, Swarnakumari N, Sivakumar M, Prakasam V, Mohan L, Ramaraj B, Samiyappan R (2001) Biological suppression of fungal nematode complex diseases of major cruciferous vegetables. South Indian Hort 49:296–300

Mishra S, Arora NK (2012) Evaluation of rhizospheric *Pseudomonas* and *Bacillus* as biocontrol tool for *Xanthomonas campestris* pv *campestris*. World J Microbiol Biotechnol 28:693–702

Mohana Sundaram A, Dhandapani N (2009) Gibberellic acid (GA_3) and *Pseudomonas fluorescens* as components of biointensive pest management strategy against *Plutella xylostella* in cauliflower. Pest Manag Hortilcult Ecosyst 15:1–8

Pereira CMF, Chaves GM, Zambolim L, Matsuoka K, Acuna RS, DoVale FXR (1996) Control of *Sclerotium cepivorum* by the use of vermicompost, solarization, *Trichoderma harzianum* and *Bacillus subtilis*. Summa Phytopathol 22:228–234

Utkhede RS, Rahe JE (1983) Effect of *Bacillus subtilis* on growth and protection of onion against white rot. Phytopathol Z 106:199–203

10 Leguminous Vegetable Crops

10.1 Pea, *Pisum sativum*

10.1.1 Damping Off, *Pythium ultimum*

10.1.1.1 Symptoms

Damping off can occur before or after crop emergence. Pre-emergence damping off results in a brown, gelatinous rotting within the seed coat. Radicals and cotyledons may become brown and soft after germination, but fail to emerge. Water-soaked, greasy lesions may form at the soil line and on roots after emergence when infected with *Pythium* spp., causing plants to collapse and wither (Fig. 10.1). Patches of seedlings can be girdled and killed.

10.1.1.2 Biomanagement

Seed treatment with *P. fluorescens* significantly reduced *P. ultimum* infection in pea (Whipps and Lumsden 1991). Similarly application of *P. fluorescens* strain 63-28 restricted the growth of *P. ultimum* in pea (Benhamou et al. 1996a, b). Two other strains of *P. fluorescens* R20 and R21 isolated from pea showed comparable results and are potentially good biocontrol agents against *Pythium* diseases in pea.

Competition for volatile organic materials derived from germinating seeds which may stimulate spore germination may be involved in the biocontrol of *P. ultimum* by *Pseudomonas putida* NIR (Paulitz 1991). Hyphal growth from sporangia of *P. ultimum* was stimulated by volatiles from germinating seeds of pea, and the stimulation was reduced when seeds were treated with *P. putida*. It was suggested that *P. putida* utilized the active moieties, ethanol and acetaldehyde, arising from the germinating seeds, thus providing biocontrol.

10.1.2 Powdery Mildew, *Erysiphe polygoni*

10.1.2.1 Symptoms

Mealy white patches appear on both sides of leaves, stems, branches, pods and tendrils (Fig. 10.2). These patches originate as minute discoloured specks, from which a powdery mass radiates on all sides and covers a large area of aerial parts. It causes considerable damage and may result in 20–30 % losses in pod number and quality. If it attacks early, plants fail to bear fruit or the pods get chaffy.

10.1.2.2 Integrated management

(a) *Two bioagents*: In organically grown pea, bio-priming of pea seeds with *Trichoderma harzianum* + *P. fluorescens* was most effective in improving the seedling stand. Foliar application of bioagents significantly reduced the incidence of powdery mildew disease. Application of *P. fluorescens* + *T. harzianum* through seed and foliar spray was most effective in increasing the yield of organically grown pea (Table 10.1).

P.P. Reddy, *Plant Growth Promoting Rhizobacteria for Horticultural Crop Protection*,
DOI 10.1007/978-81-322-1973-6_10, © Springer India 2014

Fig. 10.1 Damping off of pea seedlings

Fig. 10.2 Powdery mildew on pea leaves and pod

Table 10.1 Effect of different biocontrol agents applied through seed, FYM and/or foliar application on powdery mildew disease and yield of pea

Treatment	Seedling stand/m^2	Powdery mildew incidence (%)	Yield (kg/ha)
Seed bio-priming with *Trichoderma harzianum* + sprays with *Pseudomonas fluorescens*	56	13	355
FYM colonized with *Trichoderma harzianum* + sprays with *Pseudomonas fluorescens*	63	16	352
Control	52	36	301
CD at 5 %	6	10	53

10.1.3 Wilt, *Fusarium oxysporum* f. sp. *pisi*

10.1.3.1 Symptoms

Infection leads to drooping and wilting of plants at early stage of plant growth (Fig. 10.3). Vascular discolouration of stem and reddish appearance in the pith extends towards roots. The roots turn black and rot. Plant growth is checked, foliage turns yellow, and downward curling of stipules and leaflets takes place. The entire plant wilts and the stem shrivels. Epinasty of affected plants is a characteristic symptom.

Fig. 10.3 Pea plant showing symptoms of *Fusarium* wilt

10.1.3.2 Biomanagement

Bacillus pumilus strain SE34 protects pea roots against *Fusarium* wilt. In prebacterized roots, pathogen growth was restricted to the epidermis and the outer cortex. In these prebacterized roots, typical host reactions included strengthening of epidermal and cortical cell walls and deposition of wall appositions containing large amounts of callose as well as phenolic compounds. In non-bacterized roots, the pathogen multiplied abundantly through much of the tissue including the vascular stele.

In a field naturally infested with *Pythium* spp., inoculation of pea with strain R12 of *Rhizobium leguminosarum* bv. *viciae*, isolated from lentil in Alberta, Canada, significantly increased seedling emergence 4 weeks after planting (Bardin et al. 2004). This strain was as effective as *P. fluorescens* 708, a biological control agent of *Pythium* sp. (Bardin et al. 2003).

10.1.3.3 Integrated Management

(a) *Two bioagents*: In the greenhouse pot trial, biopesticides *Trichoderma* along with *Pseudomonas* showed significant control of *Fusarium* wilt disease of pea (50.6 % disease control) (Fig. 10.4) (Sharma 2011).

10.1.4 *Rhizoctonia* Root Rot

10.1.4.1 Symptoms

It primarily rots seeds and causes pre- and post-emergence damping off. Roots of older plants

Fig. 10.4 Treatments: *Left* – *Pseudomonas* + *Trichoderma*, *Right* – control

Fig. 10.5 *Left* – Rhizoctonia rot patches in green pea, *Right* – pea roots rotted by *R. solani*

can be rotted and stems can be infected to cause death of plants (Fig. 10.5). Mild lesions and root rot can result in stunted and stressed plants. A firm, dry, brown to reddish-brown decay or sunken lesion appears on the root and stem below or near the soil line. Plants may be stunted and may wilt or break off.

10.1.4.2 Biomanagement

B. subtilis produced an antibiotic which significantly reduced severity of *Rhizoctonia* root rot of pea in Canada when applied to soil before planting.

10.1.4.3 Integrated Management

(a) *Bioagents and chemicals*: By itself, *B. subtilis* provided more disease protection than the fungicide Anchor, but Anchor combined with *B. subtilis* was even more effective.

Fig. 10.6 Management of pea root rot. *Left* – plants treated with *Burkholderia cepacia*; *Right* – plants treated with captan fungicide

10.1.5 Root Rot, *Aphanomyces euteiches* f. sp. *pisi*

10.1.5.1 Symptoms

Infected roots often appear grey and water-soaked, eventually becoming soft and honey-brown or blackish-brown in appearance. Infection causes a reduction in root volume and function, including reduced nodulation. Symptoms in above-ground plant tissue can include chlorosis of cotyledons and necrosis of epicotyls and/or hypocotyls, stunting and wilting of foliage.

10.1.5.2 Biomanagement

Coating of pea seeds with *Burkholderia cepacia* strain AMMD gave effective control of pea root rot (Fig. 10.6).

10.1.5.3 Integrated Management

(a) *Two bioagents*: *T. harzianum* with *P. fluorescens* was able to suppress *A. euteiches* f. sp. *pisi* but did not result in better disease suppression (Dandurand and Knudsen 1993).

10.1.6 Charcoal Rot/Ashy Stem Blight, *Macrophomina phaseolina*

10.1.6.1 Symptoms

The pathogen infects seedling stems near the soil line at the base of developing cotyledons. The fungus produces black, sunken cankers that have sharp margins (Fig. 10.7) and often contains concentric rings. The plant's growing tip may be killed or the stem broken where it is weakened by the canker. Infection may continue into the hypocotyl and root region or the primary leaf petioles. Root infection causes a brown to black necrosis. If plants are grown under dryland conditions, young plants can be killed.

Infection of older seedlings and plants may cause stunting, leaf chlorosis, premature defoliation and plant death, especially during periods of high temperature and particularly following drought stress. On older plants, 'charcoal dust' often appears on the surface of the stems and is diagnostic evidence for this disease. This charcoal effect is caused by the production of small, black microsclerotia just below the epidermis and in the vascular tissue. This symptom is also called ashy stem blight.

10.1.6.2 Biomanagement

Rhizobium species were found to be effective in suppressing *M. phaseolina* in pea (Chakraborty and Purkayastha 1984).

10.2 French Bean, *Phaseolus vulgaris*

10.2.1 Rust, *Uromyces phaseoli*

10.2.1.1 Symptoms

Disease symptoms are mostly confined to leaves but young stem and branches are also attacked. The reddish-brown rust pustules appear abundantly on the lower surface of leaves (Fig. 10.8). When the leaves become thoroughly infected, these shrivel and fall off from the plant. In severe cases, most of the vines are defoliated. Uredospores of the rust pathogen are the main source of infection.

10.2.1.2 Biomanagement

Application of *B. subtilis* to bean leaves decreased the incidence of bean rust by 75 % equivalent to weekly treatments with the fungicide mancozeb (Baker et al. 1983). Application of *B. subtilis* three times at weekly interval controlled bean rust by the production of antibiotics (Baker et al. 1985).

10.2.2 Grey Mould, *Botrytis cinerea*

10.2.2.1 Symptoms

Symptoms appear as brown spots, which enlarge to give a more damaging aggressive phase

Fig. 10.7 Symptoms of charcoal rot in bean

Fig. 10.8 Rust on leaves of French bean

Fig. 10.9 Grey mould symptoms on bean pods

Fig. 10.10 Sclerotonia white rot on beans in field

(Fig. 10.9) in cool, wet or damp weather. Autumn sown beans are more likely to suffer yield losses, especially when the plant population is high and the crop becomes tall.

10.2.2.2 Biomanagement

The induction of systemic resistance in bean against *B. cinerea* by *Pseudomonas aeruginosa* 7NSK2 was found to be iron-regulated. Under iron limitation, 7NSK2 produces three siderophores: pyoverdin, pyochelin and salicylic acid (SA). Both pyoverdin-negative mutants and mutants lacking both pyoverdin and pyochelin induced resistance in bean, whereas mutants deficient in SA production did not. The resistance induced by 7NSK2 in bean is dependant on bacterially produced SA (De Meyer and Hofte 1997).

10.2.3 White Rot, *Sclerotinia sclerotiorum*

10.2.3.1 Symptoms

A watery-brown soft rot develops, followed by a white-cottony mould. Stems, leaves, flowers and pods may be affected and stem infections near soil level lead to plant collapse (Fig. 10.10). Black, seed-like resting bodies called 'sclerotia' develop in the mould or in the stem pith (tissue located in the centre of the stem).

10.2.3.2 Biomanagement

Das and Phookan (2002) reported that treatment of French bean seed with *B. subtilis* and *P. fluorescens*

Fig. 10.11 Symptoms of southern blight on beans

could significantly manage the collar rot infection with yield enhancement.

10.2.4 Southern Blight, *Sclerotium rolfsii*

10.2.4.1 Symptoms

Initial symptoms of southern blight include a yellowing of the foliage with slight darkening of the stem just above the soil line. Lesions on the stem at or near the soil line develop rapidly, girdling the stem, and result in a sudden and permanent wilt of the plant. The fungus grows downward in the stem and root, rotting the cortical tissue. White mats of mycelium develop on the stem and in adjacent soil. In a few days, tan to brown spherical sclerotia (small dormant structures) about 1.5 mm in diameter appears on the mycelial mat (Fig. 10.11). The abundant sclerotia are a good diagnostic feature of this disease.

10.2.4.2 Biomanagement

The endophytic *B. subtilis*, *B. cereus* and *B. pumilus* reduced the incidence of *S. rolfsii* in bean seedlings by 72, 79 and 26 %, respectively (Pleban et al. 1995).

Pseudomonas strains have been isolated that reduce the incidence of southern blight in bean. Inoculation with *P. putida*, *P. fluorescens* and *P. alcaligenes* reduced the incidence of disease caused by *S. rolfsii* in bean (Gamliel and Katan 1993).

Serratia marcescens was found to be the best biocontrol agent of *S. rolfsii* of bean (up to 75 % disease reduction). The drench and drip application of *S. marcescens* were more effective in controlling *S. rolfsii* than spraying, mixing in soil or seed coating (Ordentlich et al. 1987).

10.2.5 Fusarium Wilt, *Fusarium solani* f. sp. *phaseoli*

10.2.5.1 Symptoms

Symptoms include stunted growth, yellowing and wilting of the leaves; reddish discolouration of the xylem vessels, visible inside the stem as lines (if the stem is cut open lengthways) or dots (if it is cut across); and white, pink or orange fungal growth on the outside of affected stems, particularly in wet conditions.

10.2.5.2 Biomanagement

Rhizobium sp. antagonistic to *F. solani* f. sp. *phaseoli* isolated from commercial snap bean appeared to have a good potential for controlling *Fusarium* rot (Buonassisi et al. 1986).

10.2.6 Halo Blight, *Pseudomonas syringae* pv. *phaseolicola*

10.2.6.1 Symptoms

Leaf symptoms appear several days after infection as small (3 mm), water-soaked, tan brown, angular spots, each surrounded by a relatively wide, diffuse, pale green to yellow green tissue (halo) (Fig. 10.12). Infected plants may defoliate, wilt and die. Diseased plants may be dwarfed, with the upper leaves crinkled and mottled. Dark green, water-soaked areas may also be visible on the stems.

Pod infections develop rapidly under cool, wet conditions resulting in oval to circular, dark green, water-soaked 'greasy' lesions 9 mm in diameter. Halos do not develop around pod lesions (Fig. 10.12). As pods mature and turn yellow, pod lesions may remain green and may exhibit crusty bacterial ooze on the surface. Developing seed may be shrivelled or discoloured if lesions expand to involve the pod surface. With age the lesions become slightly sunken and reddish-brown.

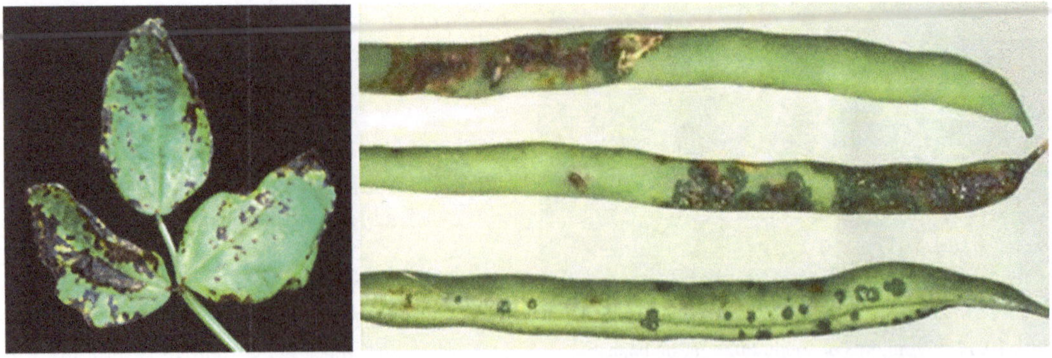

Fig. 10.12 Halo blight of beans. *Left*, lesions and bacterial crust of halo blight on leaf. *Right*, halo blight on pods

Fig. 10.13 Golden mosaic virus infection on French bean leaves

10.2.6.2 Biomanagement

Seed treated with *P. fluorescens* strain 97 protected beans against halo blight disease (Alstroem 1991).

10.2.7 Golden Mosaic Virus

As much as 70–80 % losses in yield has been observed in case of early infection.

10.2.7.1 Symptoms

The disease is characterized by scattered, slightly discoloured patches on leaves, which gradually turn bright yellow (Fig. 10.13). At times, the whole leaf turns golden yellow. The chlorophyll of the leaves is partially or completely destroyed. Affected plants are usually dwarf and pod production is greatly reduced. In nature, the disease is transmitted by the whitefly, *Bemisia tabaci*.

10.2.7.2 Biomanagement

Induction of systemic disease resistance in faba bean (*Vicia faba*) against bean yellow mosaic potyvirus (BYMV) via seed bacterization with *P. fluorescens* and *Rhizobium leguminosarum* has been investigated by Elbadry et al. (2006). They isolated PGPR strains from the roots of faba bean and examined singly or in combination the induction of resistance in faba bean against BYMV. The results established a pronounced and significant reduction in percent disease incidence (PDI) as well as in virus concentration (ELISA) in plants treated with *P. fluorescens* and *R. leguminosarum* as compared to the non-bacterized plants.

10.2.8 Root-Knot Nematodes, *Meloidogyne* spp.

10.2.8.1 Symptoms

Root-knot nematodes are known to cause significant damage. Yield reductions due to high populations of root-knot nematodes may range from 45 to 90 %. Yield losses due to root-knot infestation are typically most severe in sandy soils. These nematodes are also known to predispose plants to other soilborne pathogens that cause root rot and wilt diseases. Gall size on roots of French bean is variable (Fig. 10.14). Large egg masses are present on galls.

Fig. 10.14 Root-knot nematode on French bean roots

10.2.8.2 Integrated Management

(a) *Bioagents and botanicals*: The combination of *B. subtilis* (isolate K194) inoculum and cow manure led to a 54 % reduction in numbers of root-knot nematodes, compared to the untreated control. Consequently, damage by root-knot nematodes produced galls with galling indices 1.6 and 4.5, respectively, in plots treated in combination (*B. subtilis* and cow manure) and in untreated control plots, respectively. Compared to the other treatments, combining *B. subtilis* and organic amendments resulted in the highest nematode diversity. It can therefore be concluded that the root-knot nematodes associated with common bean can be maintained at levels below economic threshold using *B. subtilis* combined with cow manure, an integration which also demonstrated conservation of the nematode diversity (Wepuhkhulu et al. 2011).

10.3 Cowpea, *Vigna unguiculata*

10.3.1 Tomato Spotted Wilt Virus

10.3.1.1 Symptoms

Cowpeas develop purplish dead spots on the stem and leaves, followed by mottling, circular spots and wavy lines on the pods.

10.3.1.2 Biomanagement

Kandan et al. (2003) reported that the tomato spotted wilt virus disease intensity in *P. fluorescens* COT-I + CHAO and *P. fluorescens-CHAO-treated* (seed treatment) cowpea plants was reduced by 63 and 53 %, respectively. *P. fluorescens-treated* cowpea plants significantly enhanced oxidative enzymes like phenylalanine ammonia lyase and peroxidase. The phenolic content was also enhanced in the treated plants compared to the untreated plants.

10.4 Pigeonpea, *Cajanus cajan*

10.4.1 Wilt, *Fusarium udum*

10.4.1.1 Symptoms

The disease is characterized by slow wilting of the plant. The symptoms can be observed after a month of sowing or at the flowering or pod formation stage. The affected plants become yellow in colour followed by drooping and finally the whole plant dries up. The symptoms resemble as if the plant is suffering from the drought. The disease can be diagnosed whenever the affected stem is cut and opened where the browning of the xylem vessels could be clearly seen (Fig. 10.15). It has been recorded that wilting can be partial or total. However, it has been observed that partial

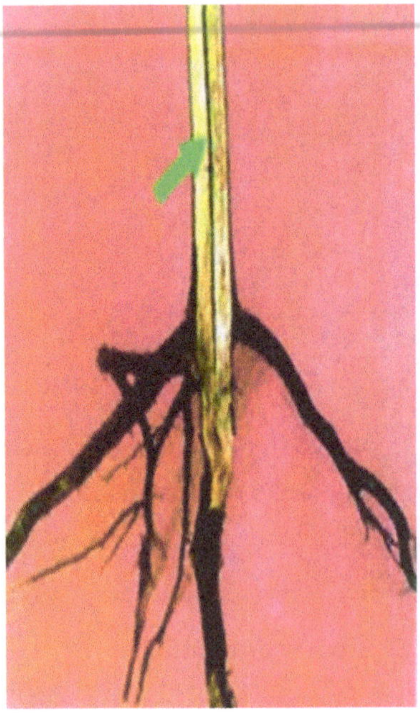

Fig. 10.15 Blackening of the xylem due to *Fusarium* wilt

wilting is mainly associated with lateral root infection, whereas tap root infection may cause complete wilting.

10.4.1.2 Biomanagement

Vasudeva and Roy (1950) observed the low incidence of *Fusarium* wilt of pigeon pea due to the suppressive effects of an antagonistic bacterium *B. subtilis* which released some antibiotic in the rhizosphere. The antibiotic was isolated and named as bulbiformin. *B. subtilis* brought about a marked reduction in the incidence of pigeon pea wilt (Vasudeva et al. 1962). Amendment of soil with roots of certain leguminous crops (sweet clover), molasses and oil cakes (groundnut cake) markedly increased the antibiotic (bulbiformin) production by *B. subtilis* and reduced the incidence of wilt by 88 % (Vasudeva et al. 1963). Seedlings gained resistance to *Fusarium* infection when the seeds were bacterized with *B. subtilis* before sowing. It was suggested that bulbiformin became systemic in the plant and provided a protective zone around the roots of pigeon pea seedlings.

Seed treatment of pigeon pea with talc-based formulation of fluorescent pseudomonads (*P. fluorescens* strain Pf1 and Pf2) at the rate of 4 g/kg of seed followed by soil application at the rate of 2.5 kg/ha at 0, 30 and 60 days after sowing controlled pigeon pea wilt incidence under field conditions. The additional soil application of talc-based formulation improved disease control and increased yield compared to seed treatment alone (Vidhyasekaran et al. 1997).

10.4.1.3 Integrated Management

(a) *Bioagents and AMF*: Application of *B. subtilis*, *Bradyrhizobium japonicum* and *Glomus fasciculatum* used either alone or in combination increased shoot dry weight, number of nodules and phosphorus content and reduced nematode multiplication and wilting index in pigeon pea (Siddiqui and Mahmood 1995a, b).
(b) *Bioagents and botanicals*: Pigeon pea seed treatment with bacterial antagonists (*Pseudomonas putida*, *B. subtilis*) at 10 g/kg seed + soil application at 2.5 kg/ha of above bioagents mixed with FYM gave 22–28 % *Fusarium* wilt control and increased the yield by 12–14 % (Ramanujam et al. 2003) (Table 10.2).

Seed treatment of pigeon pea with peat-based formulation of *B. subtilis* supplemented with 0.5 % chitin or with 0.5 % of sterilized *Aspergillus* mycelium controlled wilt of pigeon pea. It also increased growth promotion even in the presence of inoculum pressure (Manjula and Podile 2001). Chitin supplementation enhances the biocontrol efficacy of formulations.

10.4.2 Pigeonpea Cyst Nematode, *Heterodera cajani*

10.4.2.1 Symptoms

The most important characteristic symptom is the presence of cysts on the root surface. Identification of 'pearly root' caused by the presence of white females is a useful symptom of *H. cajani* infestation in pigeon pea at the vegetative stage

Table 10.2 Field performance of bacterial antagonists against *Fusarium* wilt of pigeon pea

Treatment	Formulation used and cfu/g	Disease control (%)	Increase in yield (%)
Seed treatment with *Pseudomonas fluorescens* at 10 g/kg + soil appln. at 2.5 kg/ha mixed with FYM	Talc 10^7–10^8	22	6
Seed treatment with *Pseudomonas putida* at 10 g/kg + soil appln. at 2.5 kg/ha mixed with FYM	Talc 10^6–10^7	28	12
Seed treatment with *Bacillus subtilis* at 10 g/kg + soil appln. at 2.5 kg/ha mixed with FYM	Talc 10^6–10^7	26	14

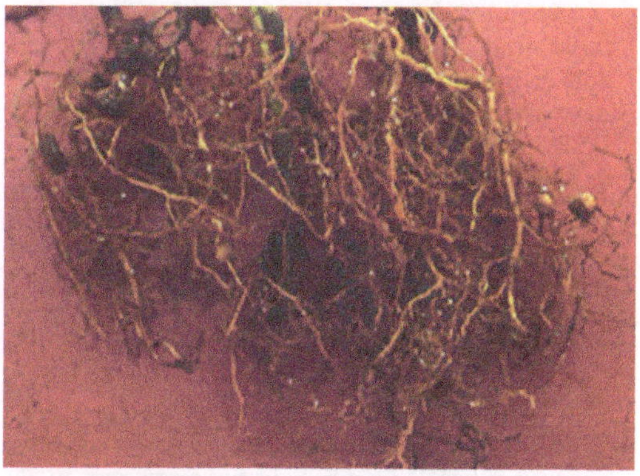

Fig. 10.16 Symptoms of pearly root caused by *Heterodera cajani* in pigeon pea

(Fig. 10.16). The symptoms of nematode injury include stunting, reduced leaf lamina size and yellowing on cotyledonary leaves. Flowers and pods are reduced in size and number and the root system may also be poorly developed. The cyst nematode retarded the emergence of leaves and reduced the number of flowering buds, flowers, growing pods and yield.

10.4.2.2 Biomanagement

Application of *P. fluorescens* at 2.5 kg/ha at sowing significantly decreased *H. cajani* population in roots at 45 DAS by 18.32 % and final nematode population by 29.55 % and thereby gave 35.93 % and 32.46 % higher pod and grain yield, respectively.

10.4.2.3 Integrated Management

(a) *Bioagents and botanicals*: Seed treatment with neem seed kernel powder at 10 % w/w + *P. fluorescens* at 10 g/kg seed was highly effective which recorded 30.11 % reduction in nematode population and 29.10 % increase in yield over control. Soil application of neem cake at 10 g/m^2 + *P. fluorescens* at 2.5 kg/ha was also highly effective which recorded 31.63 % % reduction in nematode population and 17.88 % increase in yield over control.

10.5 Cluster Bean, *Cyamopsis tetragonoloba*

10.5.1 Root Rot, *Sclerotium rolfsii*

10.5.1.1 Symptoms

The entire bark of the plant near collar region rots. The characteristic symptoms include white, cottony fungus growth on the affected portion as well as on parts in contact with soil. Gradually this hyphal mat is converted into small, mustard-like sclerotia that survive in the soil.

10.5.1.2 Biomanagement

Chaudhary and Gupta (1970) achieved control of root rot of cluster bean with *Streptomyces nigrifaciens*.

10.5.2 Root-Knot Nematode, *Meloidogyne javanica*, and Wilt, *Fusarium solani*, Disease Complex

10.5.2.1 Integrated Management

(a) *Two bioagents*: *Pseudomonas aeruginosa* and *Paecilomyces lilacinus* used alone or

Table 10.3 Effect of *P. aeruginosa* and *P. lilacinus* on plant height and control of root rot/wilt disease complex in cluster bean

			Infection %			
Treatment	Plant height (cm)	Root-knot index	*M. phaseolina*	*R. solani*	*F. solani*	*F. oxysporum*
Control	24.5	4.1	31	0	75	81
P. lilacinus (*PL*)	28.0	3.5	6	0	62	81
P. aeruginosa (*PA*)	27.7	2.0	19	0	44	50
PL + *PA*	32.0	1.4	6	0	37	44
CD (P=0.05)	2.2	0.34	6.1	6.1	6.1	6.1

together significantly ($P<0.05$) reduced infection of root-knot nematode *M. javanica* and root-infecting fungi. viz. *M. phaseolina*, *R. solani*, *F. solani* and *F. oxysporum*, on cluster bean. *P. aeruginosa* was more effective than *P. lilacinus* in reducing the *M. javanica* infection. Combined use of *P. lilacinus* and *P. aeruginosa* was more effective in reducing the infection of root-knot nematode and *F. solani* on cluster bean than either used alone. Use of *P. aeruginosa* and *P. lilacinus* significantly ($P<0.05$) increased plant height of cluster bean (Perveen et al. 1998) (Table 10.3).

References

Alstroem S (1991) Induction of disease resistance in common bean susceptible to halo blight bacterial pathogen after seed bacterization with rhizosphere pseudomonads. J Gen Appl Microbiol 37:495–501

Baker SC, Stavely JR, Thomas CA, Sasser M, Mac Fall SJ (1983) Inhibitory effect of *Bacillus subtilis* on *Uromyces phaseoli* and on development of rust pustules on bean leaves. Phytopathology 73:1148–1152

Baker CJ, Stavely JR, Mock N (1985) Biocontrol of bean rust by *Bacillus subtilis* under field conditions. Plant Dis 69:770–772

Bardin SD, Huang HC, Liu L, Yanke LJ (2003) Control of *Pythium* damping off of canola, safflower, dry pea and sugar beet by microbial seed treatment. Can J Plant Pathol 25:268–275

Bardin SD, Huang HC, Pinto J, Amundsen EJ, Erickson RS (2004) Biological control of *Pythium* damping-off of pea and sugar beet by *Rhizobium leguminosarum* bv. *viceae*. Can J Bot 82:291–296

Benhamou N, Belanger RR, Paulitz T (1996a) Ultrastructural and cytochemical aspects of the interaction between *Pseudomonas fluorescens* and Ri T-DNA transformed pea roots: host response to colonization by *Pythium ultimum* Trow. Planta 199:105–117

Benhamou N, Kloepper JW, Quadt-Hallmann A, Tuzun S (1996b) Induction of defense-related ultrastructural modifications in pea root tissues inoculated with endophytic bacteria. Plant Physiol 112:919–929

Buonassisi AJ, Copeman RJ, Pepin HS, Eaton GW (1986) Effect of *Rhizobium* spp. on *Fusarium solani* f. sp. *phaseoli*. Can J Phytopathol 8:140–146

Chakraborty U, Purkayastha RP (1984) Role of rhizobitoxine in protecting soybean roots from *Macrophomina phaseolina* infection. Can J Microbiol 30:285–289

Chaudhary KCB, Gupta S (1970) Control of root rot of guar caused by *Sclerotium rolfsii* with *Streptomyces nigrifaciens*. Neth J Plant Pathol 76:326–328

Dandurand LM, Knudsen GR (1993) Influence of *Pseudomonas fluorescens* on hyphal growth and biocontrol efficacy of *Trichoderma harzianum* in the spermosphere and rhizosphere of pea. Phytopathology 83:265–270

Das BC, Phookan AK (2002) Final project report on biological control of Sclerotinia sclerotiorum under agro-ecological conditions of Assam. Assam Agricultural University, Jorhat

De Meyer G, Hofte M (1997) Salicylic acid produced by the rhizobacterium *Pseudomonas aeruginosa* 7NSK2 induces resistance to leaf infection by *Botrytis cinerea* on bean. Phytopathology 87:588–593

Elbadry M, Taha RM, Eldougdoug KA, Gamal-Eldin H (2006) Induction of systemic resistance in faba bean (*Vicia faba* L.) to bean yellow mosaic potyvirus (BYMV) via seed bacterization with plant growth promoting rhizobacteria. J Plant Dis Protect 113:247–251

Gamliel A, Katan J (1993) Suppression of major and minor pathogens by fluorescent pseudomonads in solarized and non solarized soils. Phytopathology 81:122

Kandan A, Radjacommare R, Ramiah M, Ramanathan A, Samiyappan R (2003) PGPR induced systemic resistance in cowpea against tomato spotted wilt virus by activating defense against tomato spotted wilt virus by activating defense related enzymes and compound. In: Sarma YR (ed) Proceedings of the sixth international workshop on plant growth promoting Rhizobacteria. IISR Publishers, Calicut, pp 480–486

Manjula K, Podile AR (2001) Chitin-supplemented formulations improve biocontrol and plant growth promoting efficiency of Bacillus subtilis AF 1. Can J Microbiol. 47(7):618–625

Ordentlich A, Elad Y, Chet I (1987) Rhizosphere colonization by *Serratia marcescens* for the control of *Sclerotium rolfsii*. Soil Biol Biochem 19:747–751

Paulitz TC (1991) Effect of *Pseudomonas putida* on the stimulation of *Pythium ultimum* by seed volatiles of pea and soybean. Phytopathology 81:1282–1287

Perveen S, Ehteshamul-Haque S, Ghaffar A (1998) Efficacy of *Pseudomonas aeruginosa* and *Paecilomyces lilacinus* in the control of root rot-root knot disease complex on some vegetables. Nematol Meditter 26:209–212

Pleban S, Ingel F, Chet I (1995) Control of *Rhizoctonia solani* and *Sclerotium rolfsii* in the greenhouse using endophytic *Bacillus* spp. Eur J Plant Pathol 101:665–672

Ramanujam B, Prasad RD, Rangeswaran R (2003) Achievements in biological control of diseases with antagonistic organisms at Project Directorate of Biological Control, Bangalore. In: Ramanujam B, Rabindra RJ (eds) Current status of biological control of plant diseases using antagonistic organisms in India. Project Directorate of Biological Control, Bangalore, pp 258–271

Sharma P (2011) Evaluation of disease control and plant growth promotion potential of biocontrol agents on *Pisum sativum* and comparison of their activity with popular chemical control agent – carbendazim. J Toxicol Environ Health Sci 3(5):127–138

Siddiqui ZA, Mahmood I (1995a) Biological control of *Heterodera cajani* and *Fusarium udum* by *Bacillus subtilis*, *Bradyrhizobium japonicum* and *Glomus fasciculatum* on pigeonpea. Fund Appl Nematol 18:559–566

Siddiqui ZA, Mahmood I (1995b) Some observations on the management of the wilt disease complex of pigeonpea by treatment with a vesicular arbuscular fungus and biocontrol agents for nematodes. Biosource Technol 54:227–230

Vasudeva RS, Roy TC (1950) The effect of associated soil microflora on *Fusarium udum* Butler, the fungus causing wilt of pigeonpea (*Cajanus cajan*). Ann Appl Biol 39:229–238

Vasudeva RS, Singh GP, Iyengar MRS (1962) Biological activity of bulbiformin in soil. Ann Appl Biol 50:113–119

Vasudeva RS, Singh P, Sengupta PK, Mahmood M (1963) Further studies on the biological activity of bulbiformin. Ann Appl Biol 51:415–423

Vidhyasekaran P, Sethuraman K, Rajappan K, Vasumathi K (1997) Powder formulation of *Pseudomonas fluorescens* to control pigeonpea wilt. Biol Control 8:166–171

Wepuhkhulu M, Kimenju J, Anyango B, Wachira P, Kyallo G (2011) Effect of soil fertility management practices and *Bacillus subtilis* on plant parasitic nematodes associated with common bean, *Phaseolus vulgaris*. Trop Subtrop Agroecosyst 13:27–34

Whipps JM, Lumsden RD (1991) Biological control of *Pythium* species. Biocontrol Sci Technol 1:75–90

11 Cucurbitaceous Vegetable Crops

11.1 Pumpkin, *Cucurbita moschata*

11.1.1 Black Rot, *Didymella bryoniae*

11.1.1.1 Symptoms

Black rot is an important pre- and postharvest fruit rot of pumpkins and winter squashes. Symptoms of black rot can appear in the field before harvest and continue to develop during transit and in the storage. Small, water-soaked, dark-brown lesions develop and enlarge on fruit. As the lesions expand, they become blackish and sunken (Fig. 11.1). Fungal fruiting bodies (pycnidia) develop on infected areas. The disease gets its name from the rot, which includes blackened fungal tissue on the rind. Affected fruit may collapse during or after harvest.

11.1.1.2 Biomanagement

Disease severity on pumpkin by *D. bryoniae* is reduced by *Pseudomonas chlororaphis* and in combination with *Lysobacter gummosus*, *P. chlororaphis*, *Paenibacillus polymyxa* and *Serratia plymuthica* (Fürnkranz et al. 2012).

11.1.2 Root-Knot Nematode, *Meloidogyne javanica*, and Root Rot/Wilt, *Macrophomina phaseolina*, *Fusarium oxysporum*, *F. solani*, Disease Complex

11.1.2.1 Integrated management

(a) *Two bioagents*: *P. aeruginosa* and *P. lilacinus* used alone or together significantly ($P < 0.05$) reduced infection of root-knot nematode *M. javanica* and root-infecting fungi, namely, *M. phaseolina*, *R. solani*, *F. solani* and *F. oxysporum* on pumpkin. *P. aeruginosa* was more effective than *P. lilacinus* in reducing the *M. javanica* infection. Combined use of *P. lilacinus* and *P. aeruginosa* was more effective in reducing the infection of *M. phaseolina* and *F. oxysporum* on pumpkin than either used alone (Table 11.1) (Perveen et al. 1998).

P.P. Reddy, *Plant Growth Promoting Rhizobacteria for Horticultural Crop Protection*,
DOI 10.1007/978-81-322-1973-6_11, © Springer India 2014

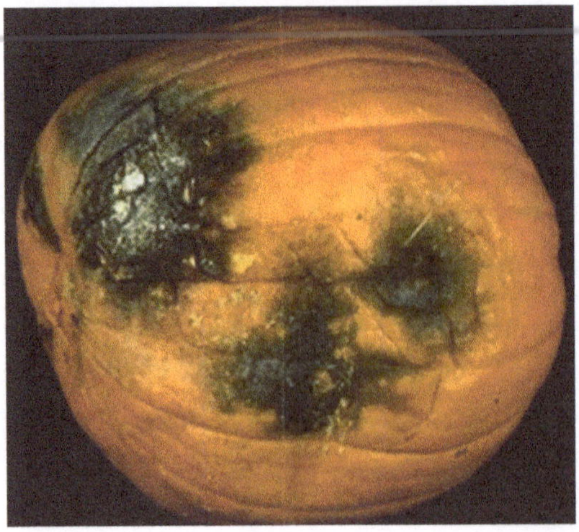

Fig. 11.1 Pumpkin fruit with black rot lesions

Table 11.1 Effect of *Pseudomonas aeruginosa* and *Paecilomyces lilacinus* on growth and control of root disease complex of pumpkin

Treatment	Plant height (cm)	Root-knot index	*M. phaseolina* infection %
Control	26.5	3.1	87
P. lilacinus (PL)	32.2	1.2	94
P. aeruginosa (PA)	30.0	0.6	50
PL + PA	28.2	0.5	19
CD (*P*=0.05)	**2.2**	**0.34**	**6.1**

11.2 Cucumber, *Cucumis sativus*

11.2.1 Damping-Off, *Pythium* sp.

11.2.1.1 Symptoms

In seedlings, a watery rot develops in the tap root and hypocotyl at or near the soil line. Damping-off or a slow decline may occur when seedling death is preceded by cotyledon and leaf chlorosis. Young seedlings wilt and die.

11.2.1.2 Biomanagement

Bacillus mycoides suppressed damping-off of cucumber seedlings caused by *P. mamillatum*. *Bacillus cereus* when sprayed or wiped on cucumbers or applied as a dip suppressed damping-off caused by *P. aphanidermatum*. Two strains of *B. subtilis* (FZB 13 and 44) tested on cucumber against *P. aphanidermatum* in greenhouse trials partially compensated for damage caused by the pathogen (Grosch et al. 1999).

Enterobacter cloacae competed for seed-derived nutrients in the spermosphere during the control of damping-off in cucumber (Maloney et al. 1994).

Pseudomonas corrugata and *P. fluorescens* induced systemic resistance to damping-off in cucumber (Zhou and Paulitz 1994). In cucumber, when *P. corrugata* 13 and *P. aureofaciens* 63-28 induced systemic resistance against *P. aphanidermatum*, salicylic acid accumulated up to sixfold in bacterized roots (Chen et al. 1999).

Formation of physical and chemical barriers at sites of potential fungal entry was detected in cucumber plants that reacted more rapidly and more efficiently to infection by pathogenic *P. ultimum* when pretreated with the endophytic biocontrol bacterium *Serratia plymuthica* strain RIGC4 (Benhamou et al. 2000).

11.2.1.3 Integrated management

(a) *Bioagents, AMF and botanicals*: Combined use of AMF *Gigaspora margarita* and 3–15 % charcoal compost (which contained antagonistic microorganisms such as *B. subtilis*, *Thermomonospora* sp. and *Thermoactinomyces* sp.) drastically reduced damping-off caused by *Pythium splendens* or *Rhizoctonia solani* in 2- and 3-week-old cucumber seedlings. Moreover, AMF and charcoal compost stimulated rooting and increased the root volume and hence plant growth (Kabayashi 1989b).

11.2.2 Anthracnose, *Colletotrichum orbiculare*, *C. lagenarium*

11.2.2.1 Symptoms

Small yellowish or water-soaked areas that enlarge rapidly and turn brown in most cucurbits appear on the foliage (Fig. 11.2). Growing leaves may be destroyed and coalescing spots may cause death of the entire leaf. On fruits, the spots are roughly circular, sunken, water soaked with dark borders. In the centre of these spots, pointwise black stromata bearing pink spore masses are produced in humid weather.

11.2.2.2 Biomanagement

PGPR as a seed treatment alone or as seed treatment plus soil drenching have protected cucumber plants against anthracnose disease (Wei et al. 1996). Liu et al. (1995c) reported that *Pseudomonas putida* strain 89B-27 and *Serratia marcescens* 90-166 have induced systemic resistance in cucumber against anthracnose caused by *C. orbiculare* in susceptible cultivars.

Fig. 11.2 Anthracnose on cucumber leaves

11.2.3 Fusarium Wilt, *Fusarium oxysporum* f. sp. *cucumerinum*

11.2.3.1 Symptoms

Yellowing of leaves progresses upward from the base of the plant. Wilting or yellowing may occur only on one side of a leaf or a branch or one side of the plant (Fig. 11.3). Yellow leaves wilt noticeably before they die. Wilting may occur at midday, when sunlight is bright and temperature is high. Infected plants are stunted and both fruit size and yield are reduced.

11.2.3.2 Biomanagement

Pseudomonas fluorescens strains 712, NIR and 346 effectively controlled *Fusarium* wilt infection in cucumber (Sneh et al. 1984).

P. putida 89B-27 and *Serratia marcescens* 90-166 induced systemic resistance to *Fusarium* wilt in cucumber (Liu et al. 1995a, b).

Seed treatment with vermiculite-based *P. putida* reduced *Fusarium* root rot of cucumber and increased the yield and growth of cucumber (Amer and Utkhede 2000).

S. marcescens strain 90-166 is known to exhibit antifungal activity against *F. oxysporum* f. sp. *cucumerinum*.

Fig. 11.3 *Fusarium* wilt of cucumber

Fig. 11.4 Cucumber wilt symptoms

11.2.3.3 Integrated management

(a) *Bioagents, AMF and botanicals*: Combined use of AMF *Gigaspora margarita* and 3–15 % charcoal compost (which contained antagonistic microorganisms such as *Bacillus subtilis*, *Thermomonospora* sp. and *Thermoactinomyces* sp.) drastically reduced *Fusarium* wilt in cucumber. Moreover, AMF and charcoal compost stimulated rooting and increased the root volume and hence plant growth (Kabayashi 1989a).

(b) *Two bioagents*: Combinations of nonpathogenic fusaria and fluorescent pseudomonads significantly reduced *Fusarium* wilt disease incidence on cucumber, although neither group alone induced significant biological control. This suggests that PGPR may be components of mixtures of biological agents in disease control strategies.

Application of a mixture of two chitinolytic bacterial strains, namely, *Paenibacillus* sp. and *Streptomyces* sp., in the ratio of 1:1 or 4:1 was more effective than when they were applied individually for the control of *Fusarium* wilt of cucumber (Singh et al. 1999).

11.2.4 Bacterial Wilt, *Erwinia tracheiphila*

11.2.4.1 Symptoms

Individual runners or whole cucumber plants wilt and die rapidly. Wilt symptoms may appear at any growth stage but are most often noticed early in the season while the plants are growing rapidly. Affected runners first appear dark green and then wilt and die. An easy way to diagnose bacterial wilt is to sever a runner close to the crown, rejoin the cut surfaces for a moment and then slowly pull them apart again. If the bacterial wilt pathogen is present, you will notice sticky strands connecting the cut surfaces as they are pulled apart (Fig. 11.4). *E. tracheiphila* is spread from one plant to another when striped cucumber beetle or the spotted cucumber beetle feed on the plant leaves.

Fig. 11.5 PGPR-treated cucumber (right) showing protection against bacterial wilt of cucumber

11.2.4.2 Biomanagement

Field experiments in cucumber demonstrated that plants grown from seed treated with PGPR sustained significantly lower populations of cucumber beetles, *Diabrotica undecimpunctata howardi* and *Acalymma vittatum*, and lower incidence of bacterial wilt disease (Fig. 11.5) compared with

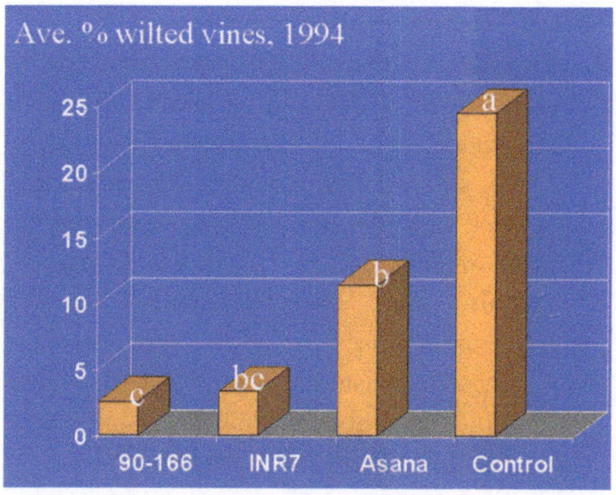

Fig. 11.6 Incidence of bacterial wilt in field cucumber with 90-166 (*Streptomyces marcescens*), INR-7 (*Bacillus pumilus*) and Asana (esfenvalerate)

nontreated control plants and plants sprayed weekly with the insecticide esfenvalerate (Zehnder et al. 1997a). Cucumber beetle feeding damage on cotyledons and stems and incidence of wilt symptoms were significantly lower on PGPR-treated plants than on nontreated plants (Fig. 11.6) (Zehnder et al. 1997a).

Treatment of cucumber seed with *P. putida* strain 89B-27 and *S. marcescens* strain 90-166 decreased the incidence of bacterial wilt disease (Kloepper et al. 1993).

Induction of systemic resistance by PGPR strains, namely, *P. putida* strain 89B-27, *S. marcescens* strain 90-166, *Flavimonas oryzihabitans* strain INR-5 and *Bacillus pumilus* strain INR-7, has significantly reduced populations of the striped cucumber beetles and the spotted cucumber beetle on cucumber. Among these strains, *S. marcescens* strain 90-166 was more effective in reducing the population of both the beetles and its efficacy was better than application of the insecticide esfenvalerate (Zehnder et al. 1999).

Delivering of *Serratia marcescens* strain 90-166 as seed dip before planting and soil application of 100 ml of the same at the rate of 10^8 cfu/ml to the sterilized soil-less planting mix after seeding reduced bacterial wilt of cucumber and controlled cucumber beetles besides increasing the fruit weight (Zehnder et al. 2001).

Fig. 11.7 Angular leaf spot on cucumber leaf

11.2.5 Angular Leaf Spot, *Pseudomonas syringae* pv. *lachrymans*

11.2.5.1 Symptoms

Leaf spots are variable in size and may be angular in shape because leaf veins limit enlargement of spots. Initial symptoms appear as brown water-soaked spots (Fig. 11.7). Initial brown spots may be surrounded by a yellow halo, become white and, as they dry, tear away from healthy tissue, producing irregular holes in affected leaves.

11.2.5.2 Biomanagement

Of all the products evaluated, the effective ones include Cease/Millstop, Citrex, KPX-B2, KleenGrow, Kocide, K-Phite, SC, MBI 10605, SUMM-IR 4 and AMV 4024 Sum. The mean disease severities for these products were in the range of 1.7–3.0 with no significant differences between them. The pathogen control plants recorded a mean disease severity of 5.0. Correspondingly, the number of bacterial leaf spots for these products ranged from 35 to 115.3 compared to the pathogen control which recorded a mean of 156.3 spots/leaf (Vijay Krishna Kumar et al. 2011).

Seed treatment of cucumber with *P. putida* strain 89B-27, *Flavimonas oryzihabitans* strain INR-5, *S. marcescens* strain 90-166 and *B. pumilus* strain INR-7 provided systemic protection against angular leaf spot by reducing total lesion diameter compared with nontreated plants (Wei et al. 1996).

Treatment of cucumber seeds with strain mixtures comprising of *B. pumilus* INR-7, *B. subtilis* GBO3 and *Curtobacterium flaccumfaciens* ME1 with a mean bacterial density of 5×10^9 cfu/seed reduced intensity of angular leaf spot and anthracnose equivalent to the synthetic elicitor Actigard and better than seed treatment with individual strains (Raupach and Kloepper 1998).

11.2.6 Cucumber Diseases

11.2.6.1 Biomanagement

ISR by *P. putida* strain 89 B-27 and *S. marcescens* strain 90-166 against anthracnose of cucumber was established by Wei et al. (1991). Later studies showed that the same PGPR strains induced systemic protection against angular leaf spot caused by *P. syringae* pv. *lachrymans* (Liu et al. 1993a), *Fusarium* wilt incited by *F. oxysporum* f. sp. *cucumerinum* (Liu et al. 1993b) and cucurbit wilt caused by *Erwinia tracheiphila* (Kloepper et al. 1993).

Seed treatment of *S. marcescens* strain 90-166 has shown ISR in cucumber against anthracnose, cucumber mosaic virus, bacterial angular leaf spot and cucurbit wilt diseases (Kloepper et al. 1993; Liu et al. 1995a, b).

11.2.6.2 Integrated management

(a) *Two bioagents*: PGPR strains INR 7 (*B. pumilus*), GBO3 (*B. subtilis*) and ME1 (*Curtobacterium flaccumfaciens*) were tested alone and in combination for biocontrol against anthracnose (*Colletotrichum orbiculare*), angular leaf spot (*Colletotrichum orbiculare*) and cucurbit wilt (*Erwinia tracheiphila*) disease. Greater suppression and enhanced consistency was observed against multiple cucumber pathogens using strain mixtures (Raupach and Kloepper 1998).

11.2.7 Cucumber Mosaic Virus (CMV)

11.2.7.1 Symptoms

The young leaves develop greenish-yellow areas and they are more translucent than those in the remaining parts of the leaf. Yellow mottle is seen on the leaves. This is followed by leaf distortion and stunting of plants. Infected fruits are mottled with yellowish-green colour and blistered (Fig. 11.8). Mosaic virus causes deformation of the cucumber plants, with small leaves and underdeveloped plants. Discolouration and spotting on the leaves are observed. This disease is spread by aphids.

Fig. 11.8 Cucumber mosaic virus on leaves and fruits

11.2.7.2 Biomanagement

Under field conditions, Jetiyanon et al. (2003) observed that a PGPR mixture containing *B. amyloliquefaciens* strain IN937a and *B. pumilus* strain IN937b induced systemic resistance against mosaic disease of cucumber caused by cucumber mosaic virus (CMV).

Induction of systemic resistance by PGPR against viral diseases has been reported in cucumber plants. Seed treatment with *P. fluorescens* strain 89B-27 and *S. marcescens* strain 90-166 has consistently reduced the number of cucumber mosaic virus-infected plants (CMV) and delayed the development of symptoms in cucumber (Raupach et al. 1996). *S. marcescens* strain Gsm01 also protects cucumber from cucumber mosaic virus.

11.2.8 Root-Knot Nematodes, *Meloidogyne* spp.

11.2.8.1 Symptoms

Root-knot nematodes cause galls or swellings on plant roots. In case of heavy attacks, galls can become very large, the root system being reduced to a swollen stump without hairs (Fig. 11.9). It restricts the uptake of nutrients from the root system to the foliage, resulting in a yellow and stunted plant.

11.2.8.2 Integrated management

(a) *Bioagents and AMF*: Combined inoculation of AMF and *P. fluorescens* had positive effect on root-knot nematode control on cucumber (Jakobsen 1999).

11.2.9 Striped Cucumber Beetle, *Acalymma vittatum*; Spotted Cucumber Beetle, *Diabrotica undecimpunctata*

11.2.9.1 Damage

Adults will attack the tender young growth of stems and leaves and the buds and petals on mature specimens (Fig. 11.10). They also carry and spread the bacterial wilt organism, *Erwinia tracheiphila*, and the cucumber mosaic virus.

11.2.9.2 Biomanagement

Several strains of PGPR like *Pseudomonas putida* strain 89 B-27, *Serratia marcescens* strain 90-166, *Flavimonas oryzihabitans* strain INR 5 and *Bacillus pumilus* strain INR 7 had induced systemic resistance in striped and spotted cucumber beetles (Zehnder et al. 1997a, b). Among these strains, *S. marcescens* strain 90-166 was more effective in reducing the population of both the beetles and its efficacy was

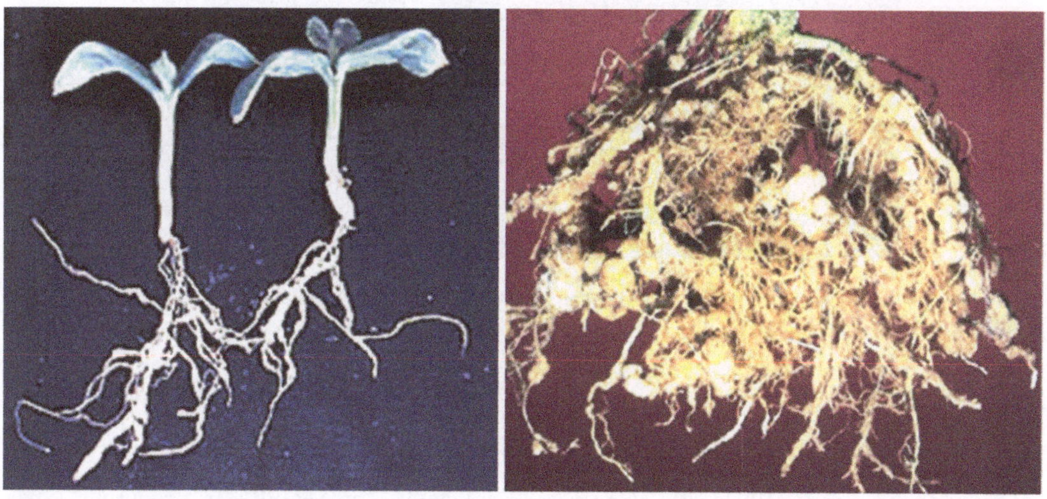

Fig. 11.9 Cucumber roots severely galled due to root-knot nematode infection

Fig. 11.10 Striped and spotted cucumber beetles

Table 11.2 Effect of PGPR on cucumber beetles under field conditions

Treatments	Mean no. of beetles/plant*
Bacillus pumilus strain INR 7	2.96[ab]
Serratia marcescens strain 90-166	2.34[a]
Fenvalerate at 56 g a.i./ha	3.62[b]
Untreated control	5.42[c]

*Means within column followed by the same letter are not significantly different

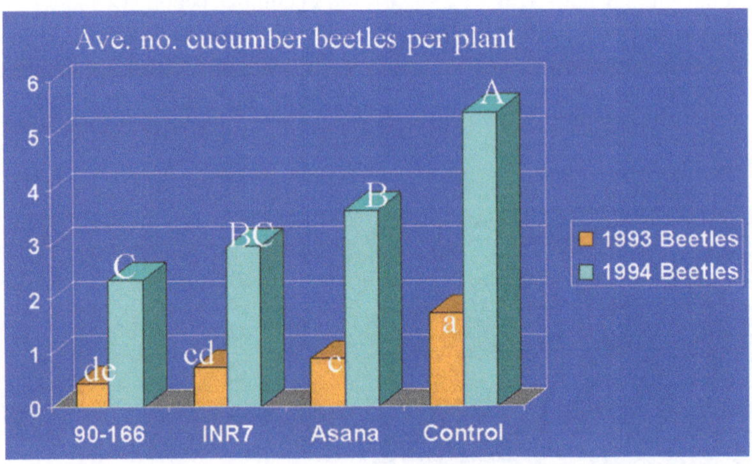

Fig. 11.11 Control of cucumber beetles in field conditions with 90-166 (*Streptomyces marcescens*), INR-7 (*Bacillus pumilus*) and Asana (esfenvalerate)

better than application of the insecticide esfenvalerate (Zehnder et al. 1997a, b). There was more reduction in the incidence of cucumber beetles (Table 11.2) (Fig. 11.11) due to the application of PGPR in cucurbits than insecticides.

11.3 Watermelon, *Citrullus lanatus*

11.3.1 Fusarium Wilt, *Fusarium oxysporum* f. sp. *niveum*

11.3.1.1 Symptoms

Symptoms include wilting and chlorosis (yellowing) of older leaves (Fig. 11.12). Wilt is most evident during the heat of the day. Plants may appear to recover by morning, only to wilt again in the afternoon. Stem cracks and brown streaks often appear near crown of plant and are associated with a red-brown exudate. The pathogens cause vascular browning that is visible in stem cross sections.

11.3.1.2 Biomanagement

The control effect of *Paenibacillus polymyxa* WY110 to watermelon *Fusarium* wilt was 52.8 % in greenhouse; it was significantly higher than the control effect of carbendazim 22.6 % (Shunhua et al. 2011).

The application in field of *Bacillus* sp. on melon crops for the management of disease caused by *F. oxysporum* showed an effect in reducing disease incidence (41 %) compared to the conventional chemical treatment (TA) and increases in yields of 26.5 % higher than TA.

B. subtilis (B1) was identified as the best biocontrol agent against *F. oxysporum* isolates of watermelon. This bioagent when coated on the seeds improved germination and could sustain the disease pressure.

Seed soaking and soil drenching with *P. aeruginosa* 231-1 suspension (10^7 or 10^8 cfu/ml) before/after/before and after pathogen inoculation reduced wilt disease infection under greenhouse conditions. The two treatments that show disease protection are seed soaking plus soil drenching with *P. aeruginosa* 231-1 (10^8 cfu/ml) before pathogen inoculation and treatment seed soaking plus soil drenching with P. *aeruginosa* 23 1-1 (10^8 cfu/ml) before and after pathogen inoculation. In field conditions, all four treatments, i.e. seed soaking plus soil drenching with bacterial suspension (10^7or 10^8 cfu/ml) with 7 or 14 days interval was found effective in the reduction of wilt disease. Especially, seed soaking plus soil drenching with *P. aeruginosa* 231-1 (10^7 cfu/ml) with 7 and 14 days intervals gave high and stable disease protection (Nguyen Thi Thu Nga et al. 2011).

11.3.2 Watermelon Mosaic Virus

11.3.2.1 Symptoms

Plants inoculated in the cotyledonary stage develop faint vein clearing symptoms in true leaves followed by systemic mottle, mosaic (Fig. 11.13) and sometimes leaf distortion.

Fig. 11.12 *Fusarium* wilt of watermelon

Fig. 11.13 Symptoms of watermelon mosaic virus

Table 11.3 Effect of *Pseudomonas aeruginosa* and *Paecilomyces lilacinus* on growth and control of root disease complex of watermelon

Treatment	Plant height (cm)	Root-knot index	*F. solani* infection %	*F. oxysporum* infection %
Control	35.1	3.8	69	12
P. lilacinus (PL)	37.5	2.8	81	0
P. aeruginosa (PA)	35.0	1.5	69	0
PL + PA	38.5	1.5	44	0
CD (P=0.05)	**2.2**	**0.34**	**6.1**	**6.1**

11.3.2.2 Biomanagement

Nallathambi et al. (2003) found that foliar spraying of *P. fluorescens* (CIAH-III) (at 1 kg talc formulation/ha) suppressed the watermelon mosaic virus. It recorded the least viral incidence (21.4 %) compared to the untreated plants (52.24 %). Seed treatment plus one foliar spray of the bacterial isolate (CIAH-11) recorded the least viral incidence (14.80 %) compared to the untreated plants (52.2 %).

11.3.3 Root-Knot Nematode, *Meloidogyne javanica* and Wilt, *Fusarium solani* Disease Complex

11.3.3.1 Integrated management

(a) *Two bioagents*: *Pseudomonas aeruginosa* and *Paecilomyces lilacinus* used alone or together significantly ($P<0.05$) reduced infection of root-knot nematode *M. javanica* and *F. solani* and *F. oxysporum* on watermelon. *P. aeruginosa* was more effective than *P. lilacinus* in reducing the *M. javanica* infection (Perveen et al. 1998). Combined use of *P. lilacinus* and *P. aeruginosa* was more effective in reducing the infection of *F. solani* on watermelon than either used alone. Use of *P. aeruginosa* and *P. lilacinus* significantly ($P<0.05$) increased plant height and fresh shoot weight on watermelon (Table 11.3).

11.4 Muskmelon, *Cucumis melo*

11.4.1 Fusarium Wilt, *Fusarium oxysporum* f. sp. *melonis*

11.4.1.1 Symptoms

Muskmelons are attacked at all stages of growth. Seeds may decay in the soil. Seedlings often wilt, collapse (damping-off) and die before or after emergence. Older plants wilt, wither and die any time during the growing season (Fig. 11.14). If

Fig. 11.14 Fusarium wilt symptoms on muskmelon

fruit forms, it is usually small and unmarketable. Necrotic lesions are often formed on the roots at sites of penetration. Wilting of runners usually progresses slowly, showing at first only during the heat of midday. Such plants recover at night, but after a few days, they wilt permanently and die. Plant wilting is promoted by a high air temperature combined with high light intensity, a high rate of evaporation and a low relative humidity. Before death the leaf margins may appear scorched. The water-conducting tissue (xylem) within the main stem is often discoloured yellow to dark brown.

In muskmelon wilt, narrow streaks of dead tissue, beginning at or near the soil line, may extend for some distance along a runner. The streaks are water-soaked at first, turning a yellowish tan to dark brown. In wet weather, a white or salmon-pink growth of the *Fusarium* fungus develops on the surface of dead muskmelon stems. Affected older plants often first show a severe stunting or yellowing of the leaves on one or more vines. The roots appear normal at first, but later turn reddish brown and finally die. Muskmelon fruit on systemically infected plants may develop sunken, irregular rot lesions due in part to the entrance of secondary fungi and bacteria.

11.4.1.2 Biomanagement

Seed treatment with wettable powder formulation of *P. putida* strain 30 and 180 suppressed the wilt of muskmelon to the extent of 63 and 50 % after 90 days of transplanting muskmelon in the field (Bora et al. 2004).

11.4.1.3 Integrated management

(a) *Bioagents and botanicals*: A bio-organic fertilizer (BIO) fortified with an antagonistic strain of *B. subtilis* Y-IVI was used to control this disease. BIO significantly reduced the disease incidence. The population of *F. oxysporum* in plant shoots of the BIO treatment was about 1,000-fold lower than the control. The population of Y-IVI remained high in muskmelon rhizosphere of the BIO treatment during the experiment. Concentration of antifungal lipopeptides, iturin A, in the BIO treatment was significantly higher than other treatments. Ten days after transplantation, the salicylic acid content in BIO-treated plant leaves was significantly higher than control. In conclusion, BIO effectively controlled muskmelon wilt, possibly because the antagonistic microbes effectively colonize the plant rhizosphere and shoots to preclude pathogen invasion. Furthermore, Y-IVI produces antifungal lipopeptides in the rhizosphere (Zhao et al. 2013).

11.5 Bottle Gourd, *Lagenaria siceraria*

11.5.1 Fusarium Wilt, *F. oxysporum* f. sp. *lagenariae*

11.5.1.1 Symptoms

The cotyledons became yellow and wilted at the early growing stage, and the foliage soon wilted from the lower position after having become yellow. The growth of wilting plants gradually declined, followed by death finally. The roots of the diseased plant were poor and their root surface became brown.

11.5.1.2 Biomanagement

Pseudomonas gladioli strain M-2196 was effective in reducing the severity of *Fusarium* wilt of bottle gourd when the roots of onion seedlings were dipped in the bacterial suspension before being transplanted alongside bottle gourd plants (Arie et al. 1987).

Bottle gourd seedlings raised from antagonist-coated seeds [*B. subtilis*-13 (8×10^7 cells/ml), *B. subtilis*-14 and *B. subtilis*-13+14] did not show any wilting symptoms and remained completely healthy. The pathogen quantum inside the host tissue was found least. The pathogen propagules per gram of host tissue were restricted and the soil population of the *F. oxysporum* was reduced very much in antagonists introduced soil (Gaikwad et al. 1987).

11.6 Bitter Gourd, *Memordica charantia*

11.6.1 Bitter Gourd Yellow Vein Mosaic Virus (BGYMV)

11.6.1.1 Symptoms

Symptoms include leaf yellowing and mosaic. The whitefly transmissibility of BGYMV was demonstrated subsequently.

11.6.1.2 Biomanagement

The bitter gourd plants treated (seed treatment) with *P. fluorescens* and *P. chlororaphis* reduced the bitter gourd yellow vein mosaic virus (BGYMV) disease incidence and increased the plant growth. It showed higher plant height and increased activity of peroxidase, polyphenol oxidase and phenol content (Rajinimala et al. 2003).

References

Amer GA, Utkhede RS (2000) Development of formulations of biological agents for management of root rot of lettuce and cucumber. Can J Microbiol 46:809–816

Arie T, Namba S, Yamashita S, Doi Y, Kijima T (1987) Biological control of Fusarium wilt of bottle gourd by mix cropping with Welsh onion or Chinese chive inoculated with *Pseudomonas gladioli*. Ann Phytopath Soc Jpn 53:531–539

Benhamou N, Gange S, Quere D, Le Dehbi L (2000) Bacterial-mediated induced resistance in cucumber: beneficial effect of the endophytic bacterium *Serratia plymuthica* on the protection against infection by *Pythium ultimum*. Phytopathology 90:45–46

Bora T, Ozaktan H, Gore E, Aslan E (2004) Biological control of *Fusarium oxysporum* f. sp. *melonis* by wettable powder formulations of the two strains of *Pseudomonas putida*. J Phytopathol 152:471–475

Chen C, Belanger RR, Benhamou N, Paulitz TC (1999) Role of salicylic acid in systemic resistance induced by *Pseudomonas* spp. against *Pythium aphanidermatum* in cucumber roots. Eur J Plant Pathol 105:477–486

Fürnkranz M, Lukesch B, Müller H, Huss H, Grube M, Berg G (2012) Microbial diversity inside pumpkins: microhabitat-specific communities display a high antagonistic potential against phytopathogens. Microb Ecol 63:418–428

Gaikwad SJ, Sen B, Meshram SU (1987) Effect of bottlegourd seedcoating with antagonists on seedlings, quantum of the pathogen inside the seedlings and population of the soil against *Fusarium oxysporum*. Plant Soil 2:205–210

Grosch R, Junge H, Kerbs B, Bochow H (1999) Use of *Bacillus subtilis* as a biocontrol agent. III. Influence of *Bacillus subtilis* on fungal root diseases and on yield in soilless culture. Z Pflanzenkr Pflanzenschutz 106:568–580

Jakobsen I (1999) Effects of *Pseudomonas fluorescens* DF 57 on growth and P uptake of two arbuscular mycorrhizal fungi in symbiosis with cucumber. Mycorrhiza 8:319–334

Jetiyanon K, Fowler WD, Kloepper JW (2003) Broad-spectrum protection against several pathogens by PGPR mixtures under field conditions in Thailand. Plant Dis 87:1390–1394

Kabayashi N (1989a) Charcoal compost fixed antagonistic microorganisms and its effect to soil-borne diseases (Abstr.). Ann Phytopath Soc Jpn 55

Kabayashi N (1989b) Suppression of *Rhizoctonia* and *Pythium* damping-off of cucumber by microorganisms in charcoal and VAM Fungi. In: Hattori H et al (eds) Research advances in microbial ecology. Japan Scientific Press, Japan, pp 242–246

Kloepper JW, Tuzun S, Liu L, Wei G (1993) Plant growth-promoting rhizobacteria as inducers of systemic disease resistance. In: Lumsden RD, Waughn JL (eds) Pest management: biologically based technologies. American Chemical Society Books, Washington, DC, pp 156–165

Liu L, Kloepper JW, Tuzun S (1993a) Induction of systemic resistance against cucumber bacterial angular leaf spot caused by *Pseudomonas syringae* pv. *lachrymans* with two plant growth promoting rhizobacteria (Abstr.). Phytopathology 82:1340

Liu L, Kloepper JW, Tuzun S (1993b) Induction of systemic resistance to *Fusarium oxysporum* by PGPR strains (Abstr.). Proc Sixth Int Congr Plant Pathol 24

Liu L, Kloepper JW, Tuzun S (1995a) Induction of systemic resistance in cucumber against Fusarium wilt by plant growth promoting rhizobacteria. Phytopathology 85:695–698

Liu L, Kloepper JW, Tuzun S (1995b) Induction of systemic resistance in cucumber against bacterial leaf spot by plant growth promoting rhizobacteria. Phytopathology 85:843–847

Liu L, Kloepper JW, Tuzun S (1995c) Induction of systemic resistance in cucumber by plant growth promoting rhizobacteria: duration of protection and effects of host resistance on protection and root colonization. Phytopathology 85:1064–1068

Maloney AP, Nelson EB, Kijk K (1994) Genetic complementation of a biocontrol- negative mutant of *Enterobacter cloacae* reveals a potential role of pathogen stimulant inactivation in the biological control of *Pythium* seed rots. In: Ryder MH, Stephens PM, Bowen GD (eds) Improving plant productivity with Rhizosphere bacteria. CSIRO Divn. of Soils, Adelide, pp 135–137

Nallathambi B, Umamaeshwari CHK, Joshi, Dharwad DG (2003) Evaluation of fluorescent pseudomonads against virus disease of watermelon. In: Proceedings of the 6th international PGPR workshop, Calicut, India, pp 181–184

Nguyen Thi Thu Nga, Nguyen Hung Vi, Tran Bach Da, Nguyen Minh Tri, Hans Jørgen Lyngs Jørgensen (2011) The effect of *Pseudomonas aeruginosa* 231-1 in protecting watermelon from wilt disease caused by *Fusarium oxysporum* f. sp. *niveum*. In: Proceedings of the 2nd Asian PGPR conference, Beijing, P.R. China, p 464

Perveen S, Ehteshamul-Haque S, Ghaffar A (1998) Efficacy of *Pseudomonas aeruginosa* and *Paecilomyces lilacinus* in the control of root rot-root knot disease complex on some vegetables. Nematol Mediterr 26:209–212

Rajinimala P, Rabindran R, Ramaiah M, Nagarajan P, Varanavasiappan S (2003) PGPR mediated disease resistance in bitter gourd against bitter gourd yellow mosaic virus. In: Proceedings of the 6th international PGPR workshop, Calicut, India

Raupach GS, Kloepper JW (1998) Mixtures of plant growth promoting rhizobacteria enhance biological control of multiple cucumber pathogens. Phytopathology 88:1158–1164

Raupach GS, Liu L, Murphy JF, Tuzun S, Kloepper JW (1996) Induced systemic resistance in cucumber and tomato against cucumber mosaic cucumo virus using plant growth promoting rhizobacteria PGPR. Plant Dis 80:891–894

Shunhua S, Ping W, Baotian X, Guoyi G (2011) Inhibitory effects and control efficacy of *Paenibacillus polymyxa* WY110 on *Fusarium oxysporum* of watermelon. In: Proceedings of the 2nd Asian PGPR conference, Beijing, P.R. China, p 393

Singh PP, Shin YC, Park CS, Chung YR (1999) Biological control of *Fusarium* wilt of cucumber by chitinolytic bacteria. Phytopathology 89:92–99

Sneh B, Dupler M, Elad Y, Baker R (1984) Chlamydospore germination of *Fusarium oxysporum* f. sp. *cucumerinum* as affected by fluorescent and lytic bacteria from *Fusarium* suppressive soil. Phytopathology 74:1115–1121

Vijay Krishna Kumar KS, Yellareddygari KR, Reddy MS, Kloepper JW, Sudini H, Zhou XG, Surendranatha Reddy EC, Yuan Zhi-lin, Yang Zhi-ling, Qi Wang, Binghai Du Shouan Zhang, Ravindra Reddy B, Bhaskara Reddy BV (2011) Over-view of commercial potential of plant growth-promoting rhizobacteria (PGPR) in the USA. In: Proceedings of the 2nd Asian PGPR conference, Beijing, P.R. China, pp 539–546

Wei G, Kloepper JW, Tuzun S (1991) Induction of systemic resistance of cucumber to *Colletotrichum orbiculare* by select strains of plant growth promoting rhizobacteria. Phytopathology 81:1508–1512

Wei L, Kloepper JW, Tuzun S (1996) Induced systemic resistance to cucumber diseases and increased plant growth by plant growth promoting rhizobacteria under field conditions. Phytopathology 86:221–224

Zehnder G, Kloepper J, Tuzun S, Yao CB, Wei G, Chambliss O, Shelby R (1997a) Insect feeding on cucumber mediated by rhizobacteria-induced plant resistance. Entomol Exp Appl 83:81–85

Zehnder G, Kloepper JW, Yao C, Wei G (1997b) Induction of systemic resistance in cucumber against cucumber beetles (Coleoptera: Chrysomelidae) by plant growth promoting rhizobacteria. J Econ Entomol 90:391–196

Zehnder GW, Murphy JF, Sikora JE, Kloepper JW (2001) Application of rhizobacteria for induced resistance. Eur J Plant Pathol 107:39–50

Zehnder GW, Yao C, Murphy JF, Sikora ER, Kloepper JW, Schuster DJ, Polston JE (1999) Microbe induced resistance against pathogens and herbivores: evidence of effectiveness in agriculture. In: Agarwal AA, Tuzun S, Bent E (eds) Induced plant defenses against pathogens and herbivores. APS Press, St. Paul, pp 335–355

Zhou T, Paulitz TC (1994) Induced resistance in the biocontrol of *Pythium aphanidermatum* by *Pseudomonas* spp. on cucumber. J Phytopathol 142:51–63

Zhao Q, Ran W, Wang H, Li X, Shen Q, Shen S, Xu Y (2013) Biocontrol of *Fusarium* wilt disease in muskmelon with *Bacillus subtilis* Y-IVI. BioControl 58:283–292

Leafy, Root and Tuber Vegetable Crops

12

12.1 Leafy Vegetable Crops

12.1.1 Lettuce, *Lactuca sativa*

12.1.1.1 Lettuce Drop, *Sclerotinia sclerotiorum*, *S. minor*

i) Symptoms

Pathogen usually invades the main stem or upper root, causing a soft, dark, watery decay. Destruction of stem tissue causes a rapid wilt, collapse and death of infected plants (Fig. 12.1). Dense masses of white fungal growth appear on the surface of rotted tissue near soil surface. Hard black irregularly shaped fungal structures called sclerotia are produced on and in rotted host tissue.

ii) Biomanagement

El-Tarabily et al. (2002) demonstrated that *Serratia marcescens* and *Streptomyces viridodiasticus* were the most promising agents of biocontrol of lettuce drop.

12.1.1.2 Bottom Rot, *Rhizoctonia solani*

i) Symptoms

Infected plants display sunken, reddish-brown lesions of varying depths and sizes on leaf petioles and midribs that touch the soil (Fig. 12.2). White to brownish mycelium grows over these lesions. Fungus can move upward from leaf to leaf until the entire head is colonized.

ii) Integrated management

(a) **Two bioagents:** The biocontrol efficacy was evaluated, combining a bacterial (*Serratia plymuthica* 3Re4-18) and a fungal (*Trichoderma viride* GB7) antagonist possessing different biocontrol activities against the bottom rot pathogen on lettuce. Both *S. plymuthica* 3Re4-18 and *T. viride* GB7 were able to suppress the bottom rot disease effectively. However, the combined application of the two antagonists improved significantly the biocontrol efficacy. Moreover, the composition of the bacterial and fungal community was more pronounced by co-inoculation of both antagonists compared to the effect of inoculation of the single agents. The colonization density of the antagonist 3Re4-18 in the rhizosphere was significantly influenced by the pathogen but not by the fungal antagonist (Grosch et al. 2011).

12.1.1.3 Root Rot, *Pythium aphanidermatum*

i) Symptoms

The pathogen causes damping-off of seedlings. The infected plants were wilting due to the decay of root and dying consequently. A brown and watery rot of the root system stunts shoot growth

P.P. Reddy, *Plant Growth Promoting Rhizobacteria for Horticultural Crop Protection*,
DOI 10.1007/978-81-322-1973-6_12, © Springer India 2014

(Fig. 12.3). Severe infections cause plants to wilt and die.

ii) Biomanagement

Seed treatment of lettuce with either vermiculite- or kaolin-based carrier of *B. subtilis* (BACT-0) significantly reduced root rot caused by *P. aphanidermatum* and it also increased the fresh weight of lettuce under greenhouse conditions.

Fig. 12.1 Lettuce drop

12.1.1.4 Root-Knot Nematodes, *Meloidogyne* spp.

i) Symptoms

Root-knot nematodes feed within the roots and cause characteristic galls on roots (Fig. 12.4). The northern root-knot nematode *M. hapla* generally occurs in cooler regions than the other three *Meloidogyne* species that prefer hot summer climates. Galls formed by *M. hapla* are spherical, distinct, and generally smaller than those caused by the three 'warm-climate' species. Plants infested as seedlings may be stunted, with patches of stunted plants becoming evident by midseason. The root-knot nematodes cause large decreases in yield of lettuce.

ii) Biomanagement

The bioagent *Streptomyces costaricus* reduced egg numbers of *M. hapla* on lettuce as compared to control (Chen et al. 2000).

iii) Integrated management

(a) *Bioagents, botanicals and physical methods*: Biofumigation with mustard under black polythene for 30 days followed by soil application of *P. fluorescens* at 2.5 kg/ha at the time of planting was on par with carbofuran at 1 kg a.i./ha in reducing the root-knot nematode population by 76.39 % and increasing the lettuce leaf yield by 23.29 %.

Fig. 12.2 Bottom rot symptoms on lettuce

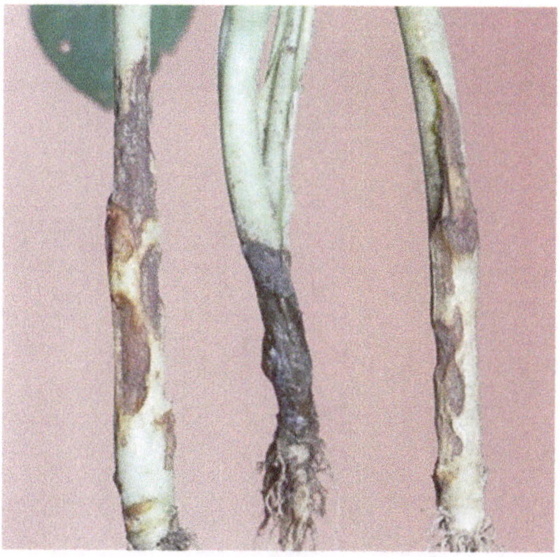

Fig. 12.3 Symptoms of root rot on lettuce

Fig. 12.4 Root-knot nematode on lettuce

12.2 Root and Tuber Vegetable Crops

12.2.1 Radish, *Raphanus sativus*

12.2.1.1 Fusarium Wilt, *Fusarium oxysporum* f. sp. *raphani*

i) Symptoms

Plants inoculated with *F. oxysporum* were dead within 24 days of inoculation. Plants inoculated with the radish isolates of *F. oxysporum* displayed 15–100 % discolouration of the vascular and cortical tissues (Fig. 12.5).

ii) Biomanagement

P. fluorescens strain WCS374 has been recorded to suppress *Fusarium* wilt in radish leading to an average increase of 40 % in yield (Bakker et al. 2007). Raajimakers et al. (1995) found that the mechanism of suppression of radish *Fusarium* wilt by *P. fluorescens* strains WCS 358 and WCS 374 was mediated by competition for iron and the level of disease incidence and population density of the bacterial strain. *P. fluorescens* strains WCS 374r and WCS 417r used commercially to control *Fusarium* wilt of radish were found to act largely through induced systemic resistance, with lipopolysaccharides of the bacterial wall strongly implicated as being involved in the induction process (Raaijimakers et al. 1995). The pseudobactin and salicylic acid produced by *P. fluorescens* strain WCS 374 induced systemic resistance to *Fusarium* wilt of radish.

Scher and Baker (1982) showed the ability of *P. putida* strain A 12 to suppress *Fusarium* wilt of radish when the soil was amended with Fe EDTA.

In the protection of radish against *Fusarium* wilt, strains of *P. putida* RE8 and *P. fluorescens* RS111a suppressed the wilting of plants by about 50 %. Both strains are able to induce systemic resistance in radish but may also possess additional antagonistic mechanisms. The combination of iron-chelating *Pseudomonas* strains (RE8 and RS111a) and inducers of systemic resistance suppressed *Fusarium* wilt of radish better than the application of individual strains (de Boer et al. 1999). Similar results were obtained in the combination of the resistance-inducing strain RE8 and *P. putida* WCS358.

Seed treatment with *P. fluorescens* strain WCS 417 has protected radish through induction of systemic resistance not only against the fungal root pathogen *F. oxysporum* f. sp. *raphani* but also against the fungal leaf pathogen *Alternaria brassicicola* (Hoffland et al. 1996).

iii) Integrated management

(a) *Two bioagents*: Strains of nonpathogenic *Verticillium lecanii, Acremonium rutilum* or

Fig. 12.5 *L* – One-sided yellowing of a radish leaf 3 weeks after inoculation of the plant roots with *Fusarium oxysporum*. *R* – Non-inoculated control and inoculated with radish isolate of *F. oxysporum*

Fusarium oxysporum with the fluorescent *Pseudomonas* spp. strains WCS358, WCS374 or WCS417 resulted in significantly better suppression of *Fusarium* wilt of radish compared to the single organism (Leeman et al. 1996).

12.2.1.2 Radish Diseases

i) Biomanagement

Seed treatment with *P. fluorescens* strain WCS 417 has protected radish through induction of systemic resistance not only against the fungal root pathogen *F. oxysporum* f. sp. *raphani* but also against the avirulent bacterial leaf pathogen *P. syringae* pv. *tomato* and fungal leaf pathogens *Alternaria brassicicola* and *F. oxysporum* (Hoffland et al. 1996). This implies that the same PGPR strain can induce resistance against multiple pathogens in the same crop.

12.2.2 Carrot, *Daucus carota*

12.2.2.1 Soft Rot, *Erwinia carotovora* s. sp. *carotovora*

i) Symptoms

The disease causes a soft, watery, slimy rot. Water-soaked irregular lesions appear on roots (Fig. 12.6). The rotted tissues are grey to brown and may have a foul odour. It decays the core of the root. Also prolonged wet weather favours disease development. It is a serious transit and storage problem if affected carrots are not discarded.

Fig. 12.6 Soft rot on carrot

In the field, tops of rotted carrots turn yellow and wilt as roots breakdown. Mostly the disease occurs after the harvest.

ii) Integrated management

(a) *Bioagents and botanicals*: Seed treatment with *Pseudomonas putida* (10^9 cfu/g) at 10 g/kg and subsequent application of 5 tons of FYM enriched with 5 kg each of *P. putida*

Fig. 12.7 Symptoms of carrot leaf blight

(10^9 cfu/g) and *T. harzianum* (10^6 cfu/g) per ha significantly reduced soft rot to 4.4 % compared to 24 % in control. Significant increase in yield (24 %) was also observed. Benefit/cost ratio calculated for marginal cost of biopesticides and returns accrued by application of biopesticides was 5.3 (Anon 2012).

12.2.2.2 Leaf Blight, *Alternaria dauci*

i) Symptoms

Alternaria leaf blight symptoms appear as dark brown to black irregularly shaped lesions on leaf blades and petioles. Spots are initially surrounded by a yellow margin and often begin on the older leaves. Leaves can be killed when spots grow together. Lesions that develop on petioles may kill the entire leaves (Fig. 12.7). Leaves weakened by blight may break off when gripped by mechanical harvesters, resulting in the roots being left in the ground. The pathogen also causes damping-off of carrot seedlings.

ii) Biomanagement

Application of *Bacillus* on carrot foliage controlled *A. dauci* incidence up to 25 %, which represents a control of two times more than the chemical treatment used for its control (Table 12.1) (Hernández et al. 2006).

Table 12.1 Product effect of *Bacillus*-based biological products and chemicals on incidence and severity of *Alternaria dauci* on carrot plants under greenhouse conditions

Treatments	Incidence (%)	Severity (%)
Strain *Bacillus* B1	25d	0.5 de
Strain *Bacillus* B3	0e	0 e
Strain *Bacillus* B9	25d	0.5 de
Strain *Bacillus* B13	25d	0.5 de
Strain *Bacillus* B15	50c	1 d
Strains *Bacillus* mix	50c	1 d
Q-L 2,000–2,000 ppm	50c	3 c
Fungicide synthetics mix[a]	75b	4.24 b
Witness	100[a]	6.75 a

[a]Chlorothalonil, iprodione, propiconazole, thiabendazole and fluazinam

12.2.2.3 Black Rot, *Alternaria radicina*

i) Symptoms

A. radicina is a seed-borne pathogen and causes carrot losses due to poor seedling establishment and damping-off. Carrots that survive early infection by *A. radicina* frequently develop a black ring of decay around the top of the stem (Fig. 12.8) and this reduces carrot quality. Older plants are particularly susceptible. Senescing leaves are often infected first, followed by infection of the crowns, which may lead to necrosis of the upper portion of the storage root. Late in the growing season, the base of the petioles turns dark brown to black and eventually the leaves are

Fig. 12.8 Black rot infected carrot roots

Fig. 12.9 Forking of carrots incited by root-knot nematodes

killed. This causes the stem tissue to break during mechanical harvesting.

ii) Biomanagement

The germination percentage, emergence percentage and the disease severity of those carrot seeds treated with *Burkholderia cepacia* no. 229 were significantly ($P<0.05$) differed from the nontreated seeds and the seed treated with other antagonists. The effects of *B. cepacia* no. 229 in promoting seed emergence and controlling disease was as good as those seeds treated with iprodione (100 ppm). Black rot lesions on carrot tap roots were significantly reduced ($P<0.05$) in size when roots were treated with *B. cepacia* no. 229 or *Bacillus amyloliquefaciens* no. 224 compared to the nontreated roots. Also, *B. cepacia* no. 229 significantly ($P<0.05$) reduced black rot on the foliage of carrot compared to check (Chen and Wu 1999).

12.2.2.4 Root-Knot Nematodes, *Meloidogyne* spp.

i) Symptoms

M. incognita was responsible for 56.64 % loss in yield of carrots. The primary symptoms include formation of galls on the side roots of tubers, formation of forked roots, split carrot tubers and spindle-shaped enlargements on tap and side roots causing yield losses. Root-knot nematodes develop characteristic forking of the roots in carrots (Fig. 12.9).

ii) Integrated management

(a) *Bioagents and botanicals*: Seed treatment with *P. fluorescens* at 10 g/kg (with 1×10^9 cfu/g) and subsequent field application of 5 MT of FYM enriched with *P. fluorescens* (with 1×10^9 cfu/g) per ha significantly reduced root-knot nematodes in carrot roots by 79 %. The yield increase was to the tune of 29.8 % with a benefit/cost ratio of 13.6.

Seed treatment with *P. putida* (10^9 cfu/g) at 10 g/kg and subsequent application of 5 tons of FYM enriched with 5 kg each of *P. putida* (10^9 cfu/g) and *T. harzianum* (10^6 cfu/g) per ha significantly reduced root-knot nematode population in roots by 77 %. Significant increase in yield (24 %) was also observed. Benefit/cost ratio calculated for marginal cost of biopesticides and returns accrued by application of biopesticides was 5.3 (Anon 2012).

12.2.2.5 Reniform Nematode, *Rotylenchulus reniformis*

i) Symptoms

General symptoms include reduced root systems, leaf chlorosis, overall stunting of host plants, and reduced yields and plant longevity. Female nematodes and their eggs are often visible when plant roots are viewed under a dissecting microscope.

ii) Biomanagement

Seed treatment of carrot with *P. fluorescens* at 20 g/kg significantly reduced the reniform nematode population both in soil and roots and increased root colonization by the bioagent and yield (Rao and Shylaja 2004) (Table 12.2).

iii) Integrated management

(a) *Bioagents and botanicals*: Seed treatment with *P. fluorescens* at 10 g/kg (with 1×10^9 cfu/g) and subsequent field application of 5 MT of FYM enriched with *P. fluorescens* (with 1×10^9 cfu/g) per ha significantly reduced reniform nematodes in carrot roots by 75 %. The yield increase was to the tune of 29.8 % with a benefit/cost ratio of 13.6.

Seed treatment with *P. putida* (10^9 cfu/g) at 10 g/kg and subsequent application of 5 tons of FYM enriched with 5 kg each of *P. putida* (10^9 cfu/g) and *T. harzianum* (10^6 cfu/g) per ha significantly reduced reniform nematode population in roots by 70 77 %. Significant increase in yield (24 %) was also observed. Benefit/cost ratio calculated for marginal cost of biopesticides and returns accrued by application of biopesticides was 5.3 (Anon 2012).

12.2.2.6 Root-Knot Nematode, *Meloidogyne incognita*, and Bacterial Wilt, *Erwinia carotovora* s. sp. *carotovora*, Disease Complex

i) Integrated management

(a) *Bioagents and botanicals*: The neem cake enriched with *P. fluorescens* applied at 10 g/m^2 increased the root colonization of the bioagent and reduced the incidence of *M. incognita* and *E. carotovora* s.sp. *carotovora* by 68 and 56 %, respectively. There was also a significant increase in the yield of carrot to the tune of 23 % (Sowmya et al. 2010).

Table 12.2 Effect of seed treatment with *Pseudomonas fluorescens* for the management of *Rotylenchulus reniformis* infecting carrot

Treatment/dose/kg seed	No. of nemas/10 g roots	No. of nemas/100 ml soil	Yield (kg)/4 m^2	Root colonization (cfu/g)
P. fluorescens – 5 g	69	88	6.7	21,745
P. fluorescens – 10 g	38	56	7.8	24,574
P. fluorescens – 20 g	35	48	8.0	23,789
Control (untreated)	92	123	5.2	–
CD (P=0.05)	**13.67**	**17.34**	**1.47**	**124,374**

12.2.3 Beetroot, *Beta vulgaris*

12.2.3.1 Damping-Off, *Pythium ultimum*

i) Biomanagement

P. fluorescens and strain R20 of *P. putida*, when inoculated on to beetroot seeds, resulted in a markedly lower incidence of colonization by *P. ultimum*. The incidence of fungal colonization of beetroot seeds treated with *P. fluorescens* and *P. putida* was 6.7 and 35.7 %, respectively, compared with 90 % of untreated seeds after planting. *P. fluorescens* inhibited both mycelial growth and sporangial germination, whereas *P. putida* inhibited only mycelial growth (Osburn et al. 1982). The viscosinamide-producing *P. fluorescens* DR54 isolate (bv. I) was shown to be an effective candidate for biological control, as tested in a pot experiment with sugar beet seedlings infested with *P. ultimum* (Nielsen et al. 1998).

ii) Integrated management

(a) *Two bioagents*: *P. fluorescens* F113 and *Stenotrophomonas maltophilia* W81 when combined in a consortium improved the level of protection (through production of the antifungal secondary metabolite 2, 4-diacetylphloroglucinol and extracellular proteolytic activity, respectively) in comparison with when used singly to protect beetroot from damping-off (Dunnea et al. 1998). The combination of W81 and F113 significantly enhanced final sugar beet stands (to the level achieved with chemical pesticides) under microcosm conditions at 28 days after sowing. In a field experiment, the only inoculation treatment capable of conferring effective protection of sugar beet was that in which W81 and F113 were co-inoculated, and this treatment proved equivalent to the use of chemical fungicides.

In conclusion, when compared with single inoculations of either biocontrol strain, the combined use of a phloroglucinol-producing *P. fluorescens* and a proteolytic *S. maltophilia* improved protection of sugar beet against *Pythium-mediated* damping-off (Dunnea et al. 1998).

12.2.3.2 Seedling Blight, *Rhizoctonia solani*

i) Symptoms

Pre-emergence damping-off results in rot of seed or seedling sprouts before emergence. Post-emergence damping-off occurs after plants have emerged from soil. A soft decay of tap root or rootlets causes collapse of seedlings. Infection by *R. solani* results in a water-soaked sunken lesion at ground level, causing plant to fall over. Surviving plants are stunted, and affected areas often show uneven growth (Fig. 12.10).

ii) Biomanagement

Streptomyces diastaticus was effective in suppressing seedling blight of beetroot.

Fig. 12.10 Seedling blight on beetroot

Fig. 12.11 Symptoms of cassava bacterial blight

12.2.3.3 Cercospora Leaf Spot, *Cercospora beticola*

i) Biomanagement

Bacillus mycoides isolate Bac J, a nonpathogenic phyllosphere-inhabiting bacterium, reduces *Cercospora* leaf spot of sugar beet by 38–91 % in both glasshouse and field experiments. Disease control was attributed to the bacterium's ability to induce systemic resistance.

12.2.4 Cassava, *Manihot esculenta*

12.2.4.1 Bacterial Blight, *Xanthomonas campestris* pv. *manihotis*

Cassava bacterial blight is a serious problem in the cassava-growing region of Nigeria.

i) Symptoms

Infected leaves show localized, angular, water-soaked areas (Fig. 12.11). Under severe disease attack, heavy defoliation occurs, leaving bare stems, referred to as 'candle sticks'. Since the disease is systemic, infected stems and roots show brownish discolouration. During periods of high humidity, bacterial exudation (appears as gum) can readily be observed on the lower leaf surfaces of infected leaves and on the petioles and stems. The disease is favoured by wet conditions.

ii) Biomanagement

Several bacterial antagonists such as *Bacillus cereus*, *B. subtilis* and *Pseudomonas* spp. were applied successfully to control the bacterial blight (Amusa and Odunbaku 2007).

Fig. 12.12 Cassava mosaic virus infection on cassava leaves

12.2.4.2 Cassava Mosaic Virus (CMV)

i) Symptoms

The virus disease causes characteristic mosaic symptoms. Young leaves have chlorotic areas and are often deformed. Leaves may also be reduced in size. When the infection is severe, leaves are small, misshapen, distorted and have bright yellow areas with normal green tissue in between (Fig. 12.12). In milder cases of infection, leaves may show only a mild chlorotic mosaic. The disease is transmitted by whitefly. Insect transmission alone does not cause much spread of the disease – man appears to be the main factor.

ii) Biomanagement

Among the four rhizobacteria [*Pseudomonas aeruginosa* (RB9), *Bacillus pumilus* (RB26), *B. subtilis* (EN16) and *Enterobacter cloacae* (EN22)] and 11 different combinations, the incidence of CMV was less in all the treated plants compared to control. About 53 % CMV incidence was noticed in the case of *E. cloacae*-treated plots compared to 90 % in control. There was significant increase in plant height, girth and tuber yield in all the treated plants compared to control. Among the combination treatments, *B. subtilis* and *E. cloacae* showed significant reduction of CMV (31 %), but it was less than *E. cloacae* (Anon 2013).

Fig. 12.13 Leaf blight damage on leaves of colocasia

12.2.5 Colocasia, *Colocasia esculenta*

12.2.5.1 Leaf Blight, *Phytophthora colocasiae*

The leaf blight of colocasia is a major and widespread disease causing 50 % loss in tuber yield.

i) Symptoms

Leaf blight is the most destructive leaf disease of colocasia. Lesions are initially small, dark and round but rapidly enlarge to 2.5–5.0 cm in diameter and become purplish to brownish in colour (Fig. 12.13). Drops of a clear liquid exude from the spots and turn yellow, orange or purple when dry. There is usually chlorotic halo round the spots. As the disease progresses, the spots coalesce and have characteristic rings of yellow or brown colour. Eventually the whole leaf may be affected and may die.

ii) Biomanagement

Prophylactic spray of *Pseudomonas* 1–2 % checked the incidence of leaf diseases like blight.

Under polyhouse conditions, the antagonistic rhizobacterial cultures S1B3, S11B4, S13B5 and S23BS when applied as seed tuber treatment reduced the *Phytophthora* blight disease severity. In these treatments there was no disease incidence compared to control where the disease severity was 2.92 on a 0–5 disease rating scale.

Soil application of rhizobacterial cultures S4BS, S13B5 and S23BS gave complete control of leaf blight disease compared to control where the disease severity was 2.83 on a 0–5 disease rating scale. Foliar application with S1B4 and S11B3 reduced the disease severity to 0–0.33 rating compared to 2.66 in control.

Under field conditions, tuber treatment with S1B3, soil application of S13B5 or foliar application with S1B4 and S11B3 reduced the disease severity and increased the yield compared to untreated pathogen-inoculated control plants. Seed treatment with S1B3 resulted in tuber yield of 255 g/plant compared to 95.42 g in control. Soil application with S13B5 resulted in 232.65 g/plant, while in foliar application with S1B4 or S11B3, yields were 274 g and 605 g per plant, respectively. These treatments promoted the plant growth also.

The field application of bacteria in combination (seed treatment, soil treatment and foliar spray) helped in reducing the leaf area damaged due to blight by 41 % during the first peak of the disease spread and by 28 % during the second peak of the disease spread. Rhizobacteria treatment also helped in reducing the storage losses. The storage loss of tubers harvested from rhizobacteria-treated plots ranged from 4.14 % to 21.24 % compared to 26.02 % in fungicide-treated plots, resulting in 18–36 % increased yield in the field trials (Sriram and Misra 2007).

12.2.6 Yam, *Dioscorea cayenensis*

12.2.6.1 Leaf Spot, *Curvularia eragrostidis*

i) Symptoms

The symptoms are necrotic leaf spots which reduce the photosynthetic area and cause losses of 35–40 % of tuber weight (yield reduction).

ii) Biomanagement

Under greenhouse conditions the antagonists IF-82 and IF-88 strains of *B. subtilis* reduced disease severity to the extent of 75 %. IF-82 strain of *B. subtilis* was the best biocontrol agent for the yam leaf spot disease (Michereff et al. 1994).

12.2.6.2 Storage Rot, *Botryodiplodia theobromae*, *Fusarium moniliforme*, *Penicillium sclerotigenum*

i) Biomanagement

Application of *Bacillus subtilis*, isolated from the epiphytic microflora of yam tuber, showed a drastic reduction in spoilage fungi of yams during 5-month storage period (Okigbo 2002, 2005).

12.2.7 Cocoyam, *Xanthosoma sagittifolium*

12.2.7.1 Root Rot, *Pythium myriotylum*

i) Biomanagement

Pseudomonas aeruginosa PNA1 strongly reduced root rot disease of tissue culture-derived cocoyam plantlets. The biocontrol activity of PNA1 against the pathogen involves phenazines. The efficacy of PNA1 to control root rot on cocoyam was improved when the strain and the pathogen were allowed to interact for 24 h prior to transplanting cocoyam plantlets (Tambong and Hofte 2001).

References

Amusa NA, Odunbaku OA (2007) Biological control of bacterial diseases of plants in Nigeria: problems and prospects. Res J Agric Biol Sci 3:979–982

Anon (2012) Annual report 2011–12. Indian Inst. of Hort. Res, Bangalore, pp 23–24

Anon (2013) Annual report 2012–13. Central Tuber Crops Research Institute, Thiruvananthapuram, p 46

Bakker PAHM, Pieterse CMJ, van Loon LC (2007) Induced systemic resistance by fluorescent *Pseudomonas* spp. Phytopathology 97:239–243

Chen J, Abawi GS, Zuckerman BM (2000) Efficacy of *Bacillus thuringiensis, Paecilomyces marguandii* and *Streptomyces costaricanus* with and without organic amendments against *Meloidogyne hapla* infecting lettuce. J Nematol 32:70–77

Chen TW, Wu WS (1999) Biological control of carrot black rot. J Phytopathol 147:99–104

de Boer M, van der Sluis I, van Loon LC, Bakker PAHM (1999) Combining fluorescent *Pseudomonas* spp. strains to enhance suppression of Fusarium wilt of radish. Eur J Plant Pathol 105:201–210

Dunnea C, Moenne-Loccoza Y, McCarthya J, Higginsa P, Powellb J, Dowlinga DN, O'Gara F (1998) Combining proteolytic and phloroglucinol-producing bacteria for improved biocontrol of *Pythium*-mediated damping-off of sugar beet. Plant Pathol 47:299–307

El-Tarabily KA, Soliman MH, Nasser AHO, Al-Hassani HA, Sivasitharam K, McKenna F, St. Hardy GE (2002) Biocontrol of *Sclerotinia minor* using a chitinolytic bacterium and actinomycetes. Plant Pathol 49:573–583

Grosch R, Dealtry S, Berg G, Mendonça-Hagler LC, Smalla K (2011) Impact of co-inoculation of bacterial and fungal antagonists on bottom rot disease and on indigenous microbial community in the lettuce rhizosphere. In: Proceedings of the 2nd Asian PGPR conference. Beijing p 392

Hernández CFD, Aguirre AA, Lira SRH, Guerrero RE, Gallegos MG (2006) Biological efficiency of organic biological and chemical products against *Alternaria dauci* Kühn and its effects on carrot crop. Int J Exp Bot 75:91–101

Hoffland E, Hakulinem J, Van Pelt JA (1996) Comparison of systemic resistance induced by avirulent and non-pathogenic *Pseudomonas* species. Phytopathology 86:757–762

Leeman M, Den Ouden FM, van Pelt JA, Dirks FPM, Steiji H (1996) Iron availability affects induction of systemic resistance to Fusarium wilt of radish by *Pseudomonas fluorescens*. Phytopathology 86:149–155

Michereff SJ, Silveira NSS, Reis A, Mariano RLR (1994) Epiphytic bacteria antagonistic to *Curvularia* leaf spot of yam. Microb Ecol 28:101–110

Nielsen MN, Sørensen J, Fels J, Pedersen HC (1998) Secondary metabolite- and endochitinase-dependent antagonism toward plant-pathogenic microfungi of *Pseudomonas fluorescens* isolates from sugar beet rhizosphere. Appl Environ Microbiol 64:3563–3569

Okigbo RN (2002) Mycoflora of tuber surface of white yam (*Dioscorea rotundata* Poir) and postharvest control of pathogens with *Bacillus subtilis*. Mycopathologia 156:81–85

Okigbo RN (2005) Biological control of postharvest fungal rot of yam (*Dioscorea* spp.) with *Bacillus subtilis*. Mycopathology 159:307–314

Osburn RM, Schroth MN, Hancock JG, Hendson M (1982) Dynamics of sugar beet seed colonization by *Pythium ultimum* and *Pseudomonas* species: effects on seed rot and damping-off. Ecol Epidemiol 79:709–716

Raaijimakers MJ, Leeman M, Mark MP, Schot V (1995) Dose response relationships in biological control of *Fusarium* wilt of radish by *Pseudomonas* sp. Phytopathology 85:1075–1081

Rao MS, Shylaja M (2004) Role of *Pseudomonas fluorescens* (Migula) in induction of systemic resistance

(ISR) and managing *Rotylenchulus reniformis* (Linford and Oliveira) on carrot (*Daucus carota* L.). Pest Manag Hortic Ecosyst 10:87–93

Scher FM, Baker R (1982) Effect of *Pseudomonas putida* and a synthetic iron chelator on induction of soil suppressiveness to *Fusarium* wilt pathogen. Phytopathology 72:1567–1573

Sowmya DS, Rao MS, Gopalakrishnan C, Ramachandran N (2010) Biomanagement of *Meloidogyne incognita* and *Erwinia carotovora* s. sp. *carotovora* infecting carrot (*Daucus carota* L.). *National Conf. on Innovations in Nematol. Res. for Agri. Sustainability – Challenges and A Roadmap Ahead*, Tamil Nadu Agri. Univ., Coimbatore, p 97

Sriram S, Misra RS (2007) Biological control of taro leaf blight caused by *Phytophthora colocasiae* (Racib.) and storage losses with rhizobacteria. J Biocontrol 21:181–188

Tambong JT, Hofte M (2001) Phenazines are involved in biocontrol of *Pythium myriotylum* on cocoyam by *Pseudomonas aeruginosa* PNA1. Eur J Plant Pathol 107:511–521

Part IV

Pest Management in Plantation and Spice Crops

Plantation Crops

13

The annual production of plantation crops is 16.358 million MT from 3.576 million ha (Tables 13.1 and 13.2.) (Tiwari et al. 2014). India has emerged as the largest producer of tea, cashew nut and areca nut.

13.1 Coffee, *Coffea arabica*, *C. canephora*

13.1.1 Lesion Nematode, *Pratylenchus coffeae*

13.1.1.1 Symptoms

In nursery, *P. coffeae* causes death of coffee plants. The diseased trees are stunted, showing thin stems and, as a result, complete destruction of secondary roots and rootlets (Fig. 13.1). In some trees, severe injury caused by *P. coffeae* results in a corky region at the base of the trunk. Complete wilting of the foliage is also a typical symptom during dry seasons. Total defoliation occurs under severe attacks.

The infected plants show loss of primary roots. The lesion nematode causes wounds in roots through which other pathogenic organisms, such as fungi and bacteria, enter the root tissues. The interaction of these agents results in the formation of lesions that finally destroy the root tissues. The root system of seedlings infected by *P. coffeae* is poor and the plants can be pulled out of the ground easily. Leaves and stems may show symptoms of mineral deficiencies.

13.1.1.2 Biomanagement

Treatments with microbial consortium [*Pseudomonas* (PFK 9), *P. fluorescens* (Pf1), arbuscular mycorrhizal fungi (AMF)] 1 month ahead of lesion nematode inoculation showed significantly higher leaf area (856.5 cm^2 $plant^{-1}$), shoot length (38.2 cm), shoot girth (3.9 mm), root length (25.8 cm), root dry weight (2.36 g $plant^{-1}$) and dry weight of plant (7.4 g $plant^{-1}$) compared to other treatments. Inoculation of coffee seedlings with microbial consortium, either 1 month ahead or along with nematode, recorded significantly higher amount of chlorophyll, total phenol and plant nutrients uptake compared to individual inoculants application. Under nursery conditions, all treatments reduced the nematode infestation and enhanced plant growth compared to infested control. The results proved that application of biocontrol agents (*P. fluorescens* (Pf1), *Pseudomonas* (PFK 9), AMF) not only has a nematicidal effect on root lesion nematode but also enhances the plant growth and nutrient uptake.

13.2 Tea, *Camellia sinensis*

13.2.1 Dieback and Canker, *Glomerella cingulata*

13.2.1.1 Symptoms

Leaves of affected branches suddenly turn yellow and wilt. Branch tips usually die. Grey blotches

P.P. Reddy, *Plant Growth Promoting Rhizobacteria for Horticultural Crop Protection*,
DOI 10.1007/978-81-322-1973-6_13, © Springer India 2014

Table 13.1 Major plantation crops producing states of India (2011–2012)

State	Area ('000 ha)	Production ('000MT)	Productivity (MT/ha)
Andaman and Nicobar	27.22	78.55	2.88
Andhra Pradesh	350.00	1396.11	3.98
Assam	94.52	268.16	2.83
Bihar	15.24	97.54	6.40
Chhattisgarh	14.29	21.32	1.49
Goa	84.93	120.62	1.42
Gujarat	28.56	240.75	8.42
Karnataka	879.68	4233.40	4.80
Kerala	958.29	4171.07	4.35
Maharashtra	206.20	339.58	1.64
Meghalaya	22.62	28.83	1.27
Odisha	211.44	354.04	1.67
Tamil Nadu	594.90	4592.28	7.71
Tripura	15.71	41.29	2.62
West Bengal	51.58	286.28	5.55
Other states	21.35	88.86	4.16
Total	3576.53	16358.68	4.57

Table 13.2 Crop-wise area, production and productivity of plantation crops in India (2011–2012)

Plantation crops	Area ('000 ha)	Production ('000MT)	Productivity (MT/ha)
Areca nut	464	681	1.46
Cashew nut	979	725	0.74
Cocoa	63	13	0.20
Coconut	2.071	14.940	7.21
Total	3.577	16.359	4.57

appear on the bark and stem, and then sunken areas (cankers) develop, eventually girdling the stem (Fig. 13.2). Parts of the plant above the stem canker lose vigour, wilt and die. Damaged plants show more symptoms during hot, dry weather.

13.2.1.2 Biomanagement

Bacillus sp. isolated from phylloplane of tea was found antagonistic to *G. cingulata*. Spraying the detached leaves with bacterial suspension and cell-free culture filtrate checked the disease (Chakraborty et al. 1996). Similarly, *Pseudomonas* sp. and *Micrococcus luteus* were found antagonistic to *G. cingulata*. Prior inoculation of leaves with these antagonists resulted in reduced disease severity (Chakraborty et al. 1994).

13.2.2 Blister Blight, *Exobasidium vexans*, *E. camelliae*

The blister blight is the major disease affecting the tender harvestable shoots of tea resulting in enormous crop loss.

13.2.2.1 Symptoms

Translucent spots appear on tender leaves. Tender stem also gets affected. Leaf galls are most often observed during the spring flush of growth. New shoots and leaves become enlarged, thickened and fleshy and appear abnormal. The colour of the affected areas turns from light green to nearly white or pink. The galls later rupture on the undersides of the leaves revealing a whitish mass of spores (Fig. 13.3). The galls eventually harden and become brown. Plants are seldom severely damaged.

13.2.2.2 Integrated Management

(a) *Bioagents and chemicals*: Field evaluation of *P. fluorescens* and *B. subtilis* revealed that they could provide moderate control of the disease, but when they were supplemented with external nutrients (salicylic acid and vermiwash), their efficacy was improved (Balasuriya and Kalaichelvan 2000) (Table 13.3).

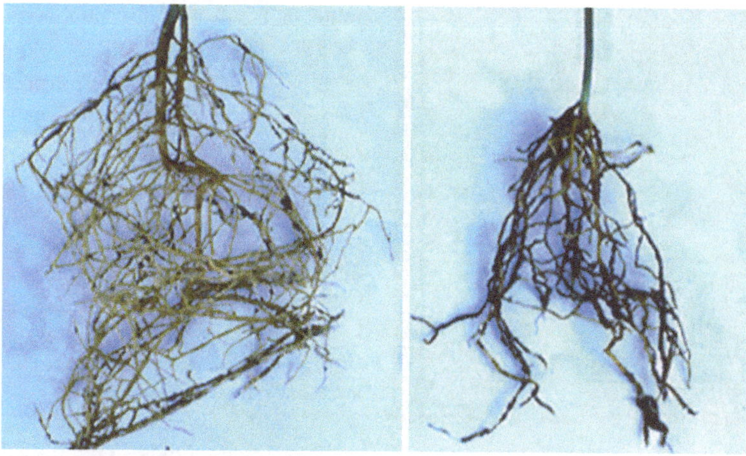

Fig. 13.1 Lesion nematode on coffee roots. *Left* – healthy. *Right* – infected

Fig. 13.2 Symptoms of dieback and canker on tea

Fig. 13.3 Symptoms of blister blight disease on tea leaves

Table 13.3 Impact of nutritional supplement on the efficacy of biocontrol agents in controlling blister blight disease of tea

Biocontrol agents	Disease incidence (%)	Protection (%)
Pseudomonas fluorescens	60.4	29.3
P. fluorescens + salicylic acid and ammonium sulphate	52.3	38.8
Bacillus subtilis	57.0	33.3
B. subtilis + salicylic acid and ammonium sulphate	47.1	44.9
Control (unsprayed)	85.5	–

Fig. 13.4 Grey blight infection on tea leaf

Table 13.4 Effect of biocontrol agents on grey blight incidence in tea

	Disease incidence (%)	
Treatment	Pretreatment	Posttreatment
Pseudomonas fluorescens	28.1	20.3 (–27.8)
P. fluorescens + vermiwash	20.5	10.6 (–48.4)
Mancozeb	20.7	8.4 (–59.7)
Carbendazim	20.6	6.6 (–68.0)
Control (unsprayed)	34.3	39.2 (+14.1)

Values in parenthesis indicate the % decrease (–) or increase (+) in disease incidence

13.2.3 Grey Blight and Dieback, *Pestalotia theae/Pestalotiopsis theae*

13.2.3.1 Symptoms

Mature leaves, young shoots and bare stalks are affected by this pathogen (Fig. 13.4). Infection on young shoot results in dieback of shoots which became a major problem mainly due to continuous shear harvesting. Grey blight adversely affects the health of the bushes, which in turn affects yield, while dieback of young shoots directly leads to substantial crop loss. Studies indicated that the disease incidence is in its peak during July to December. The crop loss due to the disease is 17 %. The economic threshold level of the disease is fixed as 18 % at a sale price of Rs.50 per kg of made tea.

13.2.3.2 Biomanagement

Field testing of biocontrol agent like *P. fluorescens* could check grey blight disease. Efficacy of this organism was improved when it was supplemented with nutrients (Table 13.4) (Chakraborty et al. 1994).

13.2.4 Brown Root Rot, *Fomes lamoensis*, and Charcoal Stump Rot, *Ustulina zonata*

13.2.4.1 Symptoms

The brown root is the most common root disease of tea. This pathogen mainly spreads by root contact. Hence, the disease occurs in distinct concentric patches in tea fields. The fungal mycelium is fawny in colour. The infected root disintegrates in advanced stages of the disease and becomes spongy. When such roots are split longitudinally, typical brownish honeycomb-like reticulations are seen. The surface of the root is heavily encrusted with soil. The fructifications are typical bracket shaped, but occur rarely.

Sudden death of bushes, white fan-shaped mycelium on the surface of wood beneath the bark and charcoal-like encrustation on bark seen in advanced stage of infection are the symptoms of charcoal stump rot of tea.

13.2.4.2 Biomanagement

Mishra et al. (2005) reported that two fluorescent *Pseudomonas* strains designated as RRLJ134 and RRLJ04 showed significant in vitro antibiosis against the brown root rot and charcoal stump rot of tea. Application of these strains resulted in significant suppression of both the diseases under gnotobiotic and nursery condition and enhanced shoot height, root length, biomass, leaf number and chlorophyll content in leaves under nursery conditions. Defence-related enzymes like phenyl alanine ammonia lyase, peroxidase and polyphenol oxidase were increased in tea cuttings treated with these bacterial strains, grown in pathogen-infested soil.

13.2.5 Red Spider Mite, *Oligonychus coffeae*

13.2.5.1 Symptoms

Leaves turn ruddy bronze due to feeding (Fig. 13.5). Crop loss studies revealed that the

Fig. 13.5 Red spider mite infestation on tea

mite can cause more than 18 % loss in crop when the infestation is severe.

13.2.5.2 Biomanagement

P. fluorescens has been identified as an entomopathogen that acts against red spider mite (Roobakkumar et al. 2011).

13.3 Coconut, *Cocos nucifera*

13.3.1 Basal Stem Rot or Thanjavur Wilt, *Ganoderma lucidum*, *G. applanatum*

In Tamil Nadu and Andhra Pradesh, Thanjavur wilt, otherwise called Ganoderma wilt, is a serious problem.

13.3.1.1 Symptoms

The symptoms are presence of bleeding patches at the stem base, premature yellowing and drooping of outer whorl of leaves and gradual drying of spindle. The other symptoms include decay of root system, flaccid spindle leaves, browning of outer leaves and appearance of bleeding patches on the basal region of the stem (Fig. 13.6). In advanced stage, bracket of fungus will be seen at the base of the tree.

13.3.1.2 Integrated Management

(a) *Bioagents, AMF and botanicals*: Field trials carried out at Veppankulam, Tamil Nadu, by soil application of *Azospirillum*, phosphobacteria and *Gigaspora calospora*, along with 5–10 kg FYM/palm/year, reduced the disease index and increased the yield (Table 13.5) (Nambiar 1996).

Fig. 13.6 Basal stem rot in coconut

(b) *Bioagents and botanicals*: Raising seedlings in *P. fluorescens*-amended compost and soil application of *P. fluorescens* along with neem cake/compost in the main field was useful in minimizing the incidence of basal stem rot of coconut.

13.3.2 Bud Rot, *Phytophthora palmivora*

13.3.2.1 Symptoms

Bud rot disease of coconut is characterized by withering, rotting of spindle leaf and foul smell (Fig. 13.7). Tender leaf and soft tissues of the crown rot into a slimy mass and emit foul odour.

13.3.2.2 Biomanagement

Bacillus subtilis, *B. macerans* and *B. amyloliquefaciens* have all been found to be checking the growth of the pathogen.

Soil drenching with *B. amyloliquefaciens* gave effective control of bud rot disease.

Table 13.5 Effect of biofertilizers on basal stem rot disease of coconut and nut yield

	Disease index		Nut yield/palm	
Treatment/dose	1990 (Initial)	1993	1990–1991	1992–1993
Azospirillum – 200 g/palm + FYM	0.4	56.5	90	90
Phosphobacteria – 200 g/palm + FYM	0.6	9.2	86	102
Control	20.8	74.3	80	76
CD (*P*=0.05)	**2.7**	**4.5**	**NS**	**3**

Fig. 13.7 Bud rot on coconut

Fig. 13.8 Stem bleeding on coconut

13.3.3 Stem Bleeding, *Thielaviopsis paradoxa*

13.3.3.1 Symptoms

The characteristic symptom of stem bleeding is the dark gummy exudation from the trunk. Exudation of reddish brown liquid is observed through longitudinal cracks in the trunk, generally at the base of the trunk (Fig. 13.8). Bleeding patches spread throughout as the disease advances. The liquid oozing out dries up and turns black. Tissues below the lesions rot and turn yellow and then black. Premature yellowing of leaves is observed in the outer whorl. Trunk gradually tapers at apex and crown size becomes reduced.

13.3.3.2 Integrated management

(a) *Bioagents and botanicals*: Raising seedlings in *P. fluorescens*-amended compost and soil application of *P. fluorescens* along with neem cake/compost in the main field was useful in minimizing the incidence of stem bleeding disease of coconut.

13.3.4 Leaf Rot, *Colletotrichum gloeosporioides*, *Exserohilum rostratum*, *Fusarium* sp.

13.3.4.1 Symptoms

Leaf rot disease of coconut is an important disease prevalent in Kerala state which disfigures the leaves and reduces yield substantially. Leaf rot

Fig. 13.9 Symptoms of leaf rot on coconut

initially appears as minute water-soaked lesions on the emerging spindle with different shapes and colours and occasionally on other young tender leaves. These lesions enlarge and coalesce leading to extensive rotting especially under favourable environmental conditions (high rainfall, RH, low maximum temperature, etc.). Mould growth (spore masses and mycelium of fungi) on the surfaces of affected leaflets is observed. The rotting may progress into the interior of the spindle also. The infected spindle gradually decays with a foul smell, attracting many insects – ants, earwigs, maggots, etc. The tips of leaflets and midribs also often become black and shrivelled. The tips of rotten spindle leaflets sometimes stick together while the base of the leaflets are still open. The rotten tissues dry up, turn black and fall off (Fig. 13.9).

13.3.4.2 Biomanagement

Spindle application of *Bacillus* spp. was found to give up to 57 % reduction in the leaf rot disease index. The newly emerged spindle leaves were completely free from any rotting symptoms. Basin + spindle application of *Bacillus* spp. was also on par with this treatment (Gunasekaran et al. 2003).

P. fluorescens applied onto the axil of the spindle region of the palms at 50 g material suspended in 1 l of water per palm or applied to the basins of the palms at 100 g/palm twice in a year during April to May and September to

Fig. 13.10 Red palm weevil damage on coconut and adult beetle

October reduced leaf rot disease index and suppressed the disease to an extent of 64 % in palms (Srinivasan 2003).

Two antagonistic bacteria (*B. subtilis* and *P. fluorescens*) were able to exert antagonism against the pathogens. *B. subtilis* (63–74 % inhibition) had an edge over *P. fluorescens* (61–69 % inhibition) in the inhibition of leaf rot pathogens.

13.3.4.3 Integrated Management

(a) *Bioagents and cultural methods*: Application of consortium of PGPR (*B. subtilis* and *P. fluorescens*) along with phytosanitation was found effective for the management of leaf rot disease.

13.3.5 Red Palm Weevil, *Rhynchophorus ferrugineus*

The pest causes serious losses in certain endemic pockets of Kerala, Karnataka, Tamil Nadu, Assam and Maharashtra particularly in areas with high water table.

13.3.5.1 Damage

Red palm weevil grubs feed on the inner tissues of trees. Infected palms show holes on trunks and a viscous brown fluid oozes out (Fig. 13.10). Chewed up fibre is extruded through the holes. Longitudinal splitting of leaf base and wilting of inner leaves can be observed. Gnawing sound produced by grubs feeding inside will be audible. Symptoms become clear at an advanced stage when the crown topples. Eventually the tree dies.

Fig. 13.11 Eryophid mite damage on coconuts

13.3.5.2 Biomanagement

Pseudomonas aeruginosa is one of the potent bacterial pathogens with an LD_{50} value in between 10^3 and 10^4.

13.3.6 Eriophyid Mite, *Aceria guerreronis*

Recently coconut gardens in some parts of Kerala, Tamil Nadu, Andhra Pradesh and Karnataka were seriously affected by a non-insect pest called eriophyid mite.

13.3.6.1 Damage

The mites infest by sucking sap from the soft tissues of buttons. In the initial stages, symptoms are seen as triangular patches close to perianth. Later, because of the continuous desapping by various stages of mites present beneath the inner bracts of perianth, brown-coloured patches are formed (Fig. 13.11). As the nuts grow in size, the injured patches become warts and then develop into longitudinal splits on the surface of nuts. The liquid oozing from these patches dries, and as a result, dried decayed matter is noticed. The damage thus caused affects the quality of husk and dehusking becomes difficult.

13.3.6.2 Biomanagement

Soil application of powder formulation of *P. fluorescens* (strain Pf1) at 100 g/tree once in 2 months reduces eriophyid mites in coconut (Table 13.6).

Table 13.6 Effect of *Pseudomonas fluorescens* on the incidence of eriophyid mite, *Aceria guerreronis*

Treatment/dose/tree	No. of mites/4 mm^2
Pseudomonas fluorescens (strain Pf1) 100 g	2.9
AMF 100 g + *Azospirillum* 100 g + phosphobacteria 100 g	22.7
Nutrient mixture	18.5
Carbofuran 250 g	11.0
Monocrotophos 15 ml + water 15 ml	13.8
Control	16.5

Fig. 13.12 Foot rot of areca nut

13.4 Areca Nut, *Areca catechu*

13.4.1 Foot Rot or Anabe Disease, *Ganoderma lucidum*

It is more prevalent in neglected gardens, causing 7 % losses.

13.4.1.1 Symptoms

The initial visible symptom is yellowing of outer whorl of leaves which gradually extends to inner whorls followed by wilting and drooping. The development of inflorescence and nuts is arrested. Nuts already formed are shed. At later stages, the weakened crown topples off, leaving a bare trunk. All around the base of the palm, brownish patches appear which exude a brown liquid (Fig. 13.12). The interior of the stem at the basal region is discoloured and rotten, emitting a foul smell. The infection extends to roots and gets discoloured, brittle and dry. The fungal invasion interrupts the uptake of water and nutrients by the palm, leading to yellowing and wilting.

13.4.1.2 Biomanagement

Antagonism of *Bacillus coagulans* and *Streptomyces* sp. on *G. lucidum* has been reported (Menon 1963).

13.5 Betel Vine, *Piper betel*

13.5.1 Foot and Root Rot, *Phytophthora nicotianae*

13.5.1.1 Symptoms

The plant loses lustre; leaves droop while still green, turn yellow and drop. Leaves turn brown and later turn black and rot (Fig. 13.13). Blackish brown marks will be visible on the stem at ground level or slightly above. The stem is discoloured and later rots. Stem becomes disintegrated.

13.5.1.2 Biomanagement

S. marcescens NBRI1213 is an important bacterium that is reported to induce systemic resistance and protects betel vine against foot and root rot disease.

13.5.2 Collar/Basal Rot, *Sclerotium rolfsii*

13.5.2.1 Symptoms

The plants are usually attacked at ground level. Dense white cotton like mass of threads (mycelium) is seen on stems. This causes rotting of the affected portion causing wilting and ultimate death of plants.

13.5.2.2 Biomanagement

Consortium of two strains of *P. fluorescens* (NBRI-N6 and NBRI-N) showed significant increase in yield of betel vine leaves. In vitro antagonism of *B. subtilis* against *S. rolfsii* was established (Agarwal et al. 1997). Field trials demonstrated that strain *P. fluorescens* NBRI-N6 was better than *P. fluorescens* NBRI-N in increasing the yield of betel vine significantly, whereas a consortium of the two strains controlled the collar rot disease more than either of the strains (Singh et al. 2003).

13.5.3 Root-Knot Nematode, *Meloidogyne incognita*, and Foot Rot, *Phytophthora capsici*, Disease Complex

13.5.3.1 Integrated Management

(a) *Two bioagents*: Combined application of *Pseudomonas* sp. (Pfbv 22) and *Bacillus* sp. (Bbv 57) gave significant reduction in nematode infestation (gall index, number of egg-laying females and soil population), wilt disease incidence and increase in leaf yield. The treatment also enhanced the biochemical markers responsible for induced systemic resistance such as peroxidase, polyphenol oxidase and phenylalanine ammonia lyase (Jonathan et al. 2006) (Table 13.7).

Fig. 13.13 Foot rot of betel vine

Table 13.7 Effect of rhizobacterial formulations on leaf yield, nematode and wilt incidence in betel vine under glasshouse conditions

Treatment/dose (2.5×10^8 cfu/g)	Plant height (cm)	No. of leaves/vine	Nema popn. in 250 ml soil	Gall index (0–5 scale)	Wilt index (0–5 scale)
Pseudomonas spp. – Pfbv 22	154	195	168	3.0	2.4
Bacillus spp. – Bbv 57	148	201	160	3.0	2.6
P. fluorescens – Pf1	135	190	158	3.5	2.4
Pfbv 22 + Bbv 57	165	225	140	2.5	1.7
Pfbv 22 + Pf1	148	197	160	3.0	2.2
Bbv 57 + Pf1	160	190	174	3.5	2.4
Pfbv 22 + Bbv 57 + Pf1	157	191	173	3.0	2.4
Metalaxyl (0.2 %) + carbofuran (2 g/vine)	160	211	149	2.4	1.9
Control	115	163	515	5.0	3.3
CD (*P*=0.05)	**8.2**	**6.5**	**6.9**	**0.5**	**0.4**

Fig. 13.14 Black pod on cocoa

13.6 Cocoa, *Theobroma cacao*

13.6.1 Black Pod Rot, *Phytophthora palmivora*

13.6.1.1 Symptoms

Black pod disease is the major disease of cocoa prevalent during monsoon season. The affected pods become pale brownish spoiling the quality of beans. Infection appears as chocolate brown spot which spreads very rapidly and soon occupies the entire surface of the pod (Fig. 13.14). As the disease advances, a whitish growth of fungus consisting of fungal sporangia is produced over the affected pod surface. Affected pods become dark brown or black.

13.6.1.2 Biomanagement

P. fluorescens an epiphytotic bacterium isolated from abandoned cocoa gardens in Costa Rica reduced infection on five cultivars under greenhouse conditions (Galindo 1992).

References

Agarwal SC, Kharu MN, Agarwal PS (1997) Biological control of *Sclerotium rolfsii* causing lentil root rot. Indian Phytopathol 30:176–179

Balasuriya A, Kalaichelvan J (2000) Is there potential in natural tea phylloplane microorganisms in the control of blister blight leaf disease of tea (*Camellia sinensis*)? Planter Kaula Lumpur 76:409–417

Chakraborty BN, Chakraborty U, Das G, Das SK (1994) Phyllosphere microflora of tea and their interaction with *Glomerella cingulata* the causal agent of brown blight disease. Tea 15:27–34

Chakraborty BN, Das G, Das SK, Chakraborty U (1996) Antagonism of *Bacillus* sp. towards *Glomerella cingulata* causing brown blight disease of tea. In: Akram I (ed) Frontiers in plant science, 6:104–128

Galindo JJ (1992) Prospects of biological control of black pod disease of cocoa. In: Cocoa pest and disease management in South-East Asia and Australia. FAO Plant Prot Bull 112:112–116

Gunasekaran M, Srinivasan N, Gupta A (2003) Biological control of leaf rot pathogen with *Pseudomonas fluorescens*. J Plantat Crops 31:52–54

Jonathan EI, Umamaheswari R, Bommaraju P (2006) Bioefficacy of native plant growth promoting rhizobacteria against *Meloidogyne incognita* and *Phytophthora capsici*. Indian J Nematol 36:230–233

Menon P (1963) Studies in *in vitro* of *Ganoderma lucidum* (Leys) Kart. Phytopathology 48:434–438

Mishra AK, Dutta S, Kumar DB (2005) Effect of fluorescent pseudomonas strains in crop enhancement and suppression of root diseases of tea. In: Proceedings of the Asian conference on emerging trends in plant-microbe interactions. University of Madras, Chennai

Nambiar KKN (ed) (1996) Basal stem rot disease of coconut. Tech. Bull., Central Plantation Crops Res. Inst, Kasargod

Roobakkumar A, Babu A, Kumar DS, Rahman VJ, Sarkar S (2011) *Pseudomonas fluorescens* as an efficient entomopathogen against *Oligonychus coffeae* Nietner (*Acari*: *Tetranychidae*) infesting tea. J Entomol Nematol 3(5):73–77

Singh US, Mishra DS, Varshney S, Zaidi NW, Khan T, Joshi D, Sharma R, Singh S, Singh N (2003) Potential and effectiveness of fungi and bacteria as biocontrol agents for plant disease management. In: Integrated pest management: principles and application (in press)

Srinivasan N (2003) Efficacy of *Pseudomonas fluorescens* against leaf rot in root (wilt) affected coconut palms. Indian Phytopathol 56:210–211

Tiwari RK, Mistry NC, Singh B, Gandhi CP (2014) Indian horticulture database-2013. National Horticulture Boad, Guragaon, 289 pp

14 Spice Crops

India has emerged as the largest producer of spices. The annual production of spices is 5.951 million MT from 3.212 million ha (Tables 14.1 and 14.2) (Tiwari et al. 2014).

14.1 Black Pepper, *Piper nigrum*

14.1.1 Quick Wilt/Foot Rot, *Phytophthora capsici*

14.1.1.1 Symptoms

The pathogen causes black spots on leaves with fine fibre-like projection from advancing margins, which rapidly enlarge and cause defoliation (Fig. 14.1). Tender leaves and succulent shoot tips of new runner shoots trailing on the soil turn black. When the main stem is affected, the entire vine wilts followed by shedding of spikes with or without black spots.

14.1.1.2 Biomanagement

Five strains of *P. fluorescens* (IISR-11, IISR-8, IISR-51, IISR-6 and IISR-13) gave protection against foot rot (above 55 % over control) to the plants and were also found to be efficient in growth promotion. Maximum disease suppression was obtained in the plants treated with strain IISR-51 (86 %).

In vitro antagonism of *Bacillus* sp. and in vivo reduction of lesions caused by *P. capsici* in black pepper has been reported (Girija and Jubina 1997).

14.1.1.3 Integrated Management

(a) *Bioagents and AMF*: Integration of *P. fluorescens* (IISR-16) with AMF was found to be highly beneficial for the growth and vigour of black pepper rooted cuttings. The rooted cuttings raised in nursery mixture fortified with the biocontrol inoculum showed field establishment as high as 90–98 % (Sarma and Saju 2004).

(b) *Two bioagents*: Integration of *T. harzianum* (IISR-1369) with *P. fluorescens* (IISR-41) gave very effective control of foot rot (10 % disease incidence as compared to 90 % in control) and increased plant height under field conditions. Integration of *T. harzianum* (IISR-1369) with *P. fluorescens* (IISR-11 or IISR-6) suppressed the root rot to the extent of 63 % over control and improved the vigour of pepper vines (Sarma and Saju 2004). Combined formulation of *T. harzianum* and *P. fluorescens* (IISR-6) was more effective for production of healthy black pepper rooted cuttings (Thankamani et al. 2005).

(c) *Bioagents, botanicals and chemicals*: Reduction in intensity of leaf infection, yellowing, defoliation, collar infection and wilting was maximum in treatment metalaxyl spray and soil drench (1.25 ml/L) + soil application of *T. harzianum* (50 g/vine), *P. fluorescens* (100 ml/vine) and neem cake (1 kg/vine) (T1) followed by potassium phosphonate spray and soil drench (3 ml/L) + soil application of *T. harzianum* (50 g/vine), *P. fluorescens*

P.P. Reddy, *Plant Growth Promoting Rhizobacteria for Horticultural Crop Protection*, DOI 10.1007/978-81-322-1973-6_14, © Springer India 2014

Table 14.1 State-wise area and production of major spice crops in India (2011–2012)

State	Area ('000 ha)	Production ('000 MT)	Productivity (MT/ha)
Andhra Pradesh	292.82	1,129.31	3.85
Assam	93.05	261.56	2.81
Gujarat	551.67	882.14	1.59
Karnataka	265.12	502.46	1.89
Kerala	254.55	112.80	0.44
Madhya Pradesh	299.91	461.17	1.53
Maharashtra	116.52	106.47	0.91
Odisha	123.92	187.50	1.51
Rajasthan	730.51	871.64	1.19
Tamil Nadu	157.33	426.38	2.71
Uttar Pradesh	58.29	201.97	3.46
West Bengal	97.12	207.70	2.13
Other states	171.66	580.36	3.38
Total	**3,212.47**	**5,951.46**	**1.85**

Table 14.2 Crop-wise area, production and productivity of spice crops in India (2011–2012)

Spices	Area ('000 ha)	Production ('000 MT)	Productivity (MT/ha)
Ajwan	35	27	0.77
Cardamom	89	16	0.17
Chillies (dried)	805	1,276	1.58
Cinnamon/Tejpatta	3	5	1.66
Celery, dill and poppy	33	33	1.00
Clove	2	1	0.50
Coriander	558	533	0.95
Cumin	594	394	0.66
Fenugreek	94	116	1.23
Fennel	100	143	1.43
Garlic	242	1,228	5.07
Ginger	155	756	4.87
Nutmeg	17	13	0.76
Pepper	200	41	0.20
Vanilla	7	1	0.14
Tamarind	58	203	3.50
Turmeric	219	1,167	5.32
Total	**3,212**	**5,951**	**1.85**

(100 ml/vine) and neem cake (1 kg/vine) (T2) (Table 14.3) (Shashidhara 2007).

14.1.2 Leaf Rot, *Rhizoctonia solani*

14.1.2.1 Symptoms

The disease is often serious in nurseries during April–May when warm humid conditions prevail. The fungus infects both leaves and stems. Greyish sunken spots and mycelial threads appear on the leaves. The infected leaves are attached to one another with the mycelial threads. On stem, the infection occurs as dark brown lesions, which spread both upwards and downwards. The new flushes subtending the points of infection gradually droop and dry up.

Fig. 14.1 Symptoms of foot rot on black pepper

Table 14.3 Effect of bioagents, botanicals and chemicals on the management of foot rot on black pepper

Treatment	Leaf infection (%)	Leaf yellowing (%)	Defoliation (%)	Collar infection (%)	Wilting (%)
T1	2.44	1.65	0.62	5.56	0.00
T2	5.97	6.58	5.56	10.07	5.56
T3 – control	52.92	62.76	52.67	77.77	50.00
CD (P=0.05)	**1.229**	**3.733**	**4.646**	**2.567**	**2.966**

T1 – metalaxyl spray and soil drench (1.25 ml/L) + soil application of *T. harzianum* (50 g/vine), *P. fluorescens* (100 ml/vine) and neem cake (1 kg/vine)

T2 – potassium phosphonate spray and soil drench (3 ml/L) + soil application of *T. harzianum* (50 g/vine), *P. fluorescens* (100 ml/vine) and neem cake (1 kg/vine)

14.1.2.2 Biomanagement

Foliar spray with liquid formulation of *P. fluorescens* and drenching at both pre- and post-monsoon treatments provided good protection. The treated plants had 35 % disease index as compared to 61 % in control and the yield was 2.34 kg/vine as compared to 0.78 kg/vine in control. However, this was on par with Bordeaux mixture spray with 29 % disease index and 3.65 kg yield/vine.

14.1.3 Anthracnose/Pollu Disease, *Colletotrichum gloeosporioides*

14.1.3.1 Symptoms

The pathogen infects the leaves causing angular to irregular yellowish brown to dark brown necrotic leaf spots with a chlorotic halo. Affected berries show brown sunken patches during early stages, and further development is affected. In later stages, the discolouration gradually increases and berries show the characteristic cross splitting. Berries turn black and dry (Fig. 14.2).

14.1.3.2 Biomanagement

Foliar spray and soil drenching with *P. fluorescens* both pre- and post-monsoon treatments provided reasonable protection. The treated plants had 35 % disease index compared to 61 % in control. The yield was 2.34 kg/plant in treated plants compared to 0.78 kg in control (Anon 2004).

14.1.4 Root-Knot Nematodes, *Meloidogyne incognita*, *M. javanica*

Up to 91 % root-knot infestation was reported from Para, Brazil, and Kerala, India. An initial

Fig. 14.2 Anthracnose on black pepper leaf and panicles

inoculum level of 10 J_2 per rooted cutting was found to reduce growth by 15 %, while at 1,00,000 J_2 level 50 % reduction in growth was observed over a 1-year period. More than 50 % death of transplants occurs in field planting of infected cuttings. *M. incognita* was responsible for a 46 % loss in yield of black pepper.

14.1.4.1 Symptoms

Prominent symptoms of root-knot infestation on black pepper are unthrifty growth and yellowing of leaves. Interveinal yellowing of the foliage is also noticeable. The leaves exhibit dense yellowish discolouration of the interveinal areas making the leaf veins prominent with deep green colour. Heavy galling of the root system is also present (Fig. 14.3).

14.1.4.2 Biomanagement

An inoculum level of 1.2×10^8 cells of *Bacillus macerans* per vine just before the monsoon period significantly reduced the root-knot nematode population in soil and roots of pepper improving the length of vine and number of leaves and weight of shoot (fresh and dry) and roots. The spores of bacteria *B. circulans*, *B. macerans*, *B. coagulans*, *B. licheniformis*, *B. subtilis* and *B. pumilus* were seen externally in the eggs and females of the root-knot nematode. Nematode count was minimum in bacterial infected samples, the percentage reduction being 35 % under natural conditions.

14.1.4.3 Integrated Management

(a) *Two bioagents*: Under field conditions, plants treated with the consortial formulation of *P. fluorescens* Pf 123 and *B. subtilis* Bs 214 significantly enhanced the yield parameters and reduced nematode infestation both in soil and roots (Table 14.4) (Devapriyaga et al. 2012).

14.2 Cardamom, *Elettaria cardamomum*

14.2.1 Capsule Rot/Azhukal, *Phytophthora meadii*, *P. nicotianae* var. *nicotianae*

14.2.1.1 Symptoms

This disease occurs during the rainy season. It affects the leaves, tender shoots, panicles and capsules. On the infected leaves, water-soaked lesions

Fig. 14.3 Symptoms of root-knot nematode infection in black pepper

Table 14.4 Efficacy of talc formulations of *Pseudomonas* and *Bacillus* isolates on root-knot nematode and yield parameters of black pepper cv. Panniyur 1 under field conditions

Treatments	No. of spikes/ vine	Wt. of spikes/ vine (g)	No. of yellow leaves/vine	No. of egg masses/5 g roots	No. of eggs/ egg mass	Gall index
Pf 123	270.30	1,928.00	13.60	30.80	155.00	2.02
Bs 214	251.00	1,898.20	17.80	37.20	169.60	2.36
Pf 123 + Bs 214	294.20	2,192.40	9.40	25.60	143.60	1.10
Carbofuran5	247.00	1,838.60	17.40	37.60	176.60	2.16
Control	172.20	906.80	26.20	48.60	260.00	4.08
CD (P=0.05)	**3.7298**	**57.9689**	**1.6746**	**1.8914**	**1.1727**	**0.2523**

appear first and rotting and shedding of leaves along the veins occur thereafter. The infected capsules become dull greenish brown and decay (Fig. 14.4). This emits a foul smell and subsequently capsules are shed. Infection spreads to the panicles also.

14.2.1.2 Biomanagement

Soil application of *B. subtilis* reduced *Phytophthora* population and suppressed the disease to the extent of 30–50 % (Suseela Bhai et al. 1993; Thomas et al. 1993).

Soil application of *B. subtilis* (PDBC) and *P. fluorescens* (PDBC) gave 98 and 96 % disease control, respectively.

Fig. 14.4 Capsule rot of cardamom

Fig. 14.5 Clump rot of cardamom

14.2.2 Rhizome Rot or Clump Rot, *Pythium vexans*, *Rhizoctonia solani*

14.2.2.1 Symptoms

The disease is prevalent during the south-west monsoon season. The symptoms include pale yellow colour on the foliage and premature death of older leaves. Collar portion of the aerial shoots become brittle and tiller breaks easily from the rhizome at the bulbous base (Fig. 14.5). Water-soaked lesions appear first on leaves. The pathogen causes rotting and shredding of leaves along the veins. Rotting develops in the collar region emitting a foul smell. Capsules turn dull greenish brown emitting a foul smell.

14.2.2.2 Biomanagement

Soil application of *B. subtilis* reduced rhizome rot caused by *P. vexans* and *R. solani*. Percentage of mortality in treated plots ranged from 8 to 10 % compared to 16.5 % in control with *P. vexans*, and it was 13–27 % with *R solani*. In combined infection, the percentage of mortality in treated plots ranged from 8 to 11 % compared to 38 % in control (Sarma and Anandraj 1998).

The application of *P. fluorescens* at the time of planting in the nursery and main field and during pre-monsoon period in established plantations is effective.

Soil application of *B. subtilis* (PDBC) and *P. fluorescens* (PDBC) gave 61–84 % disease control.

14.2.2.3 Integrated Management

(a) *Two bioagents*: Integration of *T. harzianum* (IISR-1369) with *P. fluorescens* (IISR-11 or IISR-6) suppressed the clump rot to the extent of 36 % over control.

14.3 Ginger, *Zingiber officinale*

14.3.1 Rhizome Rot, *Pythium* spp.

Soft rot is the most serious disease of ginger in India and in some other countries. It is caused by *Pythium* spp., of which *P. aphanidermatum* is the principal species in India, although *P. butleri*, *P. gracile*, *P. myriotylum*, *P. nigriotilum* and *P. vexans* have also been recorded.

14.3.1.1 Symptoms

The bases of the aerial shoots become soft and watery and then rot. The affected plants become pale; the tips of the leaves turn yellow, followed by complete yellowing and drying up of the leaves (Fig. 14.6).

The shoots fall and cease to produce rhizomes. The infection extends to the rhizomes, the inner tissues being reduced to a soft and black, putrefying mass. Losses can be high. The disease is favoured by high moisture content of the soil with insufficient drainage.

Fig. 14.6 Rhizome rot of ginger

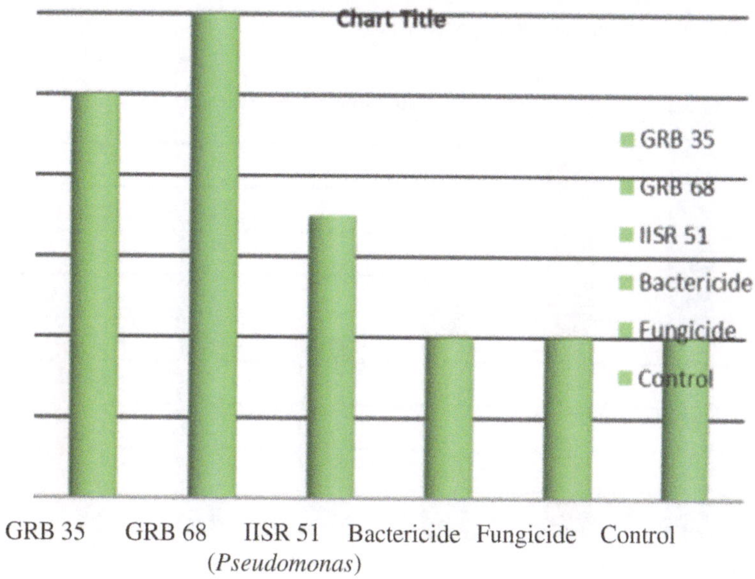

Fig. 14.7 Effect of rhizobacteria on sprouting in ginger in greenhouse

14.3.1.2 Biomanagement

The greenhouse experiments on sprouting and incidence of soft rot revealed that *B. amyloliquefaciens* (GRB 35) and *S. marcescens* (GRB 68) were superior over chemical treatments (Fig. 14.7). Both isolates (GRB 35 and GRB 68) recorded more than 75 % sprouting. Soft rot incidence was also significantly less (<10 %) compared to control (Bini et al. 2011).

The field experiments revealed that *B. amyloliquefaciens* (GRB 35) and *S. marcescens* (GRB 68) were effective for disease control and yield (Figs. 14.8 and 14.9) (Bini et al. 2011).

14.3.1.3 Integrated Management

(a) *Bioagents and botanicals*: Raising of seedlings in *P. fluorescens*-amended compost and soil application of *P. fluorescens* along with neem cake/compost in the main field was useful in minimizing the incidence of rhizome rot of ginger.

(b) *Two bioagents*: Combined application of *P. fluorescens* (IISR-11) and *T. harzianum* (IISR-1369) imparted 66.2 % survival of ginger tillers, reduced the rhizome rot infection and improved the vigour and yield of ginger plants (Sarma 2000).

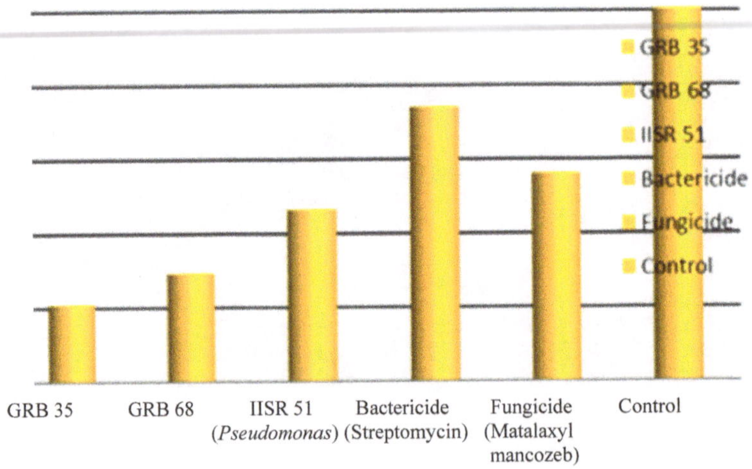

Fig. 14.8 Effect of rhizobacteria on soft rot incidence in field

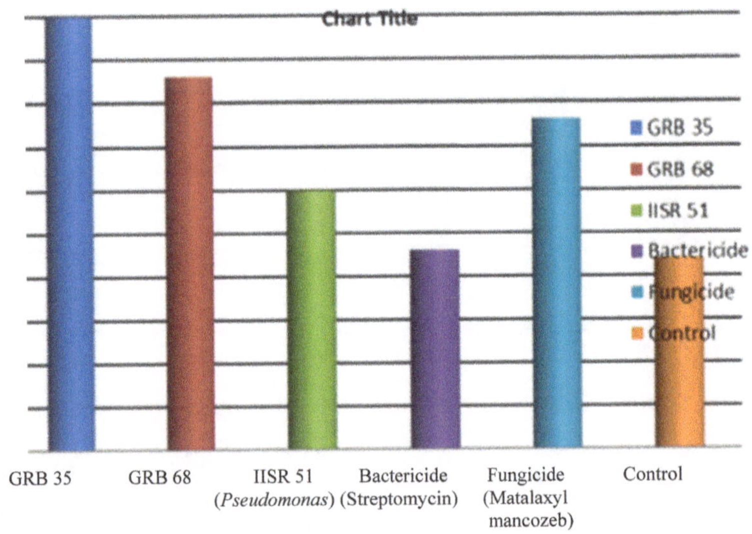

Fig. 14.9 Effect of rhizobacteria on the yield in ginger in field

Combination of *Pseudomonas* sp. and *T. harzianum* resulted in better germination and plant stand, reduced disease and increased yield. Soil application of the bioagents was more effective compared with seed treatments.

(c) *Bioagents and physical methods*: As a seed disinfection procedure, a novel technique called rhizome solarization has been devised at the Indian Institute of Spices Research, Calicut. The disinfected rhizomes when treated with *T. harzianum* and rhizobacterial strain consortia as seed treatment and soil application resulted in higher yields and growth promotion and soft rot suppression.

(d) *Bioagents and chemicals*: Integration of rhizome treatment with *Pseudomonas* sp. and fungicides [bavistin (carbendazim) + ridomil MZ (metalaxyl + mancozeb)] and soil application of *T. harzianum* was the most effective among all the tested treatments (Ram et al. 1999).

Fig. 14.10 Ginger yellows

Fig. 14.11 Performance of strain mixture of *Bacillus subtilis* and *Burkholderia cepacia* (S2BC-1 + TEPF-Sungal) (*T1*) compared to untreated control (*T2*) and pathogenic control (*T3*)

14.3.2 *Fusarium* Yellows, *Fusarium oxysporum* f. sp. *zingiberi*

14.3.2.1 Symptoms

Fusarium yellows is a very common and serious fungal disease which is specific to ginger. Infected plants are stunted and yellow; lower leaves dry out and turn brown (Fig. 14.10). Eventually all above-ground shoots dry out completely. Plant collapse is very slow (up to several weeks) compared with the rapid collapse associated with bacterial wilt infection. Diseased rhizomes show a brown internal discolouration, are normally shrivelled in appearance and eventually decay, leaving the outer shell intact with fibrous internal tissue remaining. Increased nematode infestations are usually associated with *Fusarium* rhizome rot, accentuating yield losses. *Fusarium* is also responsible for serious loss of planting pieces and poor germination. The disease spreads rapidly in the field during wet weather.

14.3.2.2 Biomanagement

B. subtilis was found effective in suppressing yellows in ginger. *B. subtilis* strain II reduced the disease incidence from 58.2 to 8.3 % and increased yield (Sharma and Jain 1978). The use of certain strains of fluorescent pseudomonads against *Fusarium* is reported.

Talc-based formulations of rhizobacterial strain XXBC-TN (*B. subtilis*) and a mixture of S2BC-1 (*B. subtilis*) with TEPF-Sungal (*Burkholderia cepacia*), which inhibits *F. oxysporum* and *F. solani*, were developed for rhizome dressing and soil application in ginger fields. The strain mixture recorded the maximum rhizome production (85.2 %) with fewer yellows and reduced rhizome rot incidences (87.8 % and 88.4 %) over the control in a polyhouse (Fig. 14.11). This was associated with an increase in the defence enzymes chitinase, β-1,3-glucanase and polyphenol oxidase. Furthermore, the strain mixture treatment promoted plant growth and enhanced rhizome production by 45.8 %. In field experiments, the PGPR strain mixture reduced yellows and rhizome rot incidences by about 50.5 %, which was comparable to that of a carbendazim and mancozeb fungicide mixture.

14.3.3 Storage Rot, *Sclerotium rolfsii*

14.3.3.1 Symptoms

This disease causes serious losses on some farms. Infection first occurs beneath the scales on the rhizomes and may progress to produce a deep brown rot over the entire surface. Affected rhizomes become enveloped in a white fungal mycelial growth which may extend up to the basal stem (Fig. 14.12). Mild infections usually produce no above-ground symptoms. Severe infections cause the stems to turn yellow and slowly dry out.

Fig. 14.12 Symptoms of storage rot of ginger

Fig. 14.13 Bacterial wilt symptoms in zinger

14.3.3.2 Integrated Management

(a) *Two bioagents*: *T. harzianum* in combination with *P. fluorescens* showed a synergistic effect in reducing the soft rot infection.

14.3.4 Bacterial Wilt, *Ralstonia solanacearum*

14.3.4.1 Symptoms

The first symptoms of the disease are yellowing and wilting of the lower leaves, which quickly spread upwards, affecting the whole plant (Fig. 14.13). In advanced stages, the base of the pseudo stem becomes water soaked, readily breaking away from the rhizome at ground level, and the plants eventually collapse. The vascular tissues become dark brown or black. When an infected stem or rhizome is cut transversely, and a little pressure applied, a milky white exudate flows freely from the cut surface. Diseased rhizomes are much darker than healthy ones. Biotype 4 was found to produce the most severe symptoms.

14.3.4.2 Biomanagement

The greenhouse experiments on sprouting, incidence of bacterial wilt and soft rot revealed that *Bacillus amyloliquefaciens* (GRB 35) and *Serratia marcescens* (GRB 68) were superior to chemical treatments. Both isolates (GRB 35 and GRB 68) recorded more than 75 % sprouting.

The field experiments revealed that two strains *B. amyloliquefaciens* (GRB 35) and *S. marcescens* (GRB 68) were effective for disease control (Fig. 14.14) and plant growth promotion. The rhizobacterial treatment recorded significantly higher yield (Bini et al. 2011).

14.3.4.3 Integrated Management

(a) *Bioagents and botanicals*: Lowest disease incidence (15.63) was recorded in the treatment of *P. fluorescens* mass cultured in the mixture of vermicompost (VC) and mustard oil cake (MOC) applied as seed treatment and soil application (Bora and Bora 2009).

14.3.5 Soft Rot, *Erwinia* sp.

14.3.5.1 Symptoms

Normally bacterial soft rot is only a storage rot. The bacteria are present in most soils but field infection usually occurs only in waterlogged areas. *Erwinia* sp. has caused severe losses of stored rhizomes on some farms, but is not considered a serious storage problem where precautions are taken. Softening of the tissue is accompanied by production of a strong odour, and the rhizome eventually collapses completely. Bacterial soft rot differs from other rhizome rots in that putrid odour is produced in this soft rot.

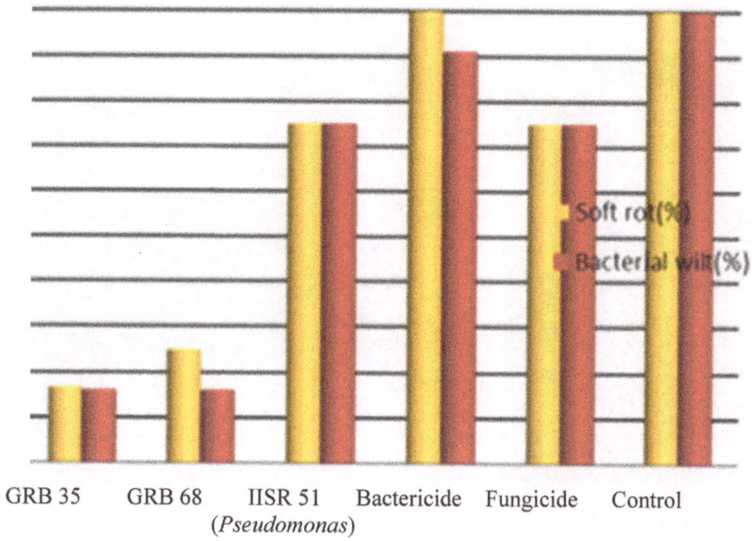

Fig. 14.14 Effect of rhizobacteria on wilt disease percentage

14.3.5.2 Integrated Management

(a) *Two bioagents*: *T. harzianum* in combination with *P. fluorescens* showed a synergistic effect in reducing the soft rot infection.

14.4 Turmeric, *Curcuma longa*

14.4.1 Rhizome Rot, *Pythium graminicolum*

14.4.1.1 Symptoms

Rhizome rot shows progressive drying up of the leaves of infected plants. The base of the aerial shoots shows water-soaked soft lesions. As the disease progresses infection gradually progress to the rhizomes, which begin to rot and become soft (Fig. 14.15). The bright orange colour of the rhizomes changes into brown. The disease may be confined to a few isolated plants or may occur in patches. In severe attacks the yield is considerably reduced.

14.4.1.2 Integrated Management

(a) *Bioagents and botanicals*: Seed treatment with *T. viride* + *P. fluorescens* at 4 g/kg seed and soil application of *T. viride* (12.5 kg/ha) and *P. fluorescens* (25 kg/ha) along with FYM at 10 MT/ha resulted in minimizing rhizome rot to 11.7 % compared to 37 % in control and higher yield of 28.6 MT/ha in treated compared to 10.3 MT/ha in control (Anon 2005).

14.4.2 Leaf Roller, *Udaspes folus*

14.4.2.1 Damage

The pest was abundant during August–October. The larvae cut and fold the leaves, remain within and feed on them. The optimum conditions for the pest's development include a temperature of 26–35 °C and relative humidity of 41–100 %.

14.4.2.2 Biomanagement

Enterobacter cloacae and *Pseudomonas* sp. were effective for the management of leaf roller larvae.

14.5 Vanilla, *Vanilla planifolia, V. andamanica*

14.5.1 Wilt, *Fusarium oxysporum* f. sp. *vanillae*

Fusarium wilt is the most serious disease of vanilla. The disease is more prevalent in younger plantation especially during the monsoon season.

Fig. 14.15 Rhizome rot of turmeric

Fig. 14.16 Vanilla *Fusarium* wilt

14.5.1.1 Symptoms

The infection starts at leaf axil and spreads to the internodal region resulting in rotting and drying of the stem above the point of infection. The fungus also causes leaf rot on the plant (Fig. 14.16).

14.5.1.2 Integrated Management

(a) *Two bioagents*: Soil application of rhizobacterial strain consortia (IISR-147 and IISR-148) with *T. harzianum* was effective in disease suppression of *F. oxysporum* f. sp. *vanillae*.

The consortia of rhizobacterial isolates, viz., (1) *P. fluorescens* isolates (HSR13, IISR51), *Bacillus* sp. (IISRIS2) and *B. polymixa* (IISR909); (2) *P. fluorescens* isolates (IISR13, IISR51), *Bacillus sp*. isolates (IISR148, IISR149, IISR152, IISR907), *B. polymixa* (IISR909) and *B. lentus* (IISR906); (3) *P. fluorescens* isolates (IISR6, IISR13, IISR51), *Bacillus* sp. isolates (IISR147, IISR151, IISR1S2, IISR153) and *B. polymixa* (IISR909); and (4) *P. fluorescens* isolates (IISR6, IISR5.1, IISR147, IISR148, IISR149 and IISR907) and *B. lentus* (IISR906), gave significant disease reduction (88.22–92.85 %) when compared to control. However, among the four rhizobacterial consortia, three showed maximum disease reduction of 92.9 % (Suseela Bhai and Kumar 2008).

Table 14.5 Evaluation of microbial antagonists against *Fusarium oxysporum* of vanilla plants

Pre inoculation with biocontrol agents	Leaf infection (%)[a]
T. virens	12.37
T. harzianum	8.49
P. fluorescens	7.18
P. putida	20.74
P. fluorescens + *T. harzianum*	7.03
Control (no biocontrol agent)	90.27
CD (*P* = 0.05 %)	**1.95**

[a]Values are means of three replicates

Combined applications of *T. harzianum* and *P. fluorescens* gave maximum reduction in percentage leaf infection (Table 14.5) (Athul Sandheep et al. 2012).

Fig. 14.17 *Phytophthora* rot symptoms on vanilla

14.5.2 *Phytophthora* Rot, *Phytophthora meadii*

14.5.2.1 Symptoms

The pathogen causes rotting of beans, leaves and stems (Fig. 14.17). In severe cases, all the beans in a bunch are completely rotten. The disease is more severe during the monsoon period, especially in shaded plantations and poorly drained soils.

14.5.2.2 Biomanagement

Spraying *P. fluorescens* during the pre-monsoon and post-monsoon period in the established plantations is effective in disease control.

14.5.2.3 Integrated Management

(a) *Two bioagents*: Soil application of rhizobacterial strain consortia (IISR-147 and IISR-148) with *T. harzianum* was effective in disease suppression of *P. meadii*.

14.6 Coriander, *Coriandrum sativum*

14.6.1 Wilt, *Fusarium* sp.

Of late, coriander wilt is becoming a serious disease which causes significant yield loss.

Table 14.6 Effect of seed treatment with biocontrol agents on wilt incidence in coriander

Treatment	Wilt incidence (%)	Yield (MT/ha)
Seed treatment with *Bacillus subtilis*	14.8	0.290
Seed treatment with *Pseudomonas fluorescens*	29.5	0.172
Seed treatment with carbendazim	14.2	0.285
Control	28.9	0.168
CD (P=0.05)	**5.8**	**26.4**

14.6.1.1 Symptoms

Symptoms include drooping of terminal portions, followed by withering and drying up of leaves, discolouration of vascular system of root, partial wilting and arrest of plant growth. Leaves become pinkish yellow to yellow. Sterility is often noticed in such plants. Seeds, if formed, are immature and light. Severe infection in early stage results in total failure of crop.

14.6.1.2 Biomanagement

Seed treatment with *P. fluorescens* at 10 g/kg gave the lowest *Fusarium* wilt incidence of 14.4 % (36.6 % in control) and increased yield (603 kg/ha as compared to 423 kg/ha in control).

Varying degrees of control were noticed on *Fusarium* isolates of coriander with the use of *Streptomyces* sp (Mathur 1963).

Table 14.7 Effect of organic amendments on wilt control potentiality of bioagents used as seed treatment in controlling cumin wilt

Treatment				
Seed treatment[a]	Soil application[b]	% disease incidence	% disease control	% increase in dry wt.
P. fluorescens	FYM	27.0	68.2	83.9
P. fluorescens	Vermicompost	26.2	68.7	83.0
P. fluorescens	Mustard cake	19.0	76.2	88.8
B. subtilis	FYM	34.7	59.1	66.7
B. subtilis	Vermicompost	28.4	66.2	65.9
B. subtilis	Mustard cake	28.5	64.4	70.8
Control	FYM	84.8	–	–
Control	Vermicompost	81.6	–	–
Control	Mustard cake	80.1	–	–

[a]Seed treatment at 6 g/kg seed
[b]FYM at 20 g/kg soil, vermicompost at 10 g/kg soil, mustard cake at 2 g/kg soil

Seed treatment and soil application with *B. subtilis* and *P. fluorescens* were also found effective in reducing coriander wilt and increased yield (Table 14.6).

14.7 Cumin, *Cuminum cyminum*

14.7.1 *Fusarium* Wilt, *Fusarium oxysporum* f. sp. *cumini*

Wilt of cumin is an endemic problem in most of the cumin growing areas of Rajasthan and usually causes substantial yield losses.

14.7.1.1 Symptoms

Infected plants show peculiar symptoms of drooping of tips and leaves, leading to mortality of the entire plant. Attack of wilt is severe in younger plants.

14.7.1.2 Integrated Management

(a) *Bioagents and botanicals*: Seed treatment with talc-based formulations of *P. fluorescens* provided excellent control of *Fusarium* wilt under greenhouse conditions. The bioagent suppressed the pathogen population in soil and enhanced the shoot and root lengths and dry weight of cumin plants. The biological potentiality of *P. fluorescens* was relatively better in the presence of FYM, vermicompost or mustard cake (Table 14.7) (Chawla and Gangopadhyay 2009).

14.8 Mustard, *Brassica nigra*

14.8.1 Sclerotinia Rot, *Sclerotinia sclerotiorum*

14.8.1.1 Symptoms

Stems develop water-soaked spots which later may be covered with a cottony white growth. As disease progresses, affected portions of the stem develop a bleached appearance, and eventually tissues shred. Girdling of the stem results in premature ripening and in lodging of plants. Hard black bodies (sclerotia) are formed inside the stem and occasionally on the stem's surface.

14.8.1.2 Biomanagement

Soil application of rhizobacterium *Azotobacter chroococcum* suppresses *Sclerotinia* rot of mustard.

References

Anon (2004) Annual report for 2003–04 of Indian Institute of Spices Research, Calicut, Kerala, India

Anon (2005) Annual report for 2003–04 of All India coordinated research project on spices, Indian Institute of Spices Research, Calicut, Kerala, India

Athul Sandheep R, Aju K, Asok Ans Jisha MS (2012) Biocontrol of *Fusarium* wilt of vanilla (*Vanilla planifolia*) using combined inoculation of *Trichoderma* sp. and *Pseudomonas* sp. Int J Pharm Bio Sci 3(B):706–716

Bini YK, Anandraj M, Dinesh R, Silna N, Kumar A (2011) Evaluation of PGPR strains for growth and disease suppression in ginger (*Zingiber officinale* Rosc). In:

Proceedings of the 2nd Asian PGPR conference, Beijing, P.R. China, pp 523–528

Bora LC, Bora P (2009) Biological approach for management of bacterial wilt of *bhoot chilli* in Assam. International conference on horticulture, Bangalore

Chawla N, Gangopadhyay S (2009) Integration of organic amendments and bioagents in suppressing cumin wilt caused by *Fusarium oxysporum* f. sp. *cumini*. Indian Phytopathol 62:209–216

Devapriyaga R, Jonathan EI, Sankarimeena K, Kavitha PG (2012) Bioefficacy of *Pseudomonas* and *Bacillus* isolates against root-knot nematode, *Meloidogyne incognita* in black pepper cv. Paniyur 1. Indian J Nematol 42:57–65

Girija VK, Jubina PA (1997) *Bacillus* sp. for management of two foliar diseases of black pepper (Abstr). In: International conference on integrated plant disease management for sustainable agriculture, New Delhi, India, p 204

Mathur BL (1963). Studies on fusarium wilt of coriander in Rajasthan. Ph.D. Thesis, University of Rajasthan, Jaipur

Ram P, Mathur K, Lodha BC (1999) Integrated management of rhizome rot of ginger involving biocontrol agents and fungicides. J Mycol Plant Pathol 29:416–417

Sarma YR (2000) Integrated management of rhizome rot of ginger. Final report of ICAR AP Cess Fund Scheme, Indian Institute of Spices Research, Calicut, Kerala, India

Sarma YR, Anandraj M (1998) Final report of the mission mode project 'Development, production and demonstration of biological agents' funded by Department of Biotechnology, Government of India, Indian Institute of Spices Research, Calicut

Sarma YR, Saju KA (2004) Biological control for the management of foot rot and slow decline diseases of black pepper. Focus on pepper (*Piper nigrum* L.) 1:25–51. International Pepper Community, Jakarta

Sharma ND, Jain AC (1978) Studies on the biological control of *Fusarium oxysporum* f. sp. *zingiberi* the causal organism of yellows disease of ginger. Indian Phytopathol 31:260–261

Shashidhara S (2007) Studies on foot rot of black pepper caused by Phytophthora capsici *Leonian, emend, Alidezeh and Tsao*. M. Sc. (Agri.) thesis in plant pathology, University of Agricultural Sciences, Dharwad, India

Suseela Bhai R, Kumar A (2008) Effect of rhizobacteria on *Phytophthora meadii*, *Fusarium oxysporum* f. sp. *vanillae* and *Colletotrichum vanillae* infecting vanilla. J Biol Control 22:33–41

Suseela Bhai R, Joseph T, Naidu R (1993) Biological control of 'Azhukal' disease of small cardamom caused by *Phytophthora meadii* Mc Rae. J Plant Crops 21(Suppl):134–139

Thankamani CK, Sreekala K, Anandraj M (2005) Effect of *Pseudomonas fluorescens* (IISR 6) and *Trichoderma harzianum* (P-26) on the growth of black pepper (*Piper nigrum* L.) in the nursery. J Spices Aromat Crops 14:112–116

Tiwari RK, Mistry NC, Singh B, Gandhi CP (2014) Indian horticulture database-2013. National Horticulture Boad, Guragaon, 289 pp

Thomas J, Suseela Bhai R, Vijayan AK, Naidu R (1993) Evaluation of antagonists and their efficacy in managing rot disease of small cardamom. J Biol Control 7:29–36

Part V

Pest Management in Ornamental, Medicinal and Aromatic Crops

15 Ornamental and Aromatic Crops

15.1 Ornamental Crops

Floriculture is estimated to cover an area of 253,660 ha with production of 1,651,610 MT of loose flowers. The area under cut flowers has increased significantly in recent years and the production has reached 7,506 million numbers (Table 15.1) (Tiwari et al. 2014). With the liberalization of the Indian economy, the domestic floricultural trade, which hitherto remained an unorganized sector, became an organized sector with the active participation of corporate giants. The establishment of state-of-the-art hi-tech floricultural units in the early 1990s for intensive cultivation of cut flowers contributes about Rs. 425 crores of valuable foreign exchange today.

15.1.1 Rose, *Rosa* spp.

15.1.1.1 Crown Gall, *Agrobacterium tumefaciens*

i) Symptoms

The bacteria enter a plant through the roots or through wounds at the root area. The bacteria cause abnormal cell growth, which produces the galls. The galls that develop on stems or just under the soil surface are round and rough textured. Round growths about 5 cm in diameter appear at the base of the plant (Fig. 15.1). Young galls are soft and white or pale green; older ones are dark and woody. Affected plants lose vigour and may become stunted, with abnormally small leaves and fewer flowers. If the disease affects the plant while it is young, the plant may be affected to the degree where it will not produce blooms. All affected plants wilt readily and grow poorly.

ii) Biomanagement

Commercial formulation of non-tumorigenic strain of *Agrobacterium rhizogenes* (K84 strain) is being used successfully to control crown gall. In Japan, strain K84 has been confirmed as an effective biocontrol agent of crown gall in rose (Makino and Morita 1985). By applying a large population of strain K84 to roots in a cell suspension at the time of transplanting rose bushes, the pathogen is inhibited before infection and biological control comes into play.

15.1.2 Carnation, *Dianthus caryophyllus*

15.1.2.1 Wilt, *Fusarium oxysporum* f. sp. *dianthi*

i) Symptoms

The initial symptoms are foliar yellowing and production of 'crook neck' or bent shoots (Fig. 15.2). The leaves and shoots wither and become brownish. Initially the symptoms are confined to a few branches or a part of the plant, but ultimately the entire plant shows the symptoms. The symptoms are first noticed during midday when the temperature is quite high inside the

P.P. Reddy, *Plant Growth Promoting Rhizobacteria for Horticultural Crop Protection*, DOI 10.1007/978-81-322-1973-6_15, © Springer India 2014

Table 15.1 Major flowers producing states of India (2011–2012)

State	Area ('000 ha)	Production ('000 MT)		Productivity of loose flowers (MT/ha)
		Loose	Cut[a]	
Andhra Pradesh	64.15	389.01	709.939	6.06
Chhattisgarh	8.41	32.85	0.00	3.90
Delhi	5.50	5.70	103.800	1.03
Gujarat	15.96	135.50	0.00	8.48
Haryana	6.34	64.15	126.947	10.11
Karnataka	29.22	211.54	1,038.800	7.23
Madhya Pradesh	15.61	150.67	0.00	9.65
Maharashtra	18.88	104.00	791.400	5.50
Odisha	7.54	26.08	602.000	3.45
Punjab	2.06	10.05	0.007	4.87
Rajasthan	2.49	2.69	0.00	1.08
Tamil Nadu	32.32	332.81	0.00	10.29
Uttar Pradesh	14.49	27.05	419.400	1.86
West Bengal	23.92	63.91	2,504.210	2.67
Other states	6.77	95.60	1,210.095	14.12
Total	253.66	1,651.61	7,506.598	6.51

[a]Cut flowers in million numbers

Fig. 15.1 Crown galls on stem of rose

polyhouses, but they are not visible when the temperature goes down. Stems when cut open show brown discolouration at the vascular region. Sometimes the pith and cortex also get discoloured. Finally the stems show shredding of the internal tissue.

ii) Biomanagement

Streptomyces griseoviridis was tested as a biocontrol agent against *Fusarium* wilt of carnations in commercial greenhouses, where it reduced the spread of the disease and increased yield (Lahdenpera 1987). Rattink (1992) reported that antagonist *S. griseoviridis* reduced the wilt of carnation by more than 50 %.

F. oxysporum f. sp. *dianthi* has been reported to be highly sensitive to inhibition by *B. subtilis* (Filippi et al. 1987).

Production of siderophores by *Pseudomonas* spp. was shown to be important in control of *Fusarium* wilt of carnation (Duijff et al. 1993). Carnation plants expressing induced systemic resistance mediated by *P. fluorescens* strain WCS417r produce significantly more phytoalexins upon stem inoculation with *F. oxysporum* f. sp. *dianthi* (van Peer et al. 1991). This enhanced phytoalexin production which may contribute to limiting ingress of the pathogen in ISR-expressing plants.

iii) Integrated Management

(a) *Two bioagents*: Increased suppression of *Fusarium* wilt of carnation was observed by combining *P. putida* WCS358 with nonpathogenic

Fig. 15.2 *Fusarium* wilt on carnation

F. oxysporum Fo47 (Lemanceau et al. 1993). The enhanced disease suppression may be due to siderophore-mediated competition for iron by WCS358, which makes the pathogenic *F. oxysporum* strain more sensitive to competition for glucose by the nonpathogenic strain Fo47.

15.1.2.2 Root-Knot Nematode, *Meloidogyne incognita*

The root-knot nematode *M. incognita* is one of the serious limiting factors in commercial cultivation of carnation under polyhouse conditions. *M. incognita* was responsible for 27 % loss in flower yield of carnation. Most of the highly fetching exotic cultivars of carnation from Europe have shown 40–60 % mortality in polyhouse beds due to root-knot nematode infection in and around Bangalore.

i) Symptoms

Plants infested with M. incognita were smaller and showed leaf chlorosis. Initially, the lower leaves turned yellow followed by the upper leaves, and, in rare cases of severe infestation, the stalk started turning yellow at the base. The stalk became weak and unable to stand erect. Flowers were rated as 'B' grade commercially. Flowering was delayed by more than a month and some nematode-infested plants never bore flowers. Only small sized galls were produced by *M. incognita* on carnation (Fig. 15.3).

ii) Biomanagement

In carnation, the application of *Pseudomonas* sp. strain WCS 417r protected plants systemically against *Fusarium* wilt disease (Van Peer et al. 1991).

iii) Integrated Management

(a) *Bioagents and AMF*: Combined inoculation of AMF and *P. fluorescens* had positive effect on root-knot nematode control on carnation (Anusuya and Vadivelu 2002).

15.1.3 Gerbera, *Gerbera jamesonii*

15.1.3.1 Root-Knot Nematode, *Meloidogyne incognita*, and Root Rot, *Phytophthora cryptogea*, Disease Complex

Sustainable production of gerbera is seriously hampered by the disease complex caused by *M. incognita* and *P. cryptogea*. Most of the highly

Fig. 15.3 Root galling on carnation due to root-knot nematodes

fetching exotic cultivars of gerbera have shown 40–60 % mortality in polyhouse beds due to root-knot nematode infection.

i) Integrated Management

(a) *Bioagents and botanicals*: Experiments were conducted for the biological control of the disease complex using *P. fluorescens* with different organic substrates (deoiled neem cake, FYM, vermicompost and coco peat) (Manoj Kumar and Rao 2011). Seedling treatment with *P. fluorescens* + soil application with neem cake enriched with *P. fluorescens* proved significantly effective in the management of the disease complex caused by *M. incognita* and *P. cryptogea* on gerbera under field conditions (Table 15.2). There was a significant reduction in the root-knot index of *M. incognita* and percentage of disease incidence by *P. cryptogea*. Reduction of *M. incognita* population to the extent of 57~64 % in roots and 60~65 % reduction in disease infestation was observed on gerbera in the beds which were treated with deoiled neem cake enriched with *P. fluorescens*. This helped in the reduction of the cost of production. The above treatment also increased plant growth parameters, flower yield of gerbera and root colonization of *P. fluorescens*. These treatments also increased yield by 20~22 %. Additionally, it was also found that the flowers harvested from the above treatment showed an increased vase life compared to control (Manoj Kumar and Rao 2011).

Combined application of neem cake enriched with *P. fluorescens* [mixing 50 g *P. fluorescens* (2×10^8 cfu/g) in 1 kg of neem cake] applied at 25 g/m^2 was found effective for the management of the disease complex and increased the flower yield by 26 % in gerbera cv. Debora (Manoj Kumar et al. 2010).

15.1.4 Gladiolus, *Gladiolus* spp.

15.1.4.1 Root-Knot Nematodes, *Meloidogyne* spp.

i) Symptoms

The root-knot nematode causes stunting of plants and reduction in leaf count. Severe galling on roots

Table 15.2 Effect of bioagents and botanicals on root-knot and root rot disease complex

Treatments	Plant height (cm)	Root galling 1–10 scale	Disease incidence (%)	% increase in yield	Root colonization of *P. fluorescens* (CFU/g × 10^6)
T1	15.0	7.7 (18.94)	38.06 (24.67)	15.12	1.03
T2	23.2	6.3 (3.68)	20.04 (60.34)	21.97	1.35
T3	19.2	7.1 (25.26)	29.93 (40.76)	20.36	1.31
T4	27.2	4.1 (56.84)	18.16 (64.06)	28.52	2.10
T5	24.6	5.3 (44.21)	25.77 (49.00)	26.92	1.97
T6	19.4	8.5 (10.52)	45.40 (10.15)	10.78	0.00
T7	18.2	9.3 (2.10)	49.49 (2.05)	9.63	0.00
T8	22.1	6.4 (32.63)	35.67 (29.40)	16.22	1.78
T9	19.3	7.4 (22.10)	35.61 (29.52)	15.97	1.65
T10	11.6	9.5	50.53	0.00	0.00
CD – 5 %	**2.26**	**0.44**	–	–	**0.74**

Values in parentheses are % decrease over control

T1 – seedling treatment with *P. fluorescens* (2 × 10^8 cfu/ml); T2 – untreated seedlings + soil application with neem cake enriched with *P. fluorescens* 20 g/m²; T3 – untreated seedlings + soil application with FYM enriched with *P. fluorescens* 200 g/m²; T4 – seedling treatment with *P. fluorescens* + soil application with neem cake enriched with *P. fluorescens*; T5 – seedling treatment with *P. fluorescens* + soil application with FYM enriched with *P. fluorescens*; T6 – untreated seedlings + soil application with neem cake at 20 g/m²; T7 – untreated seedlings + soil application with FYM at 200 g/m²; T8 – seedling treatment with *P. fluorescens* + soil application with neem cake at 20 g/m²; T9 – seedling treatment with *P. fluorescens* + soil application with FYM at 200 g/m²; T10 – control (untreated seedlings transplanted in soil without any treatment)

Fig. 15.4 Root-knot nematode symptoms on gladiolus leaves and corm

results in yellowing of leaves, which subsequently leads to stunted growth (Fig. 15.4). The nematode invades roots, daughter corms and cormels which develop after flowering.

ii) Integrated Management

(a) *Bioagents and botanicals*: Application of 5 MT of FYM per ha enriched with *P. fluorescens* (with 1 × 10^9 cfu/g) significantly reduced *M. incognita* by 61–76 %, *R. reniformis* by 65–70 % in the roots of gladiolus and increased the flower yield by 19–22 %.

Corms of gladiolus treated with neem suspension enriched with *P. fluorescens* strains 1 and 2 (5 g/l of water) and application of neem cake enriched with *P. fluorescens* strains to beds at 20 g/m² were

Table 15.3 Effect of combi-formulation of *P. fluorescens* and *P. chlamydosporia* with neem in the management of disease complex in gladioli under field conditions

Treatments	Plant height (cm)	No of *M. incognita*/5 g root[a]	Disease incidence (%)[a]	Yield (no. of spikes/plot of 4×2.5 m)[b]
T1	82	12.54 (34)	33.7 (23)	41.2 (6)
T2	107	9.88 (48)	27.2 (38)	43.6 (12)
T3	73	13.3 (30)	35.0 (20)	40.5 (4)
T4	95	11.02 (42)	29.3 (33)	42.8 (10)
T5	89	11.78 (38)	31.5 (28)	42.0 (8)
T6	121	7.22 (62)	20.6 (53)	47.1 (21)
T7	114	7.98 (58)	22.3 (49)	45.9 (18
T8	101	8.93 (53)	25.4 (42)	44.7 (15)
T9	130	6.46 (66)	18.4 (58)	47.8 (23)
T10	68	13.68 (28)	35.9 (18)	39.7 (2)
T11	53	19.0	43.8	38.9
C.D. – 5 %	**8.43**	**2.72**	–	**3.48**

T1 – corm treatment with 10 % suspension of *P. fluorescens* (2×10^8 cfu/ml); T2 – untreated corms + soil application with neem cake (20 g/m^2) enriched with *P. fluorescens*; T3 – corm treatment with 10 % suspension of *P. chlamydosporia* (2×10^6 cfu/ml); T4 – untreated corms + soil application with neem cake (20 g/m^2) enriched with *P. chlamydosporia*; T5 – corm treatment with combi-formulation of *P. fluorescens* and *P. chlamydosporia*; T6 – untreated corms + soil application with neem cake enriched with combi-formulation of *P. fluorescens* and *P. chlamydosporia* (20 g/m^2); T7 – corm treatment with *P. fluorescens* + soil application with neem cake enriched with *P. fluorescens* (20 g/m^2); T8 – corms treated with *P. chlamydosporia* + soil application with neem cake enriched with *P. chlamydosporia*; T9 – corm treatment with combi-formulation of *P. fluorescens* and *P. chlamydosporia* + soil application with neem cake enriched with combi-formulation of *P. fluorescens* and *P. chlamydosporia* (20 g/m^2); T10 – untreated corms + soil application with neem cake (20 g/sq.mt); T11 – untreated corms without any soil treatment (Control)

[a]Values in parentheses are % decrease over control

[b]Values in parentheses are % increase over control

found to be effective in reducing the root population of *M. incognita* by 68 % and increased the flower yield by 28 %.

15.1.4.2 Root-Knot Nematode, *Meloidogyne incognita*, and Wilt, *Fusarium oxysporum* f. sp. *gladioli*, Disease Complex

Productivity of gladiolus is seriously hampered by the disease complex caused by *M. incognita* and *F. oxysporum* f. sp. *gladioli*. These two pathogens reduce the productivity to the tune of 40~60 %. Infection of roots by root-knot nematodes predisposes plants to infection by soilborne root-infecting *F. oxysporum* f. sp. *gladioli* which results in the manifestation of the disease complex. The disease complex damages plants more severely and renders disease control more difficult than single pathogens.

i) Biomanagement

The application of *P. fluorescens* effectively controlled the disease complex leading to significant improvement in plant growth and flowering (Mustafa and Khan 2004).

ii) Integrated Management

Bioagents and botanicals were evaluated either singly or in combination for the management of disease complex in gladiolus (Sowmya and Rao 2011). Corm treatment with both *P. fluorescens* and *P. chlamydosporia* along with application of neem cake enriched with *P. fluorescens* and *P. chlamydosporia* proved significantly effective in the management of the disease complex caused by *F. oxysporum* f. sp. *gladioli* and *M. incognita* on gladiolus under field conditions. There was a significant reduction in the *M. incognita* population (66 %) and disease incidence by *F. oxysporum* f. sp. *gladioli* by about 58 % (Table 15.3). There was also a significant increase in plant height and yield of the crop by about 23 % (Sowmya and Rao 2011).

Rhizospheric density of *P. fluorescens* (7.3×10^6) and *P. chlamydosporia* (6.9×10^5) was found to be maximum in treatment T9 (Corm

Table 15.4 Effect of combi-formulation of *P. fluorescens* and *P. chlamydosporia* with neem on rhizospheric density and root colonization

Treatment	Density of *P. fluorescens* (cfu/g) in soil ($\times10^6$)	Root colonization (cfu/g) of *P. fluorescens* ($\times10^6$)	Density of *P. chlamydosporia* (cfu/g) in soil ($\times10^5$)	Root colonization (cfu/g) of *P. chlamydosporia* ($\times10^5$)
T1	3.8	2.4	0.0	0.0
T2	5.8	4.5	0.0	0.0
T3	0.0	0.0	2.7	2.2
T4	0.0	0.0	5.3	4.1
T5	4.0	3.3	4.3	3.3
T6	6.2	5.8	5.8	4.9
T7	6.9	6.0	0.0	0.0
T8	0.0	0.0	6.5	5.5
T9	7.3	6.6	6.9	5.9
T10	0.0	0.0	0.0	0.0
T11	0.0	0.0	0.0	0.0
C.D. – 5 %	**0.45**	**0.26**	**0.72**	**0.34**

treatment with combi-formulation of *P. fluorescens* and *P. chlamydosporia* + soil application with neem cake enriched with combi-formulation of *P. fluorescens* and *P. chlamydosporia*). Root colonization and colony-forming units (cfu) of *P. fluorescens* (6.6×10^6) and *P. chlamydosporia* (5.9×10^5) were also high in T9 (Table 15.4) (Sowmya and Rao 2011).

15.1.5 Tuberose, *Polianthes tuberosa*

15.1.5.1 Foliar Nematode, *Aphelenchoides besseyi*

Foliar nematode emerged as a serious problem in tuberose reported from Ranaghat areas of Nadia district of West Bengal. *A. besseyi* is becoming a major limiting factor for cultivation of tuberose in Ranaghat, Haringhata and Panskura areas of West Bengal. The 'Calcutta Single' cultivar of tuberose is more vulnerable to *A. besseyi* than the 'Calcutta Double' cultivar. *A. besseyi* is the primary causal agent for malformed flowers. In 'Calcutta Double' cultivar 30–40 % flower stalk renders unsalable and individual flower stalk harbours up to 45,000 nematodes. In 'Calcutta Single', the yield loss may occur to the extent of 59 %. The presence of nematode species in the cut flower and stalk is a constraint in export of flowers to other countries of the world for pest risk.

i) Symptoms

The infected flower stalk initially appears rough and then becomes crinkled, stunted and finally distorted, and in severe cases flower buds fail to bloom. Brown streaks appear on leaf bracts and petals and subsequently develop into rusty brown spots. The severely infected flower stalk becomes rotten and brittle over drying and even becomes blind, and the number of flowers per stalk is also reduced (Fig. 15.5). The nematode forms 'nematode wool' upon dark brown spots. The ovary contains a large number of nematodes. This nematode is generally more serious during the rainy season from June to September, and cent per cent loss of the second year crop of the 'Calcutta Single' cultivar of tuberose is encountered.

ii) Integrated Management

(a) *Bioagents and chemicals*: Pre-soaking of tuberose tubers overnight in plain water followed by dipping of bulbs in monocrotophos 36 SL at 700 ppm for 6 h + four sprayings with *P. fluorescens* at 2 g/L of water starting from sprouting at 15 days interval significantly reduced nematode infestation and

Fig. 15.5 Effects of foliar nematode infection on tuberose. (**a**) Healthy control stem with abundant white florets. (**b–d**) Infected plants. Note the browned stem, leaves and florets, as well as the distorted skin of the leaves on the stem

Table 15.5 Efficacy of bioagent and chemical on bud and leaf nematode infestation and spike yield of tuberose

Treatment	Nematode infestation (%)	Spike yield/ha	Benefit/cost ratio
P. fluorescens + monocrotophos	34.22	597,880	4.28
Farmers' practice[a]	45.80	584,388	2.72
Untreated control	53.39	580,952	–
CD = (*P*=0.05)	**4.94**	–	–

[a]Farmers' practice – 3 sprayings with monocrotophos 36 SL at 500 ppm when disease appeared

increased spike yield and benefit/cost ratio (Table 15.5) (Pathak and Khan 2010).

15.1.5.2 Leaf Spot/Blight, *Alternaria polyantha*

i) Integrated Management

(a) *Two bioagents*: Treatment with *Trichoderma harzianum* and *P. fluorescens* at 4 g/kg tuber and seedling dip in *T. harzianum* suspension before planting resulted in effective control of blight and increased the plant vigour.

15.1.6 Crossandra, *Crossandra undulaefolia*

15.1.6.1 Root-Knot Nematode, *Meloidogyne incognita*

The nematode is responsible for 25.62 % and 21.64 % loss in number of flowers and weight of flowers, respectively.

i) Symptoms

Infected plants are stunted with dried peripheral branches bearing smaller chlorotic leaves almost turning to white at later stages. Roots exhibit severe galling. Inflorescences are small and sometimes fail to produce flowers.

ii) Integrated Management

(a) *Two bioagents*: Treatment of nursery beds with *Pochonia chlamydosporia* and *P. fluorescens* each at 50 g/m^2 and seed treatment with the latter were significantly effective in reducing the number of nematodes in roots and soil and in increasing parasitization and suppression of eggs and flower yield of crossandra. The seedlings were colonized by both the bioagents and when transplanted in the field the bioagents reached the field soil as they were recovered from root and soil samples at harvest of the crop. Individual effect of bioagents was maximized when both these organisms were integrated in the nursery bed stage. This could be due to the combined effect of both organisms on root-knot nematode. Combined use of *P. fluorescens* and *P. chlamydosporia* did not affect the colonization of each other on root (Rao 2007).

15.2 Aromatic Crops

In recent years, there has been an increased interest in the cultivation of aromatic plants to meet the requirements of cosmetic, flavouring and perfumery industries. Aromatic crops are estimated to cover an area of 505,600 ha with production of 565, 700 MT during 2011–2012 (Table 15.6) (Tiwari et al. 2014). The aromatic plants provide raw material for the production of flavours, herbal cosmetics, perfumery, etc. Aromatic plants and their products, particularly the essential oils, are now becoming one of the more important export items from many developing countries of Asia. India has enjoyed a prominent position in the manufacture of superior perfumes and aromatics by using essential oils. The important aromatic plants grown are lemon grass, vetiver, patchouli, palmarosa, citronella, mints, geranium, lavender, basil, jasmine, etc.

15.2.1 Mints, *Mentha* spp.

15.2.1.1 Root-Knot Nematodes, *Meloidogyne incognita*, *M. javanica*, *M. hapla*

Root-knot disease of menthol/Japanese mint, spearmint, scotch spearmint, peppermint and bergamot mint caused by *M. incognita* and *M. javanica* was observed for the first time in Lucknow and the Terai region of Uttar Pradesh which reduces the herbage yield and oil content. *M. incognita* dominates over *M. javanica* in mixed infection. The quality of mint oil was also adversely affected due to nematode infection. The root-knot nematode *M. incognita* caused 40.2 % loss in herbage yield and 46.6 % loss in oil yield of menthol mint.

i) Symptoms

Initial damage occurs at 1 juvenile/g of soil with marked reduction in chlorophyll content, rate of photosynthesis and oil yield. Symptoms of the disease include occasional yellowing of leaves which in a month's time spread to a large portion of the foliage. Growth ceases soon after yellowing. Leaves turn yellow and thin, scorching easily and eventually turning brown. Initially symptoms appear in patches. Other symptoms include reduced plant growth (with smaller leaf size) and temporary wilting under slight water stress during

Table 15.6 Major aromatic crops producing states of India (2011–2012)

State	Area ('000 ha)	Production ('000 MT)	Productivity (MT/ha)
Arunachal Pradesh	5.15	109.18	21.20
Bihar	3.56	0.45	0.12
Chhattisgarh	12.12	91.41	7.54
Haryana	1.73	1.13	0.65
Karnataka	3.41	19.80	5.80
Madhya Pradesh	43.60	106.81	2.44
Odisha	1.92	0.64	0.33
Punjab	7.12	1.29	0.18
Rajasthan	280.62	149.60	0.53
Tamil Nadu	12.26	68.04	5.54
Uttar Pradesh	133.70	13.40	0.10
Other states	0.41	3.95	9.63
Total	**505.60**	**565.70**	**1.11**

hot sun. More severe symptoms such as stunted plant growth, and yellowing of leaves with veins remaining green, appear after the first harvest.

ii) Biomanagement

Haseeb et al. (2005) showed that *P. fluorescens* was effective in increasing the plant growth and oil yield as well as in suppressing *M. incognita* reproduction and root-knot index.

The below-ground symptoms are the large numbers of small sized galls with large egg masses on the roots (Fig. 15.6). Under severe infection, initiation of lateral roots and rootlets on suckers is checked. As a result, uptake of nutrients is inhibited, which in turn produces deficiency symptoms on the aerial portion of the plants.

Fig. 15.6 L – Root knots on mint

iii) Integrated Management

(a) *Bioagents and AMF*: The combination of *T. harzianum* with *G. intraradices*, *B. megaterium* and *P. fluorescens* caused 63.7 %, 45.4 % and 54.6 % reduction in root-knot indices and 67.8 %, 56.3 % and 60.4 % reduction in total *M. incognita* population, whereas the combined application of *G. intraradices* + *B. megaterium* and *G. intraradices* + *P. fluorescens* reduced only 52.7 and 54.0 % of the total nematode population, respectively. The reduction of nematode population in carbofuran-treated pots was 31.6 % (Fig. 15.7) and root-knot index was 36.3 % (Pandey et al. 2011).

The maximum enhancement of plant dry weight was recorded in *T. harzianum* + *G. intraradices* (20.0 %) followed by *T. harzianum* + *P. fluorescens* (14.4 %) as compared to untreated control (Table 15.7) (Pandey et al. 2011).

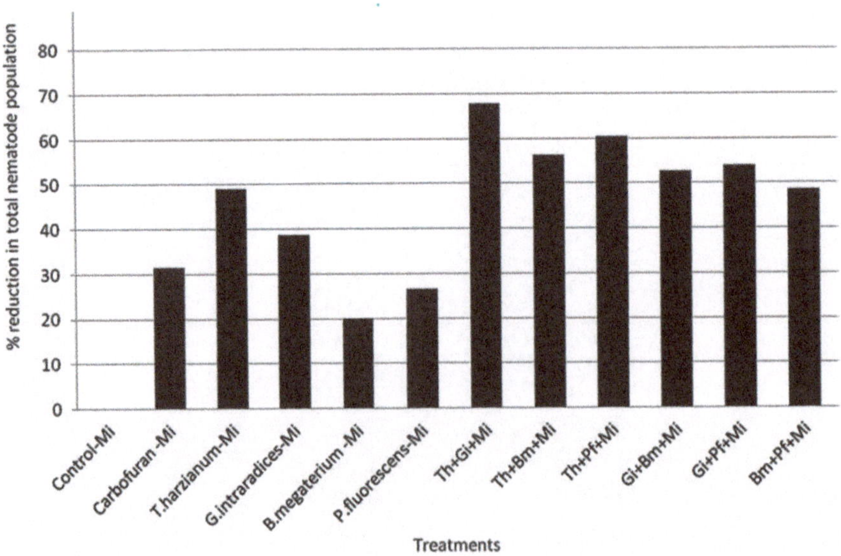

Fig. 15.7 Effect of different mutualistic fungi and PGPR on per cent reduction in total nematode population development in menthol mint

Table 15.7 Effect of mutualistic endophytes and PGPR on yield of menthol mint, root-knot index and mycorrhizal population development

Treatment	Plant dry weight (g)	Root-knot index	*G. intraradices* % root colonization
Control – uninoculated	34.0	–	–
Control – inoculated	25.7 (−24.4)[a]	3.66	–
Carbofuran	31.4 (−7.6)	2.33	–
Trichoderma harzianum (*Th*)	34.8 (−2.4)	2.00	–
Glomus intraradices (*Gi*)	33.3 (−2.1)	2.66	65
Bacillus megaterium (*Bm*)	30.9 (−9.1)	3.00	–
Pseudomonas fluorescens (*Pf*)	33.4 (−1.8)	2.66	–
Th + *Gi*	40.8 (+20.0)	1.33	78
Th + *Bm*	36.0 (+5.9)	2.00	–
Th + *Pf*	38.9 (+14.4)	1.66	–
Gi + *Bm*	34.4 (+1.2)	2.00	70
Gi + *Pf*	35.7 (+5.0)	2.00	74
Bm + *Pf*	34.5 (+1.5)	2.66	–
LSD at 5 %	**–**	**0.7433**	**4.321**

[a]Per cent increase (+) or decrease (−) over untreated–uninoculated control. Note: inoculated with 5,000 freshly hatched juveniles of *Meloidogyne incognita* (Mi)/pot

15.2.2 Patchouli, *Pogostemon patchouli*

15.2.2.1 Lesion Nematode, *Pratylenchus brachyurus*

P. brachyurus is a major problem on patchouli in the Anamalai hills of Tamil Nadu and Hessaraghatta in Karnataka.

i) Symptoms

The infected plants invariably exhibited lesions on roots and wilting symptoms.

ii) Biomanagement

Endophytic bacteria significantly reduced the population of *P. brachyurus*, and all bacterial isolates promoted growth of patchouli (shoot weight, root weight and root length). Four isolates, i.e. *Bacillus* NJ46, *Bacillus* Na22, *Bacillus* NJ2 and *Bacillus* NJ57, were among the potential biocontrol agents that reduced nematode populations as much as 68.1–73.9 %. Almost all of the isolated bacteria from patchouli roots were able to solubilize phosphate, while some of them had the ability to produce chitinase, cellulase, protease, HCN and fluorescence (Harni et al. 2007).

15.2.3 Scented Geranium, *Pelargonium graveolens*

15.2.3.1 Rust, *Uromyces pelargonii-zonalis*

i) Symptoms

Leaf rust is characterized by small yellow spots on the upper surface of the leaf and rust-coloured pustules that develop in the spots on the underside of the leaf. The pustules break open and release spores. Spores are often formed in a circular fashion around the original pustule (Fig. 15.8). When infection is severe, leaves become yellow and drop prematurely. Pustules containing the rust-coloured spores can be seen on stems, petioles and stipules. The disease is more common in greenhouse production, but it is a potentially important disease in the landscape.

Fig. 15.8 Leaf rust on geranium leaves

ii) Biomanagement

Application of *B. subtilis* three times at weekly interval controlled geranium rust by the production of antibiotics (Rytter et al. 1989).

References

Anusuya K, Vadivelu S (2002) Mutualistic symbiosis of VAM fungi on plant growth promoting rhizobacteria (PGPR) on micropropagated carnation. J Microbiol World 4:105–107

Duijff BJ, Meijer JW, Bakker P, Schippers B (1993) Siderophore-mediated competition for iron and induced resistance in the suppression of Fusarium wilt of carnation by fluorescent *Pseudomonas* spp. Neth J Plant Pathol 99:5–6

Filippi C, Baganoli G, Volterrani M, Picci G (1987) Antagonistic effects of soil bacteria on *Fusarium oxysporum* Schlecht f. sp. *dianthi* (Prilla & Del) Snyd. & Hans. Plant Soil 98:161–167

Harni R, Munif A, Supramana S, Mustika I (2007) Potential of endophytic bacteria to control lesion nematode *Pratylenchus brachyurus* on patchouli. Hayati J Biosci 14:7–12

Haseeb A, Sharma A, Abuzar S (2005) Studies on the management of *Meloidogyne incognita* on *Mentha arvensis* cv. Gomti by application of bio-agents, chemicals and organic amendments under pot conditions. In: Proceedings VIth national symposium on sustainable plant protection strategies: health and environmental concerns, Ratnagiri, p 118

Lahdenpera ML (1987) The control of Fusarium wilt of carnation with a *Streptomyces* preparation. Acta Horticult 216:85–92

Lemanceau P, Bakker PAHM, de Kogel WJ, Alabouvette C, Schippers B (1993) Antagonistic effect of non-pathogenic *Fusarium oxysporum* Fo47 and pseudobactin 358 upon pathogenic *Fusarium oxysporum* f. sp. *dianthi*. Appl Environ Microbiol 59:74–82

Makino T, Morita H (1985) Biological control of crown gall on rose seedlings by *Agrobacterium radiobacter* strain 84. Bull Shizuoka Tea Exp Stn 30:53–59

Manoj Kumar R, Rao MS (2011) Bio-management of nematode induced wilt disease complex of Gerbera using PGPRs in open field conditions. In: Proceeding of the 2nd Asian PGPR conference, Beijing, P.R. China, pp 456–463

Manoj Kumar R, Rao MS, Ramachandran N (2010) Biomanagement of disease complex caused by *Meloidogyne incognita* and *Phytophthora parasitica* Dastur in protected cultivation of gerbera using formulation of *Trichoderma harzianum* and *Pseudomonas fluorescens*. National conference on innovations in nematological research for agricultural sustainability – challenges and a roadmap ahead, Tamil Nadu Agricultural University, Coimbatore, pp 97–98

Mustafa U, Khan MR (2004) Management of disease complex caused by *Fusarium oxysporum* f. sp. *gladioli* and *Meloidogyne incognita* on gladiolus. In: National symposium on paradigms in nematological research for biodynamic farming, University of Agricultural Science, Bangalore, pp 91–92

Pandey R, Mishra AK, Tiwari S, Singh HN, Kalra A (2011) Enhanced tolerance of *Mentha arvensis* against *Meloidogyne incognita* (Kofoid and White) Chitwood through mutualistic endophytes and PGPRs. J Plant Interact 6:247–253

Pathak B, Khan MR (2010) Comparative field efficacy of chemical, botanical and biological agents against foliar nematode. Aphelenchoides besseyi infecting tuberose. Indian J Nematol 40:83–87

Rao MS (2007) Management of root-knot nematode, *Meloidogyne incognita* (Kafoid & White) Chitwood, on crossandra (*Crossandra undulaefolia* Salisb.) using *Pochonia chlamydosporia* and *Pseudomonas fluorescens*. J Orn Hortic 10:110–114

Rattink H (1992) Biological control of *Fusarium* wilt of carnation by a non-pathogenic isolate of *Fusarium oxysporum* f. sp. *dianthi*. Acta Hortic 307:37–42

Rytter JL, Lukezic FL, Craig R, Moorman GW (1989) Biological control of geranium rust by *Bacillus subtilis*. Phytopathology 79:367–370

Sowmya DS, Rao MS (2011) Bio-management of disease complex in gladiolus by using PGPR. In: Proceedings of the 2nd Asian PGPR conference, Beijing, P.R. China, pp 342–350

Tiwari RK, Mistry NC, Singh B, Gandhi CP (2014) Indian horticulture database-2013. National Horticulture Boad, Guragaon, 289 pp

Van Peer R, Niemann GJ, Schippers B (1991) Induced resistance and phytoalexin accumulation in biological control of Fusarium wilt of carnation by *Pseudomonas* sp. WCS417r. Phytopathology 81: 728–734

16 Medicinal Crops

Medicinal crops form one of the important groups due to their demand in various pharmaceutical industries and also to earn foreign exchange by way of export. The potential for earning foreign exchange by India from the exports of medicinal and aromatic plants is estimated over US$ 3,000 million per annum. The export of these plants and their products has a tremendous potential in advanced countries like Europe, the USA and Japan. The international market for medicinal plant-related trade is estimated at US$ 60 billion/year having a growth rate of 7 % per annum. Destination of foreign tourists is now aimed at India for health treatment. Kerala and Gujarat have become their favourite destinations since Ayurvedic treatment is very popular in these two states.

16.1 Coleus, *Coleus forskohlii*

16.1.1 Wilt, *Fusarium chlamydosporum*

16.1.1.1 Symptoms

In the field, the infected plants were characterized by gradual yellowing, marginal necrosis and withering of leaves followed by loss in vigour and premature death (Fig. 16.1). Such plants showed discolouration of roots and complete decaying of tap and lateral root system. The bark of such plants easily peeled off. There was extensive sloughing off and shredding of the affected bark. Such affected plants were finally killed due to severe root and collar rot. The infected tubers showed rotting and emitted a bad odour.

16.1.1.2 Biomanagement

P. fluorescens can suppress a wide range of plant pathogens including *Fusarium*. Increase in plant height, number of branches and fresh and dry weight of the plant as well as less disease severity and less root rot were observed due to fluorescent *Pseudomonas* sp. Application of PGPR strains (RB50 and RB31) recorded maximum plant height, number of branches and fresh and dry weight of the plants with less disease severity and root rot index (Mallesh 2008) (Table 16.1).

Singh et al. (2009) reported that inoculation of bioinoculant *P. fluorescens* significantly increased the forskolin content of the roots. Paramasivan et al. (2007) reported that the use of *P. fluorescens* reduced disease incidence by 20–21 %.

16.1.1.3 Integrated Methods

(a) *Two bioagents*: Inoculation with *P. fluorescens* + *T. viride* gave good results in controlling the disease. The same treatment also resulted in increased plant growth, yield and root forskolin concentration of coleus. The next best treatment was *G. mosseae* + *P. fluorescens*. The application of fungicide emisan resulted in better growth than control, but it was less when compared to single or dual inoculations of biocontrol agents (Boby and Bagyaraj 2003).

P.P. Reddy, *Plant Growth Promoting Rhizobacteria for Horticultural Crop Protection*, DOI 10.1007/978-81-322-1973-6_16, © Springer India 2014

16.1.2 Blight, *Rhizoctonia bataticola*

16.1.2.1 Symptoms

The disease was initially expressed as water-soaked areas and the affected tissues soon turned into a soft, black, watery mass at the collar region of the plant. The infection was also found on roots and caused decay, which ultimately resulted in collapse of the plant. The infected plant roots showed discolouration followed by rotting of root hairs. Extensive sloughing off and shredding of the affected bark was also observed. Under conditions of high humidity the disease was found to spread rapidly (Ramaprasad Shresti 2005). Severe infection results in defoliation and death of the plant (Fig. 16.2).

16.1.2.2 Biomanagement

The treatment *P. fluorescens* recorded significantly lowest per cent blight incidence (19.98) over control (35.52) (Ramaprasad Shresti 2005) (Table 16.2).

Fig. 16.1 Symptoms of *Fusarium* wilt of coleus

16.1.3 Bacterial Wilt, *Ralstonia solanacearum*

Bacterial wilt is the major disease of *C. forskohlii* causing heavy losses (>50 %) in south India.

16.1.3.1 Symptoms

Water-soaked patches with linear streaks were observed on the collar region of the infected plants. The leaves became flaccid, drooped quickly, wilted, showed roll-up symptoms and the whole plant dried up (Fig. 16.3). Wilted plants came off easily with a gentle pull and vascular discolourations were observed. Such tubers when pressed exhibited oozing of bacterial exudates.

16.1.3.2 Biomanagement

The application of PGPR strains (fluorescent *Pseudomonas* spp. RB50 and RB31) recorded maximum plant height, number of branches and fresh and dry weight of the plants with less disease severity and root rot index (Mallesh 2008) (Table 16.3).

16.1.3.3 Integrated Management

(a) *Bioagents and botanicals*: Application of 2 kg of *P. fluorescens* mixed with 300 kg of compost is effective.

16.1.4 Root-Knot Nematode, *Meloidogyne incognita*

The root-knot nematode infestation was reported on coleus from Kerala and Orissa. The dry

Table 16.1 Biocontrol potentiality of PGPR strains in coleus against *Fusarium* wilt

PGPR strains	Plant height (cm)	No. of branches	Fresh weight (g)	Dry weight (g)	Disease severity	Root rot index
RB01	32.5	5.0	97.5	15.6	+	1.5
RB10	29.5	4.5	93.5	14.5	++	3.0
RB13	34.0	6.0	107.5	17.0	+	1.5
RB22	28.0	4.0	94.7	15.0	++	2.5
RB31	36.5	6.5	116.5	18.5	+	1.0
RB43	30.0	5.5	95.5	15.3	++	2.0
RB50	37.0	7.0	120.0	18.9	+	1.0
Control	25.0	3.5	80.0	13.0	+++	3.5

Fig. 16.2 Blight symptoms on coleus

Table 16.2 Management of blight of *Coleus forskohlii* using biocontrol agent, organic amendments and chemicals (Ramaprasad Shresti 2005)

Treatment	% blight incidence[a]	Cfu[b] *R. bataticola*
P. fluorescens at 10 ml/plant (24×10^5 cfu/ml)	19.98 (26.51)	14.20
Pronto at 5 % soil drench	23.31 (28.84)	15.60
Neemto at 500 g/5 m^2	21.09 (27.24)	16.40
Carbofuran 3G at 15 g a.i./5 m^2	24.42 (29.57)	17.60
Farm yard manure at 5 kg/5 m^2	25.53 (30.38)	18.80
Carbendazim at 0.1 % soil drench	21.19 (27.33)	6.80
Propiconazole at 0.1 % soil drench	23.31 (28.84)	7.40
Control	35.52 (36.59)	21.60
CD at 5 %	**3.48**	**2.72**

[a]Figures in parentheses are arc sin angular transformed values

[b]Cfu – colony-forming units $\times 10^3$/g of soil (Average of 5 replications)

weight of the tubers was reduced by 20 % due to root-knot nematodes. The percentage of starch on fresh weight basis showed drastic reduction (16 %) in the infested tubers. *M. incognita* has been reported to cause a yield reduction of up to 86 %.

16.1.4.1 Symptoms

The newly inoculated *C. forskohlii* plant exhibited stunted growth, yellow patches and severe galling of roots due to *M. incognita*. Tuber formation was found commensurate with the degree of galling, and yield reduction was 65 % under field conditions.

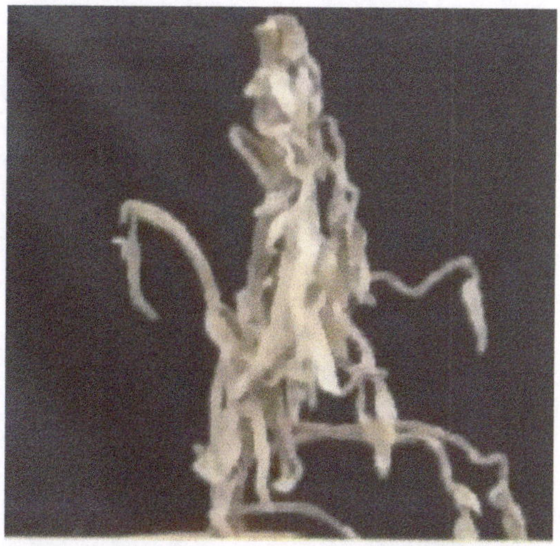

Fig. 16.3 Symptoms of bacterial wilt on coleus

The galls on coleus roots are very big and pronounced. The root-knot nematode damage often leads to crop failure. The infested tubers swell in size with irregular surface and cracking of the skin. If the infestation is severe, rotting sets in even before harvest (Fig. 16.4). Infested tubers rot after harvest and rarely reach market.

16.1.4.2 Biomanagement

Soil application of *P. fluorescens* (10^8 cfu/g) at 2.5 kg/ha recorded increased plant growth, tuber yield and propagule density in soil and reduced root galling, egg mass production and *M. incognita* population in soil (Senthamari et al. 2008) (Table 16.4).

Increase in plant height, number of branches and fresh and dry weight of the plant as well as less root-knot index were observed due to PGPR treatments (fluorescent *Pseudomonas* sp.) compared to control. Application of PGPR strains (RB50 and RB31) recorded maximum plant height, number of branches and fresh and dry weight of the plants with less root-knot index (Table 16.5) (Mallesh 2008).

Dipping of plants in 3 % solution of *Bacillus macerans* + soil drenching with 2 % solution of *B. macerans* 7 days after planting recorded maximum reduction in nematode population and root-knot count (14 galls/g of roots) and gave maximum increase in yield (Sheela et al. 2004).

Table 16.3 Biocontrol potentiality of PGPR strains in coleus against bacterial wilt

PGPR strains	Plant height (cm)	No. of branches	Fresh weight (g)	Dry weight (g)	Disease severity	Root rot index
RB01	31.5	4.5	94.0	15.00	++	2.0
RB10	27.5	4.0	92.5	14.25	+++	3.5
RB13	32.0	5.0	101.0	16.50	++	2.0
RB22	26.0	4.0	93.0	14.25	+++	3.0
RB31	34.0	5.5	107.0	16.90	++	1.5
RB43	28.0	4.5	93.5	14.50	++	2.5
RB50	35.0	6.0	110.5	17.50	+	1.5
Control	22.0	3.5	69.5	11.12	++++	4.0

Fig. 16.4 Root-knot nematode-infested plant and roots of coleus

16.1.4.3 Integrated Management

(a) *Bioagents and physical methods*: Integration of soil solarization in the nursery for 15 days with 150 gauge LDPE film and application of *Paecilomyces lilacinus* + *Bacillus macerans* in the main field are the best treatments in increasing plant height (60.3 cm compared to 40.0 cm in control), number of leaves (583.3 compared to 310.0 in control), weight of tubers/plant (546.6 g compared to 350.0 g in control) and yield (11.3 kg/plot compared to 6.9 kg/plot in control) and in reducing root galls (1.0 compared to 50.6 in control) and nematode population in soil (30.0/100 ml soil compared to 196.6/100 ml soil in control) and roots (1.6/5 g roots compared to 79.0/5 g roots in control) (Nisha and Sheela 2006).

(b) *Bioagents, botanicals and cultural methods*: Integration of dipping stem cuttings in 0.1 % *P. fluorescens* + soil application of neem cake at 400 kg/ha + growing marigold as an intercrop followed by their biomass incorporation during earthing up increased the root tuber yield by 22.7–30.0 % and reduced the root-knot nematode (*M. incognita*) population by 71.2–73.8 %. However, integration of *P. fluorescens* + marigold intercrop proved to be more economical (with benefit/cost ratio of 6.4–8.8) and effective practice for the management of root-knot nematodes in medicinal coleus (Seenivasan and Devarajan 2008) (Table 16.6).

16.1.5 Root-Knot Nematode, *Meloidogyne incognita*, and Wilt, *Fusarium chlamydosporum*, Disease Complex

16.1.5.1 Symptoms

Among the different diseases affecting coleus, root-knot and wilt disease complex caused by *M. incognita* and *F. chlamydosporum* was observed in severe form (Kumar 2008). In the interaction studies, *M. incognita* was the most aggressive pathogen compared to *F. chlamydosporum*. However, simultaneous inoculation of *M. incognita* and *F. chlamydosporum* caused greater reduc-

Table 16.4 Effect of bioagents for the management of *Meloidogyne incognita* infecting medicinal coleus

Treatment	Tuber yield (g/plant)	No. of galls/plant	No. of egg masses/g root	Spore density of *P. fluorescens* (cfu/g soil)
Control	22.71	1,680	98	–
P. fluorescens	61.57	678	31	72.05
CD (*P*=0.05)	**13.84**	**168.29**	**19.83**	–

Table 16.5 Biocontrol potentiality of PGPR strains in coleus against root-knot nematodes

PGPR strains	Plant height (cm)	No. of branches	Fresh weight (g)	Dry weight (g)	Root-knot index
RB01	31.5	4.5	94.0	15.00	2.0
RB10	27.5	4.0	92.5	14.25	3.5
RB13	32.0	5.0	101.0	16.50	2.0
RB22	26.0	4.0	93.0	14.25	3.0
RB31	34.0	5.5	107.0	16.90	1.5
RB43	28.0	4.5	93.5	14.50	2.5
RB50	35.0	6.0	110.5	17.50	1.5
Control	22.0	3.5	69.5	11.12	4.0

Table 16.6 Effect of integration of bioagents, neem cake and marigold intercrop for the management of *Meloidogyne incognita* infecting Coleus

Treatment	Gall index (1–5 scale)		Yield (MT/ha)		Benefit/cost ratio	
	Kharif	Rabi	Kharif	Rabi	Kharif	Rabi
Cutting dip in 0.1 % *P. fluorescens* + neem cake at 400 kg/ha	4.3	3.6	10.41	9.86	1.8	1.3
Cutting dip in 0.1 % *P. fluorescens* + marigold intercrop	2.3	2.1	12.08	11.07	8.8	6.4
P. fluorescens + neem cake + marigold	2.3	2.1	12.11	11.12	3.4	2.5
Carbofuran at 1 kg a. i./ha	4.3	3.6	10.36	9.85	2.1	1.6
Control	4.6	4.3	9.31	9.06	–	–
CD (*P*=0.05)	**0.36**	**0.36**	**0.83**	**0.80**	–	–

tion in plant growth as well as nematode multiplication (Fig. 16.5). Sequential inoculation of *M. incognita* 7 days prior to *F. chlamydosporum* caused reduction in plant growth parameters.

16.1.5.2 Biomanagement

Significant increase in tuber yield was recorded in all the plants treated with PGPR strains. However, the highest tuber yield was observed in the treatment RB50 (6.74 kg) followed by RB31 and RB13 (5.45 and 4.60 kg) (Mallesh 2008) (Table 16.7).

The lowest Root-knot Index (RKI) was noticed in RB50 (0.66), RB31 (0.73) and RB13 (0.80) which showed 70.00, 66.81 and 63.63 % reduction in root-knot index compared to control, respectively (Mallesh 2008).

Lowest disease incidence of 10.00 % (64.70 % reduction over control) was recorded in RB50 followed by 13.33 % in RB31 (52.94 % reduction over control). In general due to PGPR treatment, there was reduction of disease incidence in coleus which varied from 64.70 to 11.75 % reduction over control, respectively (Fig. 16.6) (Mallesh 2008).

The highest fresh and dry plant weights (1243.02 g and 212.27 g, respectively) were recorded in RB50, followed by RB31 (1094.54 g and 190.89 g, respectively) (Mallesh 2008).

16.1.5.3 Integrated Management

(a) *Bioagents and botanicals*: Combined application of plant products (neem seed kernel powder at 5 g/kg of soil) with biocontrol agents (*Trichoderma viride* + *P. fluorescens* at 10 g/kg of soil) was found effective in reducing the number of galls, nematode population, number of egg masses, root-knot index and root rot index and improving the plant growth parameters as compared to inoculated control.

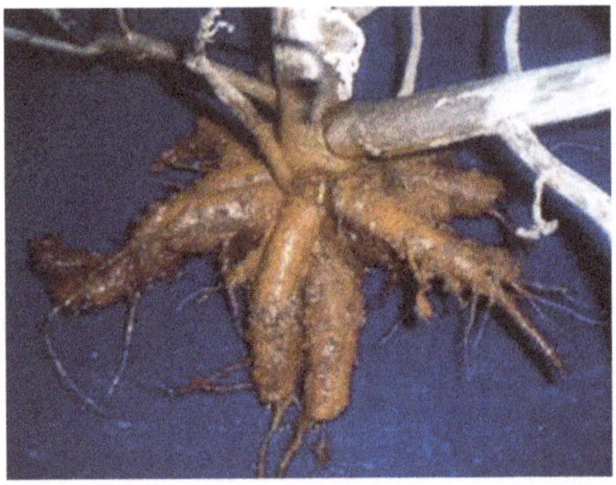

Fig. 16.5 Symptoms of root-knot and wilt disease complex on coleus

16.1.6 The Root-Knot Nematode, *Meloidogyne incognita*, and Wilt, *Fusarium chlamydosporum* and *Ralstonia solanacearum*, Disease Complex

The causal agents of wilt complex in coleus include *F. chlamydosporum*, *R. solanacearum* and *M. incognita*.

16.1.6.1 Biomanagement

Greenhouse studies using the PGPR fluorescent pseudomonads (bioformulations containing RB50 and RB31 strains) showed statistically significant increase in seedling biomass besides reduction in coleus wilt complex due to combination of pathogens (Mallesh and Lingaraju 2009).

16.1.7 Root-Knot Nematode, *Meloidogyne incognita*, and Collar Rot, *Rhizoctonia bataticola*, *Sclerotium rolfsii* and *Fusarium chlamydosporum*, Disease Complex

R. bataticola, *S. rolfsii*, *F. chlamydosporum* and *M. incognita* were found to be the most commonly

Table 16.7 Efficacy of talc formulations of PGPR strains on fresh tuber yield parameters and disease incidence in coleus under field conditions

PGPR strain	Tubers/plant	Total biomass(g) Fresh	Total biomass(g) Dry	Root-knot index	% disease incidence	% decrease over control	Tuber yield (kg/plot)
RB01	15.25	964.44	127.61	1.13	16.67	41.05	4.18
RB10	12.87	823.13	99.55	1.60	25.00	11.75	3.25
RB13	16.63	1016.74	169.85	0.80	20.00	29.40	4.60
RB22	11.87	811.84	88.60	1.53	23.33	17.64	3.36
RB31	16.50	1094.54	190.89	0.73	13.33	52.94	5.45
RB43	14.92	841.83	112.50	1.33	18.33	41.05	3.95
RB50	18.00	1243.02	212.27	0.66	10.00	64.70	6.74
Control	10.70	730.44	80.00	2.20	28.33	–	3.00
CD at 5 %	**1.52**	**14.66**	**6.55**	**0.65**	**5.02**	**–**	**0.79**

Fig. 16.6 Efficacy of talc formulation of PGPR strains against root-knot nematode on coleus under field conditions. *Left* – RB50. *Right* – Control

Fig. 16.7 Root-knot and collar rot disease complex on coleus. *Left* – Infected. *Right* – Healthy

associated fungi with collar rot disease complex (Ramaprasad Shresti 2005).

16.1.7.1 Symptoms

Wilt symptoms were first recorded at 45 days after inoculation in treatment *M. incognita* inoculated 7 days prior to inoculation of all the fungal pathogens simultaneously (*F. chlamydosporum* + *R. bataticola* + *S. rolfsii*) (Fig. 16.7) (Ramaprasad Shresti 2005).

16.1.7.2 Biomanagement

The treatment *P. fluorescens* recorded the lowest per cent wilt incidence (19.98) over control (35.52). The number of galls/5 g of root was the lowest in treatment *P. fluorescens* (18.27) (Table 16.8) (Ramaprasad Shresti 2005).

Greenhouse studies using fluorescent pseudomonads (bioformulations containing RB50 and RB31 strains) showed significant increase in seedling biomass besides reduction in coleus wilt complex due to combination of pathogens. RB50 and RB31 strains decreased root-knot index and incidence of disease complex while increasing biomass and tuber yield. Biochemical analyses in the above treatments showed elevated expression of defence enzymes (peroxidase, polyphenol oxidase and phenylala-

Table 16.8 Management of collar rot complex of *Coleus forskohlii* using different biocontrol agents, organic amendments and chemicals

Treatment[a]	% wilt incidence[b]	No. of galls/5 g roots	Cfu[c] *F. chlamydosporum*	*R. bataticola*
T1	19.98 (26.51)	18.27	8.00	14.20
T2	25.53 (30.38)	25.67	15.20	18.80
T3	21.19 (27.33)	23.33	3.60	6.80
T4	23.31 (28.84)	23.00	3.80	7.40
T5	35.52 (36.59)	28.40	19.60	21.60
CD at 5 %	**3.48**	**5.38**	**2.49**	**2.72**

[a]T1 – *Pseudomonas fluorescens* at 10 ml/plant (24 × 10^5 cfu/ml), T2 – farm yard manure at 5 kg/5 m^2, T3 – carbendazim at 0.1 % soil drench, T4 – propiconazole at 0.1 % soil drench, T5 – control

[b]Figures in parentheses are arc sin angular transformed values

[c]Cfu – colony-forming units ×10^{-3}/g of soil (average of 5 replications)

nine ammonia lyase) and higher accumulation of phenolic compounds (activity being highest in respect of RB50- and RB31-treated plants) compared to respective inoculated checks. The suppression of disease complex in coleus is largely due to aforementioned biocontrol mechanism as well as induction of systemic resistance by efficient PGPR strains (Lingaraju and Mallesh 2010).

16.1.8 Root-Knot Nematode, *Meloidogyne incognita*, and Root Rot, *Macrophomina phaseolina*, Disease Complex

The productivity of coleus has been hampered by its susceptibility to nematode and root rot disease complex. Pathogens associated with root rot are *M. phaseolina* and the root-knot nematode *M. incognita*. Severely affected areas were Salem, Attur and Rasipuram. Due to this disease complex the yield loss ranged from 50 to 60 %.

16.1.8.1 Symptoms

Simultaneous inoculation of *M. incognita* and *M. phaseolina* as well as nematode inoculation followed by fungus 15 days later caused significant reduction in tuber yield and 100 % root rot disease in medicinal coleus. The nematode multiplication was adversely affected when fungus was inoculated prior to nematode.

16.1.8.2 Integrated Management

(a) *Bioagents and cultural methods*: Integrated nematode management strategy (INMS) includes dipping of stem cuttings in *P. fluorescens* (strain Pf1) + growing marigold (*Tagetes erecta*) as intercrop and their biomass incorporation during earthing up. Biointensive disease management strategy (BDMS) include soil drenching with *P. fluorescens* (strain PfC6) at 2.5 kg/ha at planting, 30, 60, 90 and 120 days after planting. INMS along with BDMS treatment increased the tuber yield by 40.6 % and reduced nematode infestation in terms of mean number of juveniles per 100 ml soil (73.2 %), mean number of adult females per g of root (82.4 %), mean number of egg mass/g root (85.9 %) and mean number of eggs per g of root (87.9 %) with least gall index (1.6). This treatment also decreased the incidence of *M. phaseolina* root rot disease up to 50.4 %. The effect of this treatment on *M. incognita* and *M. phaseolina* and root tuber yield of medicinal coleus was almost equal to that of the treatment of INMS. But the latter is highly economical with lesser BC ratio than other treatments including standard chemical check, i.e. carbofuran 3G at 1 kg a.i./ha + soil drenching with carbendazim 0.1 %. Hence it could be concluded that treatment of INMS, i.e. dipping of stem cuttings in 0.1 % *P. fluorescens* +

Fig. 16.8 Root-knot-infested ashwagandha plant and roots

marigold intercropping, can be commercially exploited for the management of *M. incognita* and *M. phaseolina* disease complex in medicinal coleus (Seenivasan 2010).

16.2 Ashwagandha, *Withania somnifera*

16.2.1 Root-Knot Nematode, *Meloidogyne incognita*

16.2.1.1 Symptoms

The nematode-infected plants were chlorotic, stunted and less branched with fewer and smaller leaves and have poor response to fertilizers and irrigation. Such symptoms usually were not noticeable until there was severe damage to the root system by the nematodes. Roots of such plant were severely galled (Fig. 16.8) with reduced alkaloids. When the stem touches the soil, it was also found to be infested with root-knot nematode. The root-knot nematode-infected plants are more likely to be killed early than healthy non-infected plants.

16.2.1.2 Biomanagement

Soil application of *P. fluorescens* at 5 % w/w significantly increased plant height and shoot and root weight and reduced nematode population in soil, root-knot index, females in root and egg production (Ebhad and Patel 2012).

16.2.1.3 Integrated Management

(a) *Bioagents and botanicals*: Soil application of super *Pseudomonas* (consortia of *P. fluorescens* strains with chitinase, ACC deaminase and neem mixture) at 2.5 kg/ha was effective to suppress the root-knot nematode incidence by 78 % and to enhance the biomass of economic parts of ashwagandha plants by 26.87 % (dry root) and 43.19 % (seed) over untreated control.

16.2.2 Root-Knot Nematode, *Meloidogyne incognita*, and Root Rot, *Fusarium solani*, Disease Complex

A severe form of root rot disease complex in ashwagandha nurseries and regular crop was reported under Lucknow conditions. The disease was found to cause 30–50 % plant mortality.

In Karnataka, Mallesh (2008) noticed maximum disease incidence of 30.33 % in Dharwad district followed by Bijapur district (21.55 %). The pathogens, viz., *F. solani* and *M. incognita*, were isolated from disease complex plots.

16.2.2.1 Symptoms

Infected plants exhibited loss of turgidity, withering and drooping of leaves and wilting of plants. Death and decaying of roots and brown to black discolouration of roots with pulpiness were observed. White cottony growth of the fungus was observed near collar region (Fig. 16.9).

16.2.2.2 Biomanagement

Significantly higher total biomass was produced due to inoculation of PGPR strains (*P. fluorescens*). The data revealed that greater biomass accumulated on fresh and dry weight basis (41.66

Fig. 16.9 Symptoms of root-knot and root rot complex on ashwagandha. *Left* to *Right* – Rotting, decaying and disintegration of roots

Table 16.9 Efficacy of talc formulations of PGPR strains on yield parameters and disease incidence in ashwagandha under field conditions

PGPR strains	Total biomass (g)		Gall index	% disease incidence	% decrease over control	Fresh berry wt. (g/ plant)	Fresh root yield (g/plant)
	Fresh	Dry					
RB01	166.65	63.84	1.33	13.13	58.54	73.56	279.67
RB10	144.00	51.30	1.93	26.67	15.78	38.19	210.00
RB13	186.15	66.40	1.00	16.67	47.36	61.32	319.60
RB22	135.83	50.83	1.46	23.33	26.33	40.00	214.00
RB31	200.20	83.78	0.80	11.67	63.15	87.38	354.60
RB43	158.47	69.75	1.40	18.33	42.12	63.44	252.80
RB50	225.32	87.45	0.75	08.33	73.69	97.00	373.00
Control	111.86	46.92	3.06	31.67	–	35.13	185.60
CD 5 %	**19.62**	**8.91**	**0.97**	**4.19**	**–**	**9.14**	**17.92**

and 11.00 g) in RB50 which was on par with RB31 (40 and 10.50 g) (Table 16.9).

Application of PGPR strains (RB50 and RB31) recorded maximum plant height, number of branches and fresh and dry weight of the plants with less disease severity, root rot and root-knot index in individual as well as interactions of different pathogens.

The lowest root-knot index of 0.75 was observed in RB50, which was on par with RB31 (0.80). In general, there was a drastic reduction in RKI due to PGPR treatments compared to untreated.

There was significant reduction in disease incidence in PGPR-treated plots when compared to untreated plots. In PGPR treatments RB50 recorded the lowest disease incidence of 8.33% (73.69 % reduction over control), followed by RB31 11.67 % (63.15 % reduction over control).

The highest berry weight (97.00 g/plant) was recorded in RB50 followed by RB31 (87.38 g/ plant).

The highest fresh weight of ashwagandha roots was recorded in RB50 (373 g/plot) which was on par with RB31- (354.6 g/plot) (Fig. 16.10) (Table 16.9).

Biochemical analyses in the above treatments showed elevated expression of defence enzymes (peroxidase, polyphenol oxidase and phenylalanine ammonia lyase) and higher accumulation of

Fig. 16.10 Efficacy of talc formulation of PGPR strains in ashwagandha under field conditions. *Left* – RB50. *Right* – Control

phenolic compounds (activity being highest in respect of RB50- and RB31-treated plants) compared to respective inoculated checks. The suppression of disease complex in ashwagandha is largely due to the aforementioned biocontrol mechanism as well as induction of systemic resistance by efficient PGPR strains (Lingaraju and Mallesh 2010).

16.3 Sarpagandha, *Rauvolfia serpentina*

16.3.1 Foliar Blight/Spot, *Alternaria tenuis*

16.3.1.1 Symptoms

The pathogen attacks the leaves, resulting in minute, brownish or dark coloured circular spots with a yellowish margin on the ventral side of the leaves. The fungus also affects the flowers and fruits.

16.3.1.2 Integrated Management

(a) *Bioagents, botanicals and chemicals*: The application of two doses of neem cake at 50 g/plant at 30 days interval plus two sprays of *B. subtilis* in September and October along with three foliar sprays of mancozeb at 15 days interval plus two foliar sprays of carbendazim at 0.15 % at 21 days interval gave 201.5 g fresh root/plant and 75.5 % protection against foliar blight pathogen.

Fig. 16.11 Aloe rust

16.4 Aloe Vera, *Aloe indica*

16.4.1 Rust, *Phakopsora pachyrhizi*

16.4.1.1 Symptoms

It causes circular black or brown spots on the leaves of aloes. The fungal disease is especially a concern in the commercial production of aloe vera. The black spots are created by the oxidation of the phenols in plant sap and are permanent once they appear. Aloe rust is a fungus that may leave black spots on the branches of aloe plant (Fig. 16.11).

16.4.1.2 Integrated Management

(a) *Bioagents and AMF*: Four selected bio-inoculants, namely *B. subtilis, Glomus aggregatum, G. intraradices* and *Trichoderma harzianum*, were evaluated against black spot disease either alone or in different combinations. The maximum plant height attained was 56 cm in *G. intraradices*-treated plants followed by 53 cm in *B. subtilis* and *B. subtilis + G. aggregatum* treatments. Maximum plantlet production (14) was recorded in the treatments of *B. subtilis + T. harzianum + G. aggregatum* followed by 12 in others. Maximum herb yield obtained was 3.65 kg/plant in *G. aggregatum*-treated pots, followed by 3.35 kg/plant in *G. aggregatum + B. subtilis*- and *G. intraradices*-treated pots.

16.5 Soda Apple, *Solanum viarum*

16.5.1 Charcoal Rot, *Macrophomina phaseolina*

16.5.1.1 Symptoms

Leaves of the plants presented progressive necrosis, then dried out and dropped. Necrosis progressed quickly from petioles through the stems and caused entire branches to die (Fig. 16.12).

16.5.1.2 Biomanagement

P. fluorescens treatment reduced incidence and severity of charcoal rot and increased yield. Production of β-1,3-glucanase by *P. fluorescens* caused significant inhibition of *M. phaseolina.*

16.6 Kacholam, *Kaempferia galanga*

16.6.1 Root-Knot Nematode, *Meloidogyne incognita*

M. incognita was responsible for 27–38 % loss in rhizome yield of kacholam.

16.6.1.1 Biomanagement

Rhizome treatment with *P. fluorescens* at 3 % w/w significantly reduced the nematode population in soil (78 %) and roots (85 %), root galling (96 %), root-knot index (one compared to five in control), number of females in roots (90 %) and improved plant growth (number of leaves, shoot and root weight) and yield (10.93 t/ha compared to 4.88 in control) (Table 16.10) (Nisha and Sheela 2012).

Fig. 16.12 Stem lesions (**a**) and profuse production of *Macrophomina phaseolina* pycnidia on naturally infected stems and fruit of solanum plants (**b**)

Table 16.10 Effect of rhizome treatment on plant growth, root galling and yield of kacholam

Treatment	Plant weight (g)	Rhizome yield (MT/ha)	Gall index (1–5 scale)
Nimbecidine – 0.2 %	3.12	7.80	3
Glomus fasciculatum + *G. monosporum* – 3 % w/w	3.62	9.05	1
Pseudomonas fluorescens – 3 % w/w	4.37	10.93	1
Trichoderma viride – 3 % w/w	3.37	8.43	2
Hot water treatment – 55 °C	3.52	8.80	2
Dimethoate – 0.1 %	2.87	7.18	4
Untreated control	1.95	4.88	5
CD (*P*=0.05)	**0.87**	–	–

Fig. 16.13 Healthy and root-knot-infested plants and roots of henbane

16.7 Henbane, *Hyoscyamus muticus, H. niger, H. albus*

16.7.1 Root-Knot Nematode, *Meloidogyne incognita*

16.7.1.1 Symptoms

Up to 60–70 % plants were chlorotic and stunted showing a patchy appearance with fewer smaller leaves and flowers. Varying sizes of galls were found on the root system (Fig. 16.13).

16.7.1.2 Biomanagement

Significant reduction in *M. incognita* population was observed when henbane plants were inoculated with *P. fluorescens* resulting in higher herbage yield and root colonization with the bioagent.

16.7.1.3 Integrated Management

(a) *Bioagents and AMF*: Combined application of all bioinoculants (*P. fluorescens*, *G. aggregatum*, *G. fasciculatum* and *G. mosseae*) have not only enhanced the total biomass yield of *H. niger* but also significantly decreased the multiplication of nematodes. Thus, experimental evidences indicated that mixed inoculation of rhizobacterium with AM fungi could be considered for biological management instead of nematicides for reducing the deleterious effect of root-knot disease on black henbane (Pandey 1997) (Table 16.11).

16.8 Stevia, *Stevia rebaudiana*

16.8.1 Leaf Spot, *Alternaria alternata*

16.8.1.1 Symptoms

Symptoms initially appeared as small circular spots, light brown in colour. Later, many became

Table 16.11 Effect of bioagents for the management of *Meloidogyne incognita* infecting henbane

Treatment	Fresh biomass wt. (g)	% root colonization	AMF spores/100 g soil	Nema popn. (soil + roots)	Reprodn. factor	Root-knot index
Control	120.6	–	–	40,700	8.14	4.00
Pseudomonas fluorescens (*Pf*)	171.0	–	–	21,200	4.24	1.33
Glomus aggregatum (*Ga*)	172.7	65	780	22,600	4.52	1.66
G. fasciculatum (*Gf*)	144.6	38	490	28,900	5.78	2.66
G. mosseae (*Gm*)	142.0	45	580	30,260	6.05	3.00
Pf + *Ga* + *Gf* + *Gm*	202.0	68	800	20,140	4.03	1.00
CD (P=0.05)	**12.01**	**12.05**	**49.26**	**2,426.14**	**–**	**1.16**

Fig. 16.14 Alternaria leaf spot on stevia

irregular and dark brown to grey, while others remained circular with concentric rings or zones. On severely infected leaves, several spots coalesced to form large necrotic areas. On older leaves, concentric spots were more common at the tips. Leaf spots varied from 2 to 18 mm in diameter (Fig. 16.14).

16.8.1.2 Biomanagement

Foliar application of talc-based formulation of fluorescent *Pseudomonas* strain BRL-1 under field conditions showed 86–87 % reduction in disease severity.

16.9 Safed Musli, *Chlorophytum borivilianum*

16.9.1 Root-Knot Nematode, *Meloidogyne incognita*

16.9.1.1 Symptoms

The infested plants showed stunting, drying and falling of leaves and severe galling on the root system (Fig. 16.15).

16.9.1.2 Biomanagement

Plots that received vermicompost (10 t/ha), neem cake (200 kg/ha) and *Azotobacter* yielded 4.7 t/ha tubers as compared to 4.97 t/ha on plots that received full dose chemical fertilizers. The tubers from organically cultivated plots are robust and desirably whiter and are free from any infection of root-knot nematodes.

Fig. 16.15 *Left* – Healthy and root-knot-infested plants. *Right* – Root-knot-infested fingers of safed musli

References

Boby BU, Bagyaraj DJ (2003) Biological control of root rot of *Coleus forskohlii* Briq. using microbial inoculants. World J Microbiol Biotechnol 19:175–180

Ebhad D, Patel AD (2012) Management of root-knot nematodes (*Meloidogyne* spp.) in ashwagandha using bioagents in pots. Indian J Nematol 40:132–136

Kumar B (2008) Studies on root-knot and wilt complex in *Coleus forskohlii* (Wild.) Briq. caused by *Meloidogyne incognita* (Kofoid and White) Chitwood and *Fusarium chlamydosporum* (Frag. and Cif.) Booth. M. Sc. (Agri) thesis in plant pathology, University of Agricultural Sciences, Dharwad

Lingaraju S, Mallesh SB (2010) Emerging nematode diseases in two horticultural crops: their management through novel approaches. National Conference on innovations in nematological research for agricultural sustainability – challenges and a roadmap ahead, Tamil Nadu Agricultural University, Coimbatore, p 51

Mallesh SB (2008) Plant growth promoting Rhizobacteria, their characterization and mechanisms in the suppression of soil borne pathogens of Coleus and Ashwagandha. Ph.D. thesis, University of Agricultural Sciences, Dharwad, pp 206

Mallesh SB, Lingaraju S (2009) Enhancement of plant growth and suppression of wilt complex of *Coleus forskohlii* through fluorescent pseudomonads. International conference on horticulture, Bangalore

Nisha MS, Sheela MS (2006) Bio-management of *Meloidogyne incognita* on coleus, *Solenostemon rotundifolius* by integrating solarization, *Paecilomyces lilacinus, Bacillus macerans* and neem cake. Indian J Nematol 36:136–138

Nisha MS, Sheela MS (2012) Rhizome treatment – an economically feasible nematode management strategy for Kacholam. Indian J Nematol 42:125–128

Pandey R (1997) Ecofriendly approaches for the management of root-knot nematode on black henbane. Indian J Nematol 27:175–178

Paramasivan M, Mohan S, Muthukrishnan N (2007) Management of *Coleus* dry root rot pathogen *Macrophomina phaseolina* by fungal and bacterial antagonist. Indian J Plant Prot 35:133–135

Ramaprasad Shresti AY (2005) Studies on collar rot complex of Coleus forskohlii *(Wild.) Briq*. M. Sc. (Agri.) thesis, University of Agricultural Sciences, Dharwad, Karnataka, India

Seenivasan N (2010) Biointensive management of *Meloidogyne incognita* and *Macrophomina phaseolina* disease complex in medicinal coleus. Indian J Plant Prot 38:186–192

Seenivasan N, Devarajan K (2008) Integrated approach for the management of root-knot nematode, *Meloidogyne incognita* in medicinal coleus. Indian J Nematol 38:154–158

Senthamari K, Poornima K, Subramanian S, Sudheer J (2008) Nematode-fungal disease complex involving *Meloidogyne incognita* and *Macrophomina phaseolina* on medicinal coleus, *Coleus forskohlii* Briq. Indian J Nematol 38:30–33

Sheela MS, Nisha MS, Mohandas C (2004) Eco-friendly management of nematodes associated with Chinese potato (Coleus), *Solenostemon rotundifolius* (Poir) Morton. National symposium on paradigms in nematological research for biodynamic farming, University of Agricultural Sciences, Bangalore, p 90

Singh R, Parameswaran TN, Divya S, Puttanna K, Satyasrinivas KVN, Bagyaraj DJ, Kalra A (2009) Management of root-rot/wilt of *Coleus forskohlii* Briq. In: CIMAP golden jubilee national symposium on medicinal & aromatic plants "Fifty Years of Research on Medicinal & Aromatic Plants", CIMAP Regional Centre, Bangalore, p 18

Part VI

A Road Map Ahead

17 Challenges, Future Prospects and Conclusions

17.1 Challenges in PGPR Research

17.1.1 Selection and Characterization of PGPR

One of the challenges in developing PGPR for commercial application is ensuring that an effective selection and screening procedure is in place, so that the most promising organisms are identified and brought forward. In the agricultural chemical industry, thousands of prospective compounds are screened annually in efficient high-throughput assays to select the best one or two compounds for further development. Similar approaches are not yet in place for PGPR. Effective strategies for initial selection and screening of rhizobacterial isolates are required. It may be important to consider host plant specificity or adaptation to a particular soil, climatic conditions or pathogen in selecting the isolation conditions and screening assays (Bowen and Rovira 1999). One approach for the selection of organisms with the potential to control soilborne phytopathogens is to isolate from soils that are suppressive to that pathogen (Weller et al. 2002). Other approaches involve selection based on traits known to be associated with PGPR such as root colonization (Silva et al. 2003), 1-aminocyclopropane-1-carboxylate (ACC) deaminase activity (Cattelan et al. 1999; Glick 1995) and antibiotic (Giacomodonato et al. 2001) and siderophore production (Cattelan et al. 1999). The development of high-throughput assay systems and effective bioassays will facilitate selection of superior strains (McSpadden Gardener and Fravel 2002).

17.1.2 Field Application of PGPR

One of the challenges of using PGPR is natural variation. It is difficult to predict how an organism may respond when placed in the field (compared to the controlled environment of a laboratory). Another challenge is that PGPR are living organisms. They must be able to be propagated artificially and produced in a manner to optimize their viability and biological activity until field application. Like rhizobia, PGPR bacteria will not live forever in a soil, and over time growers will need to re-inoculate seeds to bring back populations.

The application of PGPR for control of fungal pathogens in greenhouse systems shows considerable promise (Paulitz and Belanger 2001), due in part to the consistent environmental conditions and high incidence of fungal disease in greenhouses. Achieving consistent performance in the field where there is heterogeneity of abiotic and biotic factors and competition with indigenous organisms is more difficult. Knowledge of these factors can aid in determination of optimal concentration, timing and placement of inoculant and of soil and crop management strategies to enhance survival and proliferation of the inoculant (McSpadden Gardener and Fravel 2002).

P.P. Reddy, *Plant Growth Promoting Rhizobacteria for Horticultural Crop Protection*,
DOI 10.1007/978-81-322-1973-6_17, © Springer India 2014

The concept of engineering or managing the rhizosphere to enhance PGPR function by manipulation of the host plant, substrates for PGPR or through agronomic practices is gaining increasing attention (Mansouri et al. 2002). Development of better formulations to ensure survival and activity in the field and compatibility with chemical and biological seed treatments is another area of focus; approaches include optimization of growth conditions prior to formulation and development of improved carriers and application technology (Yardin et al. 2000).

17.1.3 Commercialization of PGPR

Prior to registration and commercialization of PGPR products, a number of hurdles must be overcome (McSpadden Gardener and Fravel 2002). These include scaleup and production of the organism under commercial fermentation conditions while maintaining quality, stability and efficacy of the product. Formulation development must consider factors such as shelf life, compatibility with current application practices, cost and ease of application. Health and safety testing may be required to address such issues as non-target effects on other organisms including toxigenicity, allergenicity and pathogenicity, persistence in the environment and potential for horizontal gene transfer. The product claim, whether as a fertilizer supplement or for biological control, will determine to which federal agency applications for registration should be addressed. Capitalization costs and potential markets must be considered in the decision to commercialize. McSpadden Gardener and Fravel (2002) estimated that a minimum capitalization of US$1 million is required to register a biopesticide product in North America.

The success of microbial pesticides to suppress pests and diseases depends on the availability of microbes as a product or formulation, which facilitate the technology to transfer from lab to land. The constraints to biopesticide development and utilization mirror some of those factors that limit the development worldwide. Constraints include:

- Lack of suitable screening protocol for the selection of promising candidate of PGPR
- Lack of sufficient knowledge on the microbial ecology of PGPR strains and plant pathogens
- Optimization of fermentation technology and mass production of PGPR strains
- Inconsistent performance and poor shelf life
- Lack of patent protection
- Prohibitive registration cost (Fravel et al. 1999)
- Awareness, training and education shortfalls
- Lack of multidisciplinary approach
- Technology constraints (Sabitha Doraiswamy et al. 2001)

17.1.3.1 Screening and Selection of Potential PGPR Strain

Success of commercialization of PGPR strains depends on the selection of effective strains after adopting rigorous screening strategies, because any mistake during strain selection will be a costly mistake in product development (Schisler and Slininger 1997). The potentiality of the PGPR strain in the suppression of plant pathogen should be carried out at both lab and field conditions in different soil types with diversified microbial communities and climatic conditions. It would lead to the development of a viable PGPR strain.

17.1.3.2 Microbial Ecology and Interaction

Suppression of plant disease is a four-way interaction of biocontrol agents, plants, pathogens and the environment. Hence, understanding of the interaction between all these components is essential for developing a suitable biocontrol agent in disease management (Larkin et al. 1998). Extracellular metabolites produced by PGPR strains interact with microbial community (plant pathogens and other microbes) and plant in rhizosphere or spermosphere or phyllosphere and result in the suppression of pathogenic propagules either by direct action of antibiotics or through elicitation of induced systemic resistance activated by the molecular determinants (lipopolysaccharide, salicylic acid), growth regulators and siderophores of bacterial origin. However, knowledge on the influence of biotic and abiotic environment on PGPR strains to express its antimicrobial action has to be studied in depth under in vivo conditions to improve the efficacy of PGPR strains. This will facilitate to identify bacterial strains that could perform well under diverse environmental conditions around the court of infection.

17.1.3.3 Fermentation Technology and Shelf Life of Formulations

Optimization of fermentation technology (liquid or solid fermentation) with suitable medium (synthetic or semi-synthetic) for mass multiplication and identification of suitable carrier material (organic or inorganic) for formulation development with increased shelf life is a barrier in the commercial success of formulation development. The biomass production and efficacy of biocontrol agents to suppress plant pathogens varies depending on the nutrient composition of the medium (Schisler and Slininger 1997). Hence, the medium selected for biomass production should support the growth and efficacy of antagonist and the cost of medium should be economical so that the technology remains viable.

17.1.3.4 Patent Protection and Prohibitive Registration Cost

The environmental protection agency in developed and developing countries should relax the formalities and registration cost to promote registration of biocontrol agents either by universities or private companies. The patent protection rights for the effective products should be strengthened to encourage the organizations involved in identification and development of commercial biocontrol agents.

17.1.3.5 Awareness, Training and Education Shortfalls

The general level of awareness among stakeholders about the potential value of biopesticides is lacking. There is a need for the following:

- Awareness level among the policy makers of the potential for biopesticides, their efficacy and their effect in reducing the health and environmental problems should be increased.
- The opportunities offered by the commercialization in terms of generation of wealth and employment are to be promoted.
- Entrepreneurs and investors need to be informed about the opportunities that exist for establishing commercial companies to manufacture market and sell biopesticides.
- The extension workers from the state horticulture departments have to be given training on the advantages of using biopesticides for pest control. The interaction between research and extension sectors has to be improved.
- The nature and mode of action of biopesticides have to be explained to farmers who are used to chemical insecticides, which are often fast acting and are visibly effective (Sabitha Doraiswamy et al. 2001).

17.1.3.6 Lack of Multidisciplinary Approach

The process of biopesticides development to complete product requires research in areas of screening, formulation, field application, production, storage, toxicology as well as the steps necessary for commercialization, such as scaleup production, registration and regulatory matters. Most of the research efforts undertaken with the use of biopesticides are confined only to the exploration, collection, isolation and identification of biocontrol agents combined with laboratory-based bioassays. But in the process of product development the above research aspects share only a fraction of work required to develop a complete product. Product development requires a multidisciplinary approach to biopesticide research and development. Rarely, a complete range of expertise exists in a single institute or organization.

17.1.3.7 Technology Constraints

(i) *Delivery system*: Success in biocontrol depends on the understanding and use of the delivery system. The research on delivery system is well below that of chemical insecticides. The attention on the application technology can improve biopesticide performance.

(ii) *Biopesticide quality*: The major problem in the field of biopesticide production is the product quality and stability. In small-scale production, contamination of inoculum is a common problem. The long-term shelf life of the product is highly essential to attract the multinational companies to invest on a large scale.

PGPR with wide scope for commercialization include *Pseudomonas fluorescens*, *P. putida*, *P. aeruginosa*, *Bacillus subtilis* and other *Bacillus* spp. The potential PGPR isolates are formulated using different organic and inorganic carriers either through solid or liquid fermentation technologies. They are delivered either through seed treatment, bio-priming, seedling dip, soil application, foliar spray, fruit spray, hive insert, sucker treatment or sett treatment. The application of PGPR formulations with strain mixtures performs better than individual strains for the management of pests and diseases of crop plants, in addition to plant growth promotion. Supplementation of chitin in the formulation increases the efficacy of antagonists. More than 33 products of PGPR have been registered for commercial use in greenhouses and fields in North America. Though PGPR has a potential scope in commercialization, the threat of certain PGPR (*P. aeruginosa*, *P. cepacia* and *B. cereus*) to infect human beings as opportunistic pathogens has to be clarified before large-scale acceptance, registration and adoption of PGPR for pest and disease management.

17.2 Future Prospects and Conclusions

17.2.1 Future Prospects

Over the years, the PGPR have gained worldwide importance and acceptance for agricultural benefits. These microorganisms are the potential tools for sustainable agriculture and the trend for the future. Scientific researches involve multidisciplinary approaches to understand adaptation of PGPR to the rhizosphere, mechanisms of root colonization, effects on plant physiology and growth, biofertilization, induced systemic resistance, biocontrol of plant pathogens, production of determinants, etc. Biodiversity of PGPR and mechanisms of action for the different groups (diazotrophs, bacilli, pseudomonads, rhizobia, phosphate-solubilizing bacteria, lignin degrading, chitin degrading, cellulose-degrading bacteria) are demonstrated. Effects of physical, chemical and biological factors on root colonization and the proteomics perspective on biocontrol and plant defence have also shown positive results. Visualization of interactions of pathogens and biocontrol agents on plant roots using autofluorescent protein markers has provided more understanding of biocontrol processes with overall positive consequences.

As our understanding of the complex environment of the rhizosphere, of the mechanisms of action of PGPR and of the practical aspects of inoculant formulation and delivery increases, we can expect to see new PGPR products becoming available. The success of these products will depend on our ability to manage the rhizosphere to enhance survival and competitiveness of these beneficial microorganisms (Bowen and Rovira 1999). Rhizosphere management will require consideration of soil and crop cultural practices as well as inoculant formulation and delivery (McSpadden Gardener and Fravel 2002). Genetic enhancement of PGPR strains to enhance colonization and effectiveness may involve addition of one or more traits associated with plant growth promotion (Bloemberg and Lugtenberg 2001). Genetic manipulation of host crops for root-associated traits to enhance establishment and proliferation of beneficial microorganisms (Mansouri et al. 2002) is being pursued. However, regulatory issues and public acceptance of genetically engineered organisms may delay their commercialization. The use of multi-strain inocula of PGPR with known functions is of interest as these formulations may increase consistency in the field (Jetiyanon and Kloepper 2002; Siddiqui and Shaukat 2002). They offer the potential to address multiple modes of action, multiple pathogens and temporal or spatial variability.

PGPR inoculants can fulfil diverse beneficial interactions in plants leading to promising solutions for sustainable and environment-friendly agriculture. The applications of rhizosphere soil of agricultural crops with desirable bacterial populations have established considerable promises in both the laboratory and greenhouse experiment. Further, improved understanding on the way by which PGPR promote plant growth can lead to expanded exploitation of these 'biofertilizers' to reduce the potential negative environmental effects associated with the food and fibre production. Recent progress of molecular biology and biotechnology in the understanding of rhizobacterial interactions with the nodules of crop plants will encourage a suitable area of research in PGPR mechanisms relating to rhizosphere colonization. Farwell et al. (2007) compared the growth of a transgenic canola, *Brassica napus* inoculated with PGPR strain, *Pseudomonas putida* UW4, to normal canola and found a significant growth in the former. Thus, biotechnology can be applied to improve the efficacy of PGPR strains through transgenics for agricultural improvement. However, modern technology based on the transformations of 1-aminocyclopropane-1-carboxylic acid deaminase gene, which directly stimulates plant growth by cleaving the immediate precursor of plant ethylene into *Pseudomonas fluorescens* CHAO, not only increased the plant growth but also accelerated biocontrol properties of PGPR species (Holguin and Glick 2001). Genomic tinkering of naturally occurring PGPR strains with effective genes (Nakkeeran et al. 2005) could lead to accentuated expression of genomic products, thereby alleviating the attack of both pests and diseases on field crops that would further facilitate for better introduction of a single bacterium with multiple modes of action to benefit the growers. Thus, future success of industries producing microbial inoculants, especially PGPR, will depend on innovative business management, product marketing, extension education and extensive research. Further optimization is required for better fermentation and formulation processes of effective PGPR strains to introduce in agriculture.

Scientists across the world have been researching on suitable and effective PGPR as among the strategies to raising crop productivity levels. In Asian countries, PGPR technologies are now at various stages of development and utilization. For successful sustainable agriculture, basic, applied and strategic research must focus on the identification of potential PGPR screening, characterization and cataloguing the strains for their desirable traits, enhancement of potential strains and understanding the mechanism of action of these PGPR to harness their optimum potential. Thrust also must be laid on developing cheap and viable mass multiplication protocols, increasing shelf life, identifying suitable carrier systems, and working out the economics to demonstrate the usefulness of the PGPR. There is a need for the development of effective communication methods to disseminate the technology among the end users to get feedback in order to refine the technology. To bridge these gaps, strategies should be thought of by forming consortia to work in a collaborative mode and build new partnerships. Also concerted approach among various stakeholders is required on research and development aspects and positive policy interventions to promote these inputs.

17.2.2 Conclusions

The present review indicates the development and formulations of PGPR in biological promotion of different characteristics of plant growth. Most of the PGPR isolates significantly increased plant height, root length and dry matter production in various agricultural crops like potato, tomato, maize, wheat, etc. The development of stable formulations of antagonistic PGPR for sustainable agricultural systems has established another promising approach for replacing the use of chemical fertilizers and pesticides. Besides, PGPR are protecting natural environments as well as biological resources by playing a significant role in integrated pest management system (IPM). In addition to this, certain PGPR strains have also the ability to activate octadecanoid, shikimate and terpenoid pathways (Gouws 2009) which in turn assist in the alterations of VOC production in host plants. In accordance with their mode of action, PGPR can be classified as biofertilizers, phytostimulators and biopesticides with certain bacteria having overlapping applications. However, screening strategies for selecting the best rhizobacterial strain for rhizosphere competence and studies on the ecology of the introduced PGPR with the resident PGPR and other microbial species in the plant rhizosphere would require more comprehensive knowledge. The involvement of ACC deaminase gene; siderophore; phosphate; phytohormones like IAA, cytokinin, gibberellins, etc.; nodulation; disease suppression; and their coordinated expression seemed to play a role in enhancing the plant growth, yield and nutrient uptake of various crop plants in different agroecosystems. However, carefully controlled field trials of crop plants inoculated along with rhizobacteria are necessary for maximum commercial exploitation of PGPR strains.

Identifying different mechanisms of action facilitates the combination of strains, bacteria with bacteria or bacteria with fungi, to hit pathogens with a broader spectrum of microbial weapons. Along this same line, biotechnology can be applied to further improve strains that have prized qualities (e.g. formulation ease, stability or otherwise exceptionally suited to plant colonization) by creating transgenic strains that combine multiple mechanisms of action. Current and future progress in our understanding of PGPR diversity, colonization ability, mechanisms of action, formulation and application could facilitate their development as reliable components in the management of sustainable agricultural systems.

References

Bloemberg GV, Lugtenberg BJJ (2001) Molecular basis of plant growth promotion and biocontrol by rhizobacteria. Curr Opin Plant Biol 4:343–350

Bowen GD, Rovira AD (1999) The rhizosphere and its management to improve plant growth. Adv Agron 66:1–102

Cattelan AJ, Hartel PG, Fuhrmann JJ (1999) Screening for plant growth-promoting rhizobacteria to promote early soybean growth. Soil Sci Soc Am J 63:1670–1680

Farwell AJ et al (2007) Tolerance of transgenic canola plants (*Brassica napus*) amended with plant growth-promoting bacteria to flooding stress at a metal-contaminated field site. Environ Pollut 147:540–545

Fravel DR, Rhodes DJ, Larkin RP (1999) Production and commercialization of biocontrol products. In: Albajes R, Lodovica Gullino M, Van Lenteren JC, Elad Y (eds) Integrated pest and disease management in greenhouse crops. Kluwer Academic Publishers, Boston, pp 365–376

Giacomodonato MN, Pettinari MJ, Souto GI, Mendez BS, Lopez NI (2001) A PCR-based method for the screening of bacterial strains with antifungal activity in suppressive soybean rhizosphere. World J Microbiol Biotechnol 17:51–55

Glick BR (1995) The enhancement of plant growth by free-living bacteria. Can J Microbiol 41:109–117

Gouws LM (2009) The molecular analysis of the effects of lumichrome as a plant growth promoting substance. Dissertation, University of Stellenbosch, South Africa

Holguin G, Glick BR (2001) Expression of the ACC deaminase gene from *Enterobacter cloacae* UW4 in *Azospirillum brasilense*. Microbial Ecol 41:281–288

Jetiyanon J, Kloepper JW (2002) Mixtures of plant growth-promoting rhizobacteria for induction of systemic resistance against multiple plant diseases. Biol Control 24:285–291

Larkin RP, Roberts DP, Gracia-Garza JA (1998) Biological control of fungal diseases. In: Hutson D, Miyamoto Y (eds) Fungicidal activity, chemical and biological approaches. Wiley, New York, pp 149–191

Mansouri H, Petit A, Oger P, Dessaux Y (2002) Engineered rhizosphere: the trophic bias generated by opine-producing plants is independent of the opine type, the soil origin, and the plant species. Appl Environ Microbiol 68:2562–2566

McSpadden Gardener BB, Fravel DR (2002) Biological control of plant pathogens: research, commercialization, and application in the USA. Plant Health Prog. doi:10.1094/PHP-2002-0510-01-RV, Online

Nakkeeran S, Fernando WGD, Siddiqui ZA (2005) Plant growth promoting rhizobacteria formulations and its scope in commercialization for the management of pests and diseases. In: Siddiqui ZA (ed) PGPR: biocontrol and biofertilization. Springer, Dordrecht, pp 257–296

Paulitz TC, Belanger RB (2001) Biological control in greenhouse systems. Annu Rev Phytopathol 39:103–133

Sabitha D, Nakkeeran S, Chandrasekar, G (2001) *Trichoderma* – bioarsenal in plant disease management and its scope for commercialization. In: Proceedings of the Indian Phytopathological Society, Southern Zone meeting, 10–12 Dec 2001, Indian Institute of Spice Research, Calicut, pp 43–55

Schisler DA, Slininger PJ (1997) Microbial selection strategies that enhance the likelihood of developing commercial biological control products. J Ind Microbiol Biotechnol 19:172–179

Siddiqui IA, Shaukat SS (2002) Resistance against damping-off fungus *Rhizoctonia solani* systematically induced by the plant growth-promoting rhizobacteria *Pseudomonas aeruginosa* (1E-6S(+)) and *P. fluorescens* (CHAO). J Phytopathol 150:500–506

Silva HSA, Romeiro RS, Mounteer A (2003) Development of a root colonization bioassay for rapid screening of rhizobacteria for potential biocontrol agents. J Phytopathol 151:42–46

Weller DM, Raaijmakers JM, McSpadden Gardener BB, Thomashow LS (2002) Microbial populations responsible for specific soil suppressiveness to plant pathogens. Annu Rev Phytopathol 40:309–348

Yardin MR, Kennedy IR, Thies JE (2000) Development of high quality carrier materials for field delivery of key microorganisms used as bio-fertilisers and biopesticides. Radiat Phys Chem 57:565–568

Index

P.P. Reddy, *Plant Growth Promoting Rhizobacteria for Horticultural Crop Protection*, DOI 10.1007/978-81-322-1973-6, © Springer India 2014

O

P

R